AF598127

Methods in Molecular Biology™

Series Editor
John M. Walker
School of Life Sciences
University of Hertfordshire
Hatfield, Hertfordshire, AL10 9AB, UK

For further volumes, go to
http://www.springer.com/series/7651

Plant Immunity

Methods and Protocols

Edited by

John M. McDowell

Department of Plant Pathology, Physiology, and Weed Science, Virginia Tech, Blacksburg, VA, USA

Editor
John M. McDowell, Ph.D.
Department of Plant Pathology,
Physiology, and Weed Science
Virginia Tech
Latham Hall (0390) 550
Blacksburg, VA 24061, USA
johnmcd@vt.edu

ISSN 1064-3745 e-ISSN 1940-6029
ISBN 978-1-61737-997-0 e-ISBN 978-1-61737-998-7
DOI 10.1007/978-1-61737-998-7

Library of Congress Control Number: 2011921723

Printed on acid-free paper

Humana Press is part of Springer Science+Business Media (www.springer.com)

Preface

Examples of How New Experimental Technologies Have Enabled Landmark Advances in Understanding of Plant Immunity Over the Last Half-Century

This volume of Methods in Molecular Biology was designed to emphasize emerging technologies that can be applied to outstanding questions in plant immunity. The content is complementary to another recent, excellent volume in the series with a similar focus (1). Below, I provide a brief historical overview highlighting major conceptual advances in molecular plant–microbe interactions that would not have been possible without exploitation of new technologies. Additionally, I outline current conceptual challenges in our field that can be addressed with methods described in this volume. Finally, I speculate on technological advances in the near term that enable deeper understanding of plant immunity and support rational strategies for durable disease control.

As all readers of this volume know, much effort has been invested in understanding the molecular mechanisms through which plants and microbes interact. Much of the progress in this field has been fueled by timely, thoughtful exploitation of new methodologies. For example, H.H. Flor's use of classical genetics clearly demonstrated that the outcome of encounters between flax and flax rust can be dictated by single genes on both sides of the interaction (2). Equally important, his methodology revealed striking specificity in these interactions, which led to formulation of the seminal "gene-for-gene" model. This genetics-driven model provided a conceptual framework for the plant immunity that proved generally applicable and remains relevant today (3, 4).

Subsequent emergence of molecular biology tools enabled the gene-for-gene model to be elaborated in molecular terms. For example, gene cloning technologies were used to isolate avirulence (*avr*) genes, resistance (*R*) genes, and additional components of pathogenicity and immunity. Molecular approaches, along with judicious biochemistry, provided for critical examination of "receptor-ligand" models that predicted direct interaction between the products of *R* and *avr* genes (e.g., (5)). Three important themes emerged from these efforts: First, the majority of plant resistance genes encode proteins from a single superfamily, defined by a nucleotide-binding site and leucine-rich repeats (NB-LRR) (6). Second, pathogen Avr proteins are, in many cases, translocated into plant cells where they act as effectors to reprogram plant cells for susceptibility (7). Third, NB-LRR proteins often do not interact directly with corresponding Avr proteins but instead monitor guardees or decoys that are modified by the Avr protein (8–10). In addition, experiments with pathogen "elicitor" molecules revealed a second branch of the plant immune system, which directly recognizes pathogen-associated molecular patterns (PAMPs) that are evolutionarily conserved among diverse pathogens (11). The two branches of the plant immune system have been connected by recent models predicting that pathogen effectors may have evolved to interfere with PAMP-triggered immunity (12). NB-LRR receptors thereby provide a second line of defense by recognizing the molecular signatures of effector activity. In sum, the adoption of molecular biology methods led to major advances in understanding of plant immunity that could not have been anticipated by models based (however logically) on genetic data alone.

At present, "omics" tools are being used to build on molecular advances and provide new insights. For example, it is now possible to survey a plant genome and identify all of the potential immune receptors using queries based on conserved motifs (e.g., (13)). From this, it has become clear that plants maintain hundreds of probable immune receptors, which in many cases appear to be evolving dynamically to cope with ever-changing pathogen populations (14). From an applied perspective, these inventories can greatly accelerate the process of resistance gene identification (e.g., for cloning and/or marker-assisted breeding of *R* genes from wild relatives into crops).

Similar advances are underway in pathogen genomics. For example, molecular signatures are being developed for pathogen effector proteins that enable comprehensive effector gene inventories to be predicted *in silico*. Genome level comparisons have revealed that pathogen genomes contain dozens (in prokaryotes) to hundreds (in eukaryotes) of effector genes (15, 16). Like the cognate surveillance genes in plants, these genes are often variable and subject to rapid turnover. Large-scale characterization of effector functions is a major focus of effort in the field of plant immunity that is discussed further below.

Transcript profiling is also impacting understanding of plant immunity. For example, early experiments with microarrays revealed massive transcriptional changes that accompany the activation of the immune system and illuminated molecular distinctions between different resistance mechanisms (e.g., (17)). Subsequent studies that combine transcript profiling with immune response mutants have provided insight into the structure of the defense hierarchy and have identified previously unknown components of the network (e.g., (18)). Analyses of transcript profiles have also provided important insights into the molecular mechanisms through which pathogens manipulate the environment inside plant tissue (e.g., (19)).

At present, it is inarguable that our current, exciting level of understanding of plant immunity (and pathogen evasion thereof) owes much to the timely adoption of new methodologies in genetics and molecular biology, as well as genomics. However, we remain far from a complete understanding of how the plant immune system functions, or how its functionality is perturbed by adapted pathogens. Many questions remain that will require new methodologies to be developed, optimized, and widely adopted. This volume of Methods in Molecular Biology was designed to emphasize emerging technologies that can be applied to outstanding questions in plant immunity.

For example, although NB-LRR immune surveillance proteins have now been known of for 1.5 decades, we still do not understand exactly how they function, and it is not clear whether all NB-LRR proteins function in a similar manner (20, 21). Moreover, we still lack a complete inventory of downstream signaling components, and we do not understand how these components interact. Methods that can be applied for new insights into molecular functionality of NB-LRR proteins and other immune signaling components are described in Chaps. 1–4. Chapter 1 addresses the understudied question of exactly where in the cell NB-LRR proteins exercise their functions of surveillance and downstream activation. In particular, the approaches therein can be applied to study dynamic relocalization of NB-LRRs in response to pathogen invasion (e.g., (22)). Chapter 2 describes a very innovative "fragment complementation" approach for understanding the functions of intramolecular interactions between different NB-LRR functional domains (e.g., (23)). Chapters 3 and 4 provide new protocols for the oft-vexing process of purifying low-abundance protein complexes. These protocols were developed with the immediate goal of identifying the components within immune surveillance complexes (e.g., (24, 25)), but are also potentially applicable to any protein expressed *in planta*.

Chapters 5–7 have similarly broad applicability. Chapter 5 describes chromatin immunoprecipitation, which is being widely used to characterize protein–DNA interactions in vivo and identify targets of transcription factors in a variety of organisms (e.g., (26)). This chapter was written in reference to WRKY transcription factors, which are ubiquitous in plants and are key regulators of immunity and other plant processes. The procedures could be adapted for other plant proteins (e.g., NB-LRR proteins that function inside the nucleus) or for pathogen effector proteins that mimic plant transcription factors. Chapter 6 provides new information on an inducible system for plant transgene expression that is frequently used in studies of plant immunity (e.g., for expressing effector proteins *in planta* (27)). This chapter helps researchers maximize the versatility of this system and clearly understand its limitations. Chapter 7 describes a creative method for detection and quantification of alternatively spliced transcripts (28). Alternative splicing is important for the regulation of some NB-LRR resistance gene regulation and is currently understudied with respect to immune system function (29).

Chapters 8–17 describe methods used to identify and functionally characterize pathogen effector proteins. As mentioned above, pathogen genomics have revealed a plethora of candidate effectors. Understanding how they function is one of the most active areas in plant–pathogen research at present (15, 16). One emergent generality is that almost all types of pathogens deploy moderate to large batallions of secreted effectors, many of which operate inside plant cells. Chapters 9 and 10 provide approaches to isolate plant cells that are in intimate contact with fungi and nematodes, respectively. These cells can serve as sources for cDNA libraries that are enriched for transcripts encoding effectors (e.g., (30, 31)). This is a proven approach toward effector gene discovery for pathogens with no reference genome sequence.

The bacterium *Pseudomonas syringae* has been at the forefront of effector characterization, and Chap. 10 describes methods whereby single or multiple gene knockouts can be constructed. This approach is vital to establish loss-of-function phenotypes, deconvolute the redundancy in effector repertoires, and evaluate the contribution of effectors to bacterial host range (e.g., (32)). In the eukaryotic kingdom, oomycetes from the *Phytophthora* genus have been at the forefront of effector identification; however, transformation of *Phytophthora* is often challenging even for experienced labs (33). Chapter 11 provides procedures for transformation of *P. capcisi*, which appears more amenable to genetic manipulation and can infect *N. benthamiana* and defense-compromised *Arabidopsis* mutants. Chapter 12 describes procedures pertaining to a second oomycete, *Hyaloperonsopora arabidopsidis*, that has long been used as a model pathogen of *Arabidopsis* and is becoming even more widely used for oomycete comparative genomes and investigation of oomycete effector proteins (34).

Bacteria deploy dozens of effectors, and oomycetes, fungi, and nematodes likely produce many-fold more (15, 16). To facilitate functional characterization of large collections of effectors, several high-throughput assays have been recently developed. Two such assays, presented in Chaps. 13 and 14, can be used to estimate immune-suppressive capacity of effectors from almost any pathogen (e.g., (35–37)). Chapter 15 describes a transient expression system optimized for protein complex purification, similar to Chaps. 3 and 4, that can be applied at medium-throughput to identify plant proteins which interact with pathogen effectors (or other types of protein interactions *in planta*). Chapters 16 and 17 describe approaches for visualizing subcellular localization of effectors in plant cells, which is of key importance for understanding effector function (e.g., (38)).

In the final chapters, the focus returns to the plant at a fine spatial scale. One aspect of plant–pathogen interactions that has not been adequately addressed relates to spatial differences in the molecular responses of plant cells in different locations of the infected organ, relative to pathogen infection structures. Chapters 18–20 provide information on laser microdissection, which is one of the most promising technologies for addressing questions relating to spatio-temporal differences in different cell types in infected organs (e.g., (39)). Finally, Chap. 21 zooms in even further (completing a spatial circle with Chap. 1) to describe exciting approaches to visualize subcellular dynamics in infected cells (e.g., (40)). This is undoubtedly one of the major emerging areas in plant–microbe interactions in the upcoming years (41).

The authors of these chapters sincerely hope that our contributions are of use, and we wish readers the best of success in applying these methods to their favorite pathosystems. We also look forward to the next volume(s) in this series that address plant–microbe interactions. Perhaps the next volume describes new technologies for structural studies of immune receptor complexes, along with advanced proteomic and metabolomic surveys of infected tissue at fine spatial scales. A major challenge will be to integrate data from disparate approaches, with the potential payoff being holistic models of infected cells, tissues, and organs. It would be particularly valuable to understand regulatory connections between immunity and other plant processes that might predict undesirable side effects of engineered resistance strategies (e.g., yield loss, reduced resistance to abiotic stress). It is exciting to imagine that such depth of understanding might even prompt a subsequent Methods volume focusing on "translational" approaches; for example, bioinformatic approaches to efficiently identify durable resistance genes for breeding or transgenics, or even surveillance genes that are custom-designed to detect PAMPs or indispensable pathogen effectors. Is this far-fetched? Perhaps...but if we were plant breeders in the 1950s could we have anticipated the depth of understanding that has already been achieved in only five short decades?

Blackburg, VA ***John M. McDowell***

References

1. Ronald, P. (Ed.) (2007) Plant–Pathogen Interactions, Springer, New York.
2. Flor, H. H. (1956) The complementary genetic systems in flax and flax rust, *Adv Genet* **8**, 29–54.
3. Flor, H. H. (1971) Current status of the gene-for-gene concept, *Annu Rev Phytopathol* **9**, 275–96.
4. Dangl, J. L., and McDowell, J. M. (2006) Two modes of pathogen recognition by plants, *Proc Natl Acad Sci U S A* **103**, 8575–6.
5. Keen, N. T. (1990) Gene-for-gene complementarity in plant–pathogen interactions, *Annu Rev Genet* **24**, 447–63.
6. Baker, B., Zambryski, P., Staskawicz, B., and Dinesh-Kumar, S. P. (1997) Signaling in plant–microbe interactions, *Science* **276**, 726–33.
7. da Cunha, L., Sreerekha, M. V., and Mackey, D. (2007) Defense suppression by virulence effectors of bacterial phytopathogens, *Curr Opin Plant Biol* **10**, 349–57.
8. Dangl, J. L., and Jones, J. D. (2001) Plant pathogens and integrated defence responses to infection, *Nature* **411**, 826–33.
9. Jones, J. D., and Dangl, J. L. (2006) The plant immune system, *Nature* **444**, 323–9.
10. van der Hoorn, R. A., and Kamoun, S. (2008) From Guard to Decoy: a new model for perception of plant pathogen effectors, *Plant Cell* **20**, 2009–17.
11. Boller, T., and Felix, G. (2009) A renaissance of elicitors: perception of microbe-associated

molecular patterns and danger signals by pattern-recognition receptors, *Annu Rev Plant Biol* **60**, 379–406.

12. Chisholm, S. T., Coaker, G., Day, B., and Staskawicz, B. J. (2006) Host–microbe interactions: shaping the evolution of the plant immune response, *Cell* **124**, 803–14.
13. Meyers, B. C., Kozik, A., Griego, A., et al. (2003) Genome-wide analysis of NBS-LRR-encoding genes in Arabidopsis, *Plant Cell* **15**, 809–34.
14. McDowell, J., and Simon, S. (2006) Recent insights into R gene evolution, *Mol Plant Pathol* 7, 437–48.
15. Hogenhout, S. A., Van der Hoorn, R. A., Terauchi, R., and Kamoun, S. (2009) Emerging concepts in effector biology of plant-associated organisms, *Mol Plant Microbe Interact* **22**, 115–22.
16. Collmer, A., Schneider, D. J., and Lindeberg, M. (2009) Lifestyles of the effector rich: genome-enabled characterization of bacterial plant pathogens, *Plant Physiol* **150**, 1623–30.
17. Tao, Y., Xie, Z., Chen, W., et al. (2003) Quantitative nature of Arabidopsis responses during compatible and incompatible interactions with the bacterial pathogen *Pseudomonas syringae*, *Plant Cell* **15**, 317–30.
18. Wang, L., Tsuda, K., Sato, M., et al. (2009) Arabidopsis CaM binding protein CBP60g contributes to MAMP-induced SA accumulation and is involved in disease resistance against *Pseudomonas syringae*, *PLoS Pathog* **5**, e1000301.
19. de Torres-Zabala, M., Truman, W., Bennett, M. H., et al. (2007) *Pseudomonas syringae* pv. tomato hijacks the Arabidopsis abscisic acid signalling pathway to cause disease, *EMBO J* **26**, 1434–43.
20. Caplan, J., Padmanabhan, M., and Dinesh-Kumar, S. P. (2008) Plant NB-LRR immune receptors: from recognition to transcriptional reprogramming, *Cell Host Microbe* **3**, 126–35.
21. Collier, S. M., and Moffett, P. (2009) NB-LRRs work a "bait and switch" on pathogens, *Trends Plant Sci* **14**, 521–9.
22. Burch-Smith, T. M., Schiff, M., Caplan, J. L., et al. (2007) A novel role for the TIR domain in association with pathogen-derived elicitors, *PLoS Biol* **5**, e68.
23. Rairdan, G. J., Collier, S. M., Sacco, M. A., et al. (2008) The coiled-coil and nucleotide binding domains of the Potato Rx disease resistance protein function in pathogen recognition and signaling, *Plant Cell* **20**, 739–51.
24. Liu, J., Elmore, J., Fuglsang, A., et al. (2009) RIN4 functions with plasma membrane H-ATPases to regulate stomatal apertures during pathogen attack, *PLoS Biol* 7, e1000139.
25. Qi, Y., and Katagiri, F. (2009) Purification of low-abundance Arabidopsis plasma-membrane protein complexes and identification of candidate components, *Plant J* **57**, 932–44.
26. Turck, F., Zhou, A., and Somssich, I. E. (2004) Stimulus-dependent, promoter-specific binding of transcription factor WRKY1 to Its native promoter and the defense-related gene PcPR1-1 in Parsley, *Plant Cell* **16**, 2573–85.
27. Kim, M. G., da Cunha, L., McFall, A. J., et al. (2005) Two *Pseudomonas syringae* type III effectors inhibit RIN4-regulated basal defense in Arabidopsis, *Cell* **121**, 749–59.
28. Zhang, X., and Gassmann, W. (2007) Alternative splicing and mRNA levels of the disease resistance gene RPS4 are induced during defense responses, *Plant Physiol* **145**, 1577–87.
29. Gassmann, W. (2008) Alternative splicing in plant defense, *Curr Top Microbiol Immunol* **326**, 219–34.
30. Catanzariti, A.-M., Dodds, P. N., Lawrence, G. J., et al. (2006) Haustorially expressed secreted proteins from flax rust are highly enriched for avirulence elicitors, *Plant Cell* **18**, 243–56.
31. Gao, B., Allen, R., Maier, T., et al. (2003) The parasitome of the phytonematode *Heterodera glycines*, *Mol Plant Microbe Interact* **16**, 720–26.
32. Kvitko, B. H., Park, D. H., Velasquez, A. C., et al. (2009) Deletions in the repertoire of *Pseudomonas syringae* pv. tomato DC3000 type III secretion effector genes reveal functional overlap among effectors, *PLoS Pathog* **5**, e1000388.
33. Fry, W. (2008) *Phytophthora infestans*: the plant (and R gene) destroyer, *Mol Plant Pathol* **9**, 385–402.
34. Holub, E. B. (2008) Natural history of *Arabidopsis thaliana* and oomycete symbioses, *Eur J Plant Pathol* **122**, 91–109.
35. Abramovitch, R. B., Kim, Y. J., Chen, S., et al. (2003) Pseudomonas type III effector AvrPtoB induces plant disease susceptibility by inhibition of host programmed cell death, *EMBO J* **22**, 60–9.
36. Dou, D., Kale, S. D., Wang, X., et al. (2008) Conserved C-terminal motifs required for avirulence and suppression of cell death by *Phytophthora sojae* effector Avr1b, *Plant Cell* **20**, 1118–33.

37. Jamir, Y., Guo, M., Oh, H. S., et al. (2004) Identification of *Pseudomonas syringae* type III effectors that can suppress programmed cell death in plants and yeast, *Plant J* **37**, 554–65.
38. Whisson, S. C., Boevink, P. C., Moleleki, L., et al. (2007) A translocation signal for delivery of oomycete effector proteins into host plant cells, *Nature* **450**, 115–8.
39. Ithal, N., Recknor, J., Nettleton, D., et al. (2007) Parallel genome-wide expression profiling of host and pathogen during soybean cyst nematode infection of soybean, *Mol Plant Microbe Interact* **20**, 293–305.
40. Koh, S., Andre, A., Edwards, H., et al. (2005) *Arabidopsis thaliana* subcellular responses to compatible *Erysiphe cichoracearum* infections, *Plant J* **44**, 516–29.
41. Koh, S., and Somerville, S. (2006) Show and tell: cell biology of pathogen invasion, *Curr Opin Plant Biol* **9**, 406–13.

Contents

Contributors

REX ALLEN • *Department of Plant Pathology, University of Georgia, Athens, GA, USA*
RYAN G. ANDERSON • *Department of Plant Pathology, Physiology, and Weed Science, Virginia Tech, Blacksburg, VA, USA*
PAUL R. J. BIRCH • *Division of Plant Sciences, College of Life Sciences, University of Dundee at SCRI, Invergowrie, Dundee, UK*
PETRA C. BOEVINK • *Plant Pathology Programme, Scottish Crop Research Institute, Invergowrie, Dundee, UK*
ANN-MAREE CATANZARITI • *Department of Plant and Microbial Biology, University of California, Berkeley, CA, USA*
DIVYA CHANDRAN • *Department of Plant and Microbial Biology, University of California, Berkeley, CA, USA*
GITTA COAKER • *Department of Plant Pathology, University of California, Davis, CA, USA*
ALAN COLLMER • *Department of Plant Pathology and Plant-Microbe Biology, Cornell University, Ithaca, NY, USA*
PATRICK COURNOYER • *Department of Molecular, Cellular, and Developmental Biology, Yale University, New Haven, CT, USA*
DANIEL DEEGAN • *Department of Plant Pathology, Physiology, and Weed Science, Virginia Tech, Blacksburg, VA, USA*
S. P. DINESH-KUMAR • *Department of Molecular, Cellular, and Developmental Biology, Yale University, New Haven, CT, USA*
PETER DODDS • *Division of Plant Industry, Commonwealth Scientific and Industrial Research Organisation, Canberra, ACT, Australia*
JEFF ELLIS • *Division of Plant Industry, Commonwealth Scientific and Industrial Research Organisation, Canberra, ACT, Australia*
JAMES M. ELMORE • *Department of Plant Pathology, University of California, Davis, CA, USA*
WALTER GASSMANN • *Division of Plant Sciences, Christopher S. Bond Life Sciences Center, University of Missouri, Columbia, MO, USA*
XUEQING GENG • *Department of Horticulture and Crop Science, The Ohio State University, Columbus, OH, USA; Department of Plant Cellular and Molecular Biology, The Ohio State University, Columbus, OH, USA*
GREG HATHER • *Department of Plant and Microbial Biology, University of California, Berkeley, CA, USA; Department of Statistics, University of California, Berkeley, CA, USA*
TROY HOFF • *Department of Plant Pathology, Physiology, and Weed Science, Virginia Tech, Blacksburg, VA, USA*

GUOZHONG HUANG • *Department of Plant Pathology, University of Georgia, Athens, GA, USA*
QIAN HUANG • *Plant Pathology Department, Nanjing Agricultural University, Nanjing, China*
EDGAR HUITEMA • *The Sainsbury Laboratory, John Innes Centre, Norwich, UK*
RICHARD S. HUSSEY • *Department of Plant Pathology, University of Georgia, Athens, GA, USA*
NORIKO INADA • *Department of Plant and Microbial Biology, University of California, Berkeley, CA, USA*
NAGABHUSHANA ITHAL • *Division of Plant Sciences, Bond Life Sciences Center, University of Missouri, Columbia, MO, USA*
ALEXANDRA M. E. JONES • *The Sainsbury Laboratory, John Innes Centre, Norwich, UK*
SHIV D. KALE • *Virginia Bioinformatics Institute, Virginia Tech, Blacksburg, VA, USA*
SOPHIEN KAMOUN • *The Sainsbury Laboratory, John Innes Centre, Norwich, UK*
FUMIAKI KATAGIRI • *Department of Plant Biology, University of Minnesota, St. Paul, MN, USA*
ERIC KEMEN • *The Sainsbury Laboratory, John Innes Centre, Norwich, UK*
SERRY KOH • *Plant Systems Engineering Research Center, Korea Research Institute of Bioscience and Biotechnology, Daejeon, Republic of Korea*
BRIAN H. KVITKO • *Department of Plant Pathology and Plant-Microbe Biology, Cornell University, Ithaca, NY, USA*
DAVID MACKEY • *Department of Horticulture and Crop Science, The Ohio State University, Columbus, OH, USA*
Department of Plant Cellular and Molecular Biology, The Ohio State University, Columbus, OH, USA
ROHIT MAGO • *Division of Plant Industry, Commonwealth Scientific and Industrial Research Organisation, Canberra, ACT, Australia*
JOHN M. MCDOWELL • *Department of Plant Pathology, Physiology, and Weed Science, Virginia Tech, Blacksburg, VA, USA*
KURT MENDGEN • *Phytopathologie, Fachbereich Biologie, Universität Konstanz, Konstanz, Germany*
MELISSA G. MITCHUM • *Division of Plant Sciences, Bond Life Sciences Center, University of Missouri, Columbia, MO, USA*
PETER MOFFETT • *Boyce Thompson Institute for Plant Research, Ithaca, NY, USA*
YIPING QI • *Department of Plant Biology, University of Minnesota, St. Paul, MN, USA*
MARIO ROCCARO • *Department of Plant Microbe Interaction, Max Planck Institute for Plant Breeding, Cologne, Germany*
MATTHEW SMOKER • *The Sainsbury Laboratory, John Innes Centre, Norwich, UK*
SHAUNA C. SOMERVILLE • *Department of Plant and Microbial Biology, Energy Biosciences Institute, University of California, Berkeley, CA, USA*
IMRE E. SOMSSICH • *Department of Plant Microbe Interaction, Max Planck Institute for Plant Breeding, Cologne, Germany*
BRETT M. TYLER • *Virginia Bioinformatics Institute, Virginia Tech, Blacksburg, VA, USA*

WILLIAM UNDERWOOD • *Energy Biosciences Institute, University of California, Berkeley, CA, USA*
RALF T. VOEGELE • *Phytopathologie, Fachbereich Biologie, Universität Konstanz, Konstanz, Germany*
YUANCHAO WANG • *Plant Pathology Department, Nanjing Agricultural University, Nanjing, China*
STEPHEN C. WHISSON • *Plant Pathology Programme, Scottish Crop Research Institute, Invergowrie, Dundee, UK*
MARY C. WILDERMUTH • *Department of Plant and Microbial Biology, University of California, Berkeley, CA, USA*
JOE WIN • *The Sainsbury Laboratory, John Innes Centre, Norwich, UK*
XUE-CHENG ZHANG • *Division of Plant Sciences, Christopher S. Bond Life Sciences Center, University of Missouri, Columbia, MO, USA*

Chapter 1

Studying NB-LRR Immune Receptor Localization by Agroinfiltration Transient Expression

Patrick Cournoyer and S.P. Dinesh-Kumar

Abstract

NB-LRR immune receptors in plants play dual roles as sentries and as activators of defense. The site in the cell where these activities take place can be different for different NB-LRRs. Furthermore, recognition and defense activation can occur in distinct subcellular compartments. Therefore, determining the subcellular localization of NB-LRRs is a key step toward understanding how they function. Recent advances in confocal microscopy enable high-resolution imaging of proteins in live cells. Agroinfiltration in the *Nicotiana benthamiana* model plant system is a convenient way of expressing proteins for localization studies. This chapter explains how to use *N. benthamiana* to transiently express NB-LRRs for confocal fluorescence microscopy.

Key words: Localization, NB-LRR, Confocal microscopy, R gene, Agroinfiltration, *Nicotiana*, Transient expression

1. Introduction

The plant innate immune system uses two tiers of defenses to protect plants from pathogens. In the first, cell-surface receptors recognize conserved molecular patterns shared by many microbes and induce basal immune responses. When pathogens circumvent or block basal immunity, intracellular NB-LRR immune receptors act as a second line of defense. NB-LRRs recognize specific pathogen effectors present in the cell. To be effective, NB-LRRs must reside where pathogen effectors localize or where they exert their influence. They must also reside at sites where they can promote signals that lead to defense activation. NB-LRR localization is therefore an area of research that is critical to understand how these receptors function. Biochemical fractionation has been used for investigating subcellular localization of NB-LRRs. Although

John M. McDowell (ed.), *Plant Immunity: Methods and Protocols*, Methods in Molecular Biology, vol. 712,
DOI 10.1007/978-1-61737-998-7_1, © Springer Science+Business Media, LLC 2011

biochemical fractionation is informative, fluorescent protein tagging combined with confocal fluorescence microscopy provides the advantage of seeing the protein's location and movement in live cells, in real time, and in three dimensions.

Microscopy has been used to study the localization of several NB-LRRs to-date. The N immune receptor from *Nicotiana* was shown to be nuclear and cytoplasmic, even though it does not possess a canonical nuclear localization signal (NLS). N's presence in the nucleus was shown to be required for function (1). The barley immune receptor MLA10 also lacks a canonical NLS and was shown to be nuclear and cytoplasmic. Like N, MLA10 nuclear localization is required for its function (2). Biochemical fractionation experiments showed that MLA1, a related NB-LRR, is present in the nucleus and appears to accumulate there at higher levels during a defense response (2).

RRS1-R from Arabidopsis is an unusual NB-LRR because it has a WRKY transcription factor domain and a canonical NLS. In the presence of its cognate pathogen effector, it was observed in the nucleus (3). Arabidopsis RPS4 is another NB-LRR that has a canonical NLS. Microscopy analyses have shown that RPS4 localizes to the nucleus and to areas of the cytoplasm (4). Its NLS is required for its nuclear localization and for its function. Biochemical fractionation showed that RPS4 associates with endomembranes and co-fractionates with an endoplasmic reticulum marker (4). RPP1-A and RPM1 are other NB-LRRs from Arabidopsis that are known to be membrane associated from biochemical fractionation experiments (5, 6). Microscopy could be used to examine the localization of these membrane-associated NB-LRRs more closely.

Based on the NB-LRRs examined to date, it is clear that different NB-LRRs can have different localization patterns. It is also notable that NB-LRRs sometimes have localizations that could not be predicted based on their sequence. Furthermore, it is plausible that NB-LRRs move between different compartments in the cell either constitutively or upon activation. Careful examination using confocal fluorescence microscopy has great potential to inform us about how NB-LRRs detect pathogens and trigger defense.

Expressing a protein of interest fused to a fluorescent protein in live cells is now routinely used in the study of protein localization. Many factors, however, must be considered when deciding on a strategy. The goal of any localization study is to replicate as closely as possible the protein's natural state. The best strategy depends on the transformation capabilities for the plant system under investigation. *Agrobacterium*-mediated transfection methods are favored for most dicot plants while particle bombardment is commonly used to transfect monocot cells.

1.1. Agroinfiltration

Agroinfiltration permits the rapid transformation of leaf cells without the need to recover transgenic lines. Agroinfiltrated leaf cells are ready for imaging 30–48 h after infiltration, compared to months required for generating transgenic lines. Imaging is possible before the onset of RNA silencing, which can be an obstacle for protein expression in transgenic lines. It is also possible to co-express multiple proteins in the same cells without difficulty. These advantages have made agroinfiltration a popular choice for localization studies.

Agroinfiltration is most commonly used to transfect leaves of *Nicotiana tabacum* and *Nicotiana benthamiana*. Some plants are not amenable to agroinfiltration because of unfavorable leaf architecture or because of defense mechanisms that protect them from Agrobacterium in leaves. *Nicotiana* is often used to study the localization of immune receptors from other taxa because of its convenience. While still informative, there are caveats to heterologous expression. NB-LRRs rarely function outside of their native plant family and therefore may or may not exhibit native localization patterns. Heterologous expression may also lead to spurious activation of programed cell death (PCD) as in the case of RPS4 expression in *Nicotiana* (7). In Arabidopsis leaves, Agrobacterium activates the EF-Tu receptor (EFR), resulting in basal defense responses that prevent efficient transfection (8). Arabidopsis *efr* mutants are amenable to agroinfiltration. Thus, localization of Arabidopsis NB-LRRs can be studied by transient expression without the need for a heterologous system.

1.2. Other Methods of Expressing Proteins for Localization

Generating transgenic lines expressing fluorescent protein fusions is the best way to test protein localization when agroinfiltration is not possible. Transgenic lines are also useful for confirming observations in agroinfiltrated leaves. Monocots are generally recalcitrant to Agrobacterium-mediated transformation and are often transformed by particle bombardment. For localization studies, transient expression by particle bombardment is possible (see ref. 2 for an example). Below, we focus on expression using agroinfiltration in *Nicotiana benthamiana*.

2. Materials

2.1. Creating an Expression Vector

1. Cloned NB-LRR of interest.
2. Fluorescent protein fusion expression vector.

2.2. Growing Nicotiana

1. *Nicotiana benthamiana* seeds.
2. 0.1% agarose, autoclaved.
3. Growing materials: Potting mix, pots, growth facilities.

2.3. Agroinfiltration

1. *N. benthamiana* seeds.
2. *Agrobacterium tumefaciens* strain GV2260.
3. Expression vector.
4. Infiltration medium: 10 mM $MgCl_2$, 10 mM MES (2-(*N*-Morpholino) ethane sulfonic acid), 200 μM acetosyringone. Prepare fresh. Use a 1 M stock solution of acetosyringone in dimethyl sulfoxide (DMSO), which can be stored in aliquots at −20°C.
5. 1-ml syringe.

2.4. Confocal Microscopy

1. Razor blade to excise leaf sample.
2. 10-ml syringe for imbibing sample.
3. Forceps.
4. Chamber with cover glass base (Lab-Tek II Chamber #1.5 coverglass system, Nalge Nunc International).
5. Glass or acrylic block to weigh leaf sample down against cover glass base.
6. Inverted confocal laser scanning fluorescence microscope. We use a Zeiss LSM 510 meta.

3. Methods

3.1. Expression Vector

1. Select expression vector to create a fusion protein with preferred fluorescent protein (see Note 1).
2. Insert NB-LRR gene sequence in frame with fluorescent protein. The strong, constitutive 35S promoter is commonly used to drive protein expression. It is preferable, however, to express proteins in their genomic regulatory context, which better replicates native expression (see Note 2).
3. Transform *A. tumefaciens* strain GV2260 with vector (see Note 3).
4. Show that the tagged NB-LRR retains functionality. One way to do this is to show that transient co-expression of the tagged NB-LRR with its cognate pathogen effector specifically results in hypersensitive response PCD. Another way is to show that leaf tissue transiently expressing the tagged NB-LRR has resistance to the pathogen. The most stringent test for functionality is to create a stable transgenic line in a background lacking the functional NB-LRR allele and demonstrate resistance.
5. To confirm the NB-LRR fluorescent protein fusion is stable and is expressed properly, analyze agroinfiltrated tissue by western blot. If low molecular weight proteins are detected,

free fluorescent protein may be present. Free fluorescent protein makes microscopy images uninterpretable, since they cannot be distinguished from the labeled NB-LRR.

3.2. Plant Growth

1. Suspend *N. benthamiana* seeds in 0.1% agarose. Incubate at room temperature to facilitate germination.
2. Germinate seeds by pipetting onto moist potting mix in 4" pots. Use a clear cover over to maintain high humidity until seedlings are well established.
3. Keep plants at 23–25°C with at least 30% humidity. *Nicotiana* usually benefits from fertilizer, especially after the four-leaf stage.
4. Plants are ready for agroinfiltration when they are 4–5 weeks old and have six to eight true leaves (see Note 4).

3.3. Agroinfiltration

1. Inoculate 25 ml culture of LB containing appropriate selection in a 125-ml Erlenmeyer flask with a single colony of *Agrobacterium* harboring the expression vector. Incubate culture at 26–27°C overnight with shaking.
2. Centrifuge culture 1,800 × *g* for 10–15 min to collect cells.
3. Resuspend cells in 1-ml agroinfiltration medium. Adjust OD_{600} to 1.0 using agroinfiltration medium (see Note 5).
4. Induce cells by incubating at room temperature with gentle agitation for 2 h.
5. If multiple constructs are to be co-expressed in the same tissue, Agrobacterium suspensions harboring the desired constructs may be mixed (see Note 6).
6. Mark areas to be infiltrated by drawing a circle on the uppermost expanded leaves of healthy *Nicotiana* plants. Use a 1-ml needleless syringe to draw up Agrobacterium cell suspension and infiltrate into the abaxial leaf surface to fill the marked area. Keep firm, gentle pressure between the leaf and syringe by placing a finger on the leaf opposite of the syringe. Lightly touching the abaxial leaf surface with the corner of a razor blade facilitates infiltration (see Note 7).
7. Keep plants in the light for 12 h after agroinfiltration at temperatures near, but not exceeding 22°C (see Note 8).
8. Place plants under normal growing conditions until microscopy.

3.4. Preparing Samples for Microscopy

1. 30–48 h after infiltration, prepare for microscopy by excising an area of infiltrated leaf (see Note 9).
2. Imbibe the sample to fill intercellular spaces with water. To do this, place the sample in a 10-ml syringe partially filled with water. Seal the syringe opening with a finger, and create a vacuum by extending the plunger. Quickly release the opening while ensuring that the leaf remains submerged. You will notice the leaf become darker green.

3. Place the leaf sample in the chamber (see Note 10). Cover the sample with water to ensure that no air remains trapped beneath the sample (see Note 11).
4. Weigh down sample to ensure that it remains pressed down with a clean, flat object such as a small block of glass or acrylic.

3.5. Interpreting Localization Observations

1. Co-expressing the NB-LRR of interest with a fluorescent protein-tagged organellar marker is often the best way to confirm organellar localization. Counterstaining with cell-permeable fluorescent stains is usually more technically difficult and produces poorer quality images. For example, a fluorescent protein-NLS fusion is a good nuclear marker. Visualizing chloroplasts is particularly easy because they autofluoresce (650–800 nm) when excited with a wide range of wavelengths.
2. Interfering with NB-LRR subcellular targeting can show the functional significance of localization to a particular compartment or organelle. This can be done by mutating a known targeting motif like an NLS, chloroplast transit peptide, or ER retention signal and observing a change in localization. If the mechanism of localization is unknown, a more creative approach is needed. One example is to append an NES to minimize nuclear accumulation or an ER retention signal to prevent progression to the Golgi or plasma membrane.
3. Photobleaching experiments can show the extent to which NB-LRR diffusion within a particular cellular compartment is impaired by binding. Fluorescence recovery after photobleaching (FRAP) can show the proportion of protein that is immobile, determine if the diffusion of protein is slowed by participation in complexes, and get an impression of transient associations. For guidance on performing FRAP experiments, consult (9) (see Note 12).
4. If an NB-LRR is localized to more than one compartment, shuttling between them may occur. Whole-compartment FRAP is a useful technique for showing this. For guidance with whole-compartment FRAP, see ref. 10. A similar technique, fluorescence loss in photobleaching (FLIP), can show exchange between different compartments (11).

4. Notes

1. We have compared the suitability of several fluorescent proteins for microscopy in *Nicotiana*, including EGFP, citrine, and cerulean. We obtained the best results from citrine (12). We have had more success with C-terminal fluorescent

protein fusions with NB-LRRs than with N-terminal fusions. N-terminal fusions are more likely to disrupt localization than C-terminal fusions because some canonical signal peptides are located at the N terminus of proteins.

2. Preserving NB-LRR genomic context has several advantages over expression with a strong constitutive promoter. Sometimes, protein localization is mediated by transport machinery or by retention that is saturable. Therefore, overexpression has the potential to result in mislocalization. Overexpression can also result in protein misfolding and accumulation in insoluble aggregates. In addition, some NB-LRR transcripts undergo alternative splicing that is necessary for function and depends on the genomic context of the gene. Low expression, however, can make imaging difficult. If expression is too low a strong promoter may be needed.
3. *Agrobacterium* strain GV2260 is best for *N. benthamiana*. Strain GV3101 can also be used in *Nicotiana* and is the preferred strain for agroinfiltration into *Arabidopsis*. GV2260 is derived from C58C1-RS and has resistance to rifampicin and streptomycin. It harbors a helper plasmid that carries the *vir* genes and confers carbenicillin resistance.
4. Use only very healthy plants for expression. Plants with any signs of stress or chlorosis have lower expression than healthy plants.
5. The OD_{600} can be optimized depending on the expression level of the NB-LRR. Because expression of the N immune receptor is relatively low, we use $OD_{600} = 1.8$.
6. Optimal starting OD_{600} and mixing ratios for co-expression should be determined empirically.
7. Young leaves provide the best expression. We use the two uppermost fully expanded leaves for agroinfiltration.
8. Agrobacterium-mediated transformation of *Nicotiana* is most efficient at 22°C, with efficiency declining rapidly with increasing temperature (13). Darkness is inhibitory for *Agrobacterium*-mediated transformation, so keep plants in light for at least 12 h after infiltration (14).
9. Different proteins have varying rates of maturation. Therefore, we advise observing protein accumulation over a time course. Expression levels can be estimated by direct observation at the microscope or by quantitative western blotting. After 48 h, RNA silencing may reduce expression.
10. There may be differences between expression levels on the upper and lower epidermis. It is best to look at both and decide which is better on a case-by-case basis.

11. It is advantageous to exclude air from the path of light. Using a water immersion objective lens is best because the refractive index of fluids between the lens and the cells is nearly constant.
12. Expression of some NB-LRRs can be too low for FRAP when driven by genomic regulatory sequences.

Acknowledgments

An NIH grant supports the work on NB-LRR immune receptors in the S.P.D.-K., laboratory.

References

1. Burch-Smith, T.M., Schiff, M., Caplan, J.L., Tsao, J., Czymmek, K. and Dinesh-Kumar, S.P. (2007) A novel role for the TIR domain in association with pathogen-derived elicitors. *PLoS Biol.* **5**, e68.
2. Shen, Q.H., Saijo, Y., Mauch, S., Biskup, C., Bieri, S., Keller, B., Seki, H., Ulker, B., Somssich, I.E. and Schulze-Lefert, P. (2007) Nuclear activity of MLA immune receptors links isolate-specific and basal disease-resistance responses. *Science.* **315**, 1098–1103.
3. Deslandes, L., Olivier, J., Peeters, N., Feng, D.X., Khounlotham, M., Boucher, C., Somssich, I., Genin, S. and Marco, Y. (2003) Physical interaction between RRS1-R, a protein conferring resistance to bacterial wilt, and PopP2, a type III effector targeted to the plant nucleus. *Proc. Natl Acad. Sci. U. S. A.* **100**, 8024–8029.
4. Wirthmueller, L., Zhang, Y., Jones, J.D. and Parker, J.E. (2007) Nuclear accumulation of the arabidopsis immune receptor RPS4 is necessary for triggering EDS1-dependent defense. *Curr. Biol.* **17**, 2023–2029.
5. Weaver, L.M., Swiderski, M.R., Li, Y. and Jones, J.D. (2006) The *Arabidopsis thaliana* TIR-NB-LRR R-protein, RPP1A; protein localization and constitutive activation of defence by truncated alleles in tobacco and arabidopsis. *Plant J.* **47**, 829–840.
6. Boyes, D.C., Nam, J. and Dangl, J.L. (1998) The *Arabidopsis thaliana RPM1* disease resistance gene product is a peripheral plasma membrane protein that is degraded coincident with the hypersensitive response. *Proc. Natl Acad. Sci. U. S. A.* **95**(26), 15849–15854.
7. Zhang, Y., Dorey, S., Swiderski, M. and Jones, J.D. (2004) Expression of RPS4 in tobacco induces an AvrRps4-independent HR that requires EDS1, SGT1 and HSP90. *Plant J.* **40**, 213–224.
8. Zipfel, C., Kunze, G., Chinchilla, D., Caniard, A., Jones, J.D., Boller, T. and Felix, G. (2006) Perception of the bacterial PAMP EF-tu by the receptor EFR restricts agrobacterium-mediated transformation. *Cell.* **125**, 749–760.
9. van Royen, M.E., Farla, P., Mattern, K.A., Geverts, B., Trapman, J. and Houtsmuller, A.B. (2009) Fluorescence recovery after photobleaching (FRAP) to study nuclear protein dynamics in living cells. *Methods Mol. Biol.* **464**, 363–385.
10. Koster, M., Frahm, T. and Hauser, H. (2005) Nucleocytoplasmic shuttling revealed by FRAP and FLIP technologies. *Curr. Opin. Biotechnol.* **16**, 28–34.
11. van Drogen, F. and Peter, M. (2004) Revealing protein dynamics by photobleaching techniques. *Methods Mol. Biol.* **284**, 287–306.
12. Griesbeck, O., Baird, G.S., Campbell, R.E., Zacharias, D.A. and Tsien, R.Y. (2001) Reducing the environmental sensitivity of yellow fluorescent protein. Mechanism and applications. *J. Biol. Chem.* **276**, 29188–29194.
13. Dillen, W., DeClercq, J., Kapila, J., Zambre, M., VanMontagu, M. and Angenon, G. (1997) The effect of temperature on agrobacterium tumefaciens-mediated gene transfer to plants. *Plant J.* **12**, 1459–1463.
14. Zambre, M., Terryn, N., De Clercq, J., De Buck, S., Dillen, W., Van Montagu, M., Van Der Straeten, D. and Angenon, G. (2003) Light strongly promotes gene transfer from agrobacterium tumefaciens to plant cells. *Planta.* **216**, 580–586.

Chapter 2

Fragment Complementation and Co-immunoprecipitation Assays for Understanding R Protein Structure and Function

Peter Moffett

Abstract

Plant disease resistance (R) proteins confer protection against specific pathogens or pathogen isolates. R proteins function by recognizing pathogen-encoded avirulence (Avr) proteins and translating this recognition event into an initiation of downstream signaling pathways. Key to understanding this process is the study of the protein–protein interactions involving R proteins. Recognition and signaling mechanisms are mediated by both intramolecular interactions that take place between different domains of R proteins as well as intermolecular interactions between R proteins and additional plant proteins. These processes have been studied in part by using *Agrobacterium*-mediated transient expression of R protein fragments in *Nicotiana benthamiana* which allows for the rapid assessment of functionality. Furthermore, pairs of proteins or protein fragments can be transiently expressed as fusions with different epitope tags. One putative protein partner is subjected to immunoprecipitation. Subsequent immunoblotting is performed to determine whether the second protein has remained associated (or co-immunoprecipitated) with the first, indicating a protein–protein interaction. This technique has contributed substantially to structure–function analyses of R proteins and to the characterization of interactions between R proteins and other plant proteins.

Key words: *Agrobacterium*, Agroinfiltration, NB-LRR, Immunoprecipitation, Transient expression, *Nicotiana benthamiana*

1. Introduction

Agrobacterium tumefaciens transfers T-DNA to plant cells whereupon it migrates to the nucleus and becomes transcriptionally competent (1). In some species, this process is so efficient that most or all cells become transiently transformed within a leaf patch that has been syringe-infiltrated with a suspension of *Agrobacterium*. The use of *Agrobacterium*-mediated transient expression (agroexpression or agroinfiltration) has become a

John M. McDowell (ed.), *Plant Immunity: Methods and Protocols*, Methods in Molecular Biology, vol. 712,
DOI 10.1007/978-1-61737-998-7_2, © Springer Science+Business Media, LLC 2011

widely adopted technique for functional genomics in plants. Agroinfiltration is highly efficient in the model tobacco species *Nicotiana benthamiana* and *N. tabacum* as well as most other *Nicotiana* species we have tested. Agroinfiltration can also be applied to a number of other plant species (2), albeit often with much lower efficiency that may preclude biochemical analyses of the transiently expressed proteins. The high efficiency of agroinfiltration in *N. benthamiana* allows for the ability to co-express multiple proteins in the same infiltrated patch by mixing *Agrobacterium* suspensions. Furthermore, protein expression levels can be modulated by using native promoters versus strong constitutive or inducible promoters, or by co-expression with suppressors of gene silencing (3). Agroinfiltration has been very informative in the study of the recognition of pathogen-encoded Avirulence (Avr) proteins by plant disease resistance (R) proteins, particularly those encoding NB-LRR proteins (4). Since co-expression of matching R and Avr proteins is sufficient to induce a form of cell death known as the hypersensitive response (HR) agroinfiltration allows for rapid functional analysis of R and Avr protein derivatives (5–7).

1.1. NB-LRR Protein Fragment Complementation

One of the first insights into the molecular mechanisms by which plant NB-LRR proteins function was the observation that the Rx and Bs2 CC-NB-LRR proteins could undergo fragment complementation. Agroinfiltration-mediated co-expression of fragments encoding different domains of these proteins, either CC-NB plus LRR or CC plus NB-LRR, recapitulated the function of the full-length protein in conferring an Avr-dependent HR (8). The ability of these fragments to undergo functional complementation was subsequently shown to be due to the fact that these fragments physically interacted as demonstrated by co-immunoprecipitation (co-IP) of the transiently expressed fragments (8, 9). Subsequent use of these techniques has allowed researchers to fine-map the intramolecular interactions that take place between the different domains of CC-NB-LRR proteins and propose molecular models of NB-LRR protein function (10, 11).

1.2. Co-immuno-precipitation

Structure–function analyses of protein–protein interactions between known proteins or protein fragments can be undertaken rapidly by generating fusion proteins with epitope tags and making use of the commercially available antibodies that recognize them. Due to their small size, a single epitope tag can often be simply incorporated into a PCR primer allowing for rapid construction of protein derivatives for study. For example, we have co-expressed a series of differentially tagged fragments of the Rx protein and derivatives thereof in *N. benthamiana* leaves to undertake structure–function analyses of the physical interactions that take place within the Rx protein (8, 10, 11). In this method, protein is extracted from leaves co-expressing two differentially tagged fragments and antibodies recognizing one epitope are used to immunoprecipitate one of the fragments. Subsequent immunoblotting with antibodies recognizing the other epitope tag can then be used to probe for the presence of the second fragment in the immunoprecipitate, indicating a physical association between the two fragments. Similar immunoprecipitation methods can be adapted to the identification of unknown proteins using transiently expressed R protein fragments and candidates can be validated using epitope-tagged versions of such candidates (12, 13). The efficiency of agroinfiltration in *N. benthamiana* allows for the co-expression of multiple proteins (up to five in our experience). As such, the co-IP of multiprotein complexes is possible as well as the investigation of how a given protein–protein interaction is affected by the presence of a third protein. Below is a method for investigating the interaction between HA-tagged and FLAG-tagged proteins or protein fragments.

2. Materials

2.1. Agroinfiltration

1. *A. tumefaciens* strains containing a binary vector (see Note 1) encoding epitope-tagged (see Note 2) versions of the proteins or protein fragments of interest.
2. Luria–Bertani (LB) broth and agar plates supplemented with appropriate antibiotics for binary vector selection in *Agrobacterium*.
3. Incubators and shakers at 28°C.
4. High-speed table top or floor model centrifuge.
5. Infiltration buffer (see Note 3): 10 mM $MgCl_2$, 10 mM MES pH 5.6, 100 μM acetosyringone (added immediately prior to use from a 100 mM stock in DMSO).
6. Spectrophotometer for measuring optical density of *Agrobacterium* cultures.
7. Slip tip syringes, 1 or 2 mL according to preference.
8. Four- to six-week-old *N. benthamiana* plants, grown at 20–23°C, 50% humidity, 16 h/8 h light/dark cycle.

2.2. Protein Extraction

1. Pre-chilled mortars and pestles.
2. Extraction buffer: GTEN (10% (v/v)glycerol, 25 mM Tris pH 7.5, 1 mM EDTA, 150 mM NaCl), 10 mM DTT, 2% (w/v) PVPP (polyvinylpolypyrrolidone), 1× protease inhibitor cocktail (see Note 4).
3. Refrigerated microcentrifuge.

4. Desalting matrix (Bio-Gel P6 DG; Bio-Rad, 150-0738), screening columns (Fisher, 11-387-50) and 15-mL snap-cap tubes (optional; see Note 5).

2.3. Immunoprecipitation

1. 10% (v/v) Nonidet P-40 (NP-40, sold as Igepal CA-630 by Sigma; should be less than 1 week old). Should be prepared at least 1 h beforehand as NP-40 takes time to mix with water.
2. IP buffer: GTEN, 0.15% (v/v) NP-40, 0.5 mM DTT.
3. Agarose conjugated Goat IgG (Rockland, 005-0050; see Note 6).
4. EZview™ Red anti-HA and anti-FLAG affinity gels (Sigma, E6779 and F2426; see Note 6).
5. Rotating microtube mixer.
6. 1-mL syringe with 25 Ga needle.
7. 1× SDS–PAGE loading buffer: 100 mM DTT, 2% (w/v) SDS, 10% (v/v) glycerol, 50 mM Tris–HCL pH 6.8, 0.02% (w/v) bromophenol blue.

2.4. Immunoblotting

1. Any suitable SDS–PAGE and proteins transfer blotting apparatus and appropriate buffers (see Note 7).
2. 100% methanol.
3. PVDF membrane (Bio-Rad, 162-0177).
4. ANTI-FLAG® M2 monoclonal antibody peroxidase conjugate (Sigma, A8592) and Anti-HA (3F10) antibody peroxidase conjugate (Roche, 12013819001).
5. TBST: 25 mM Tris–HCL pH 7.5, 150 mM NaCl, 0.1% (v/v) Tween-20.
6. Blocking buffer: TBST, 5% (w/v) skim milk powder.
7. Amersham ECL Plus™ western blotting detection reagents (GE healthcare).
8. Saran Wrap, X-ray film, and developing machine.

3. Methods

Prior to co-IP the proteins or protein fragments of interest should be functionally validated. For R protein fragments, this may involve their co-expression with their cognate Avr protein. For example, co-expression of the Rx CC and NB-LRR fragments plus the coat protein of potato virus X results in a macroscopic HR in infiltrated patches within 2–3 days (8). Protein accumulation should be verified and an optimal time point determined for protein extraction. Likewise, optimal extraction buffers should be determined before initiating co-IPs (see Note 4). In addition, appropriate controls for the co-IP should be identified. Ideally, these should be proteins similar to those under study but that do not interact with the protein(s) of interest. For example, the Rx CC domain interacts with the Rx NB-LRR fragment, whereas the Bs2 CC domain does not (8) make the latter an appropriate negative control when demonstrating an interaction between the Rx CC and NB-LRR fragments. Likewise, the RanGAP2 protein interacts with the Rx CC domain, whereas the similar RanGAP1 protein does not, and neither RanGAP1 nor RanGAP2 interact with the Bs2 CC domain (12, 13). In the absence of a highly similar noninteracting protein, it should at least be demonstrated that the protein of interest does not simply interact with the antibody-conjugated beads or any randomly tagged protein. As such, a tagged version of a protein that would not be expected to interact, such as GUS or GFP, can be used as a negative control.

Ideally, it is desirable to perform a co-IP in both "directions." That is, to immunoprecipitate an HA-tagged protein followed by investigating the presence of the putative FLAG-tagged partner as well as performing an anti-FLAG immunoprecipitation followed by investigating the presence of the HA-tagged protein. When comparing the interaction properties of different proteins or protein derivatives, this also allows for the demonstration that equal amounts of proteins have been expressed and immunoprecipitated. Once the initial interaction is well established, however, it may be sufficient to perform the co-IP in only one direction to investigate the interaction under varying conditions or using protein derivatives.

3.1. Agroinfiltration

1. Germinate *N. benthamiana* seeds by spreading on top of wetted Metro-Mix 360 (Sun Gro) topped with a layer of vermiculite. Seeds must remain exposed to the light for germination. Two weeks later, transplant seedlings to individual 10 cm pots containing Metro-Mix 360 with growth conditions as outlined above. Plants should be fertilized weekly with Peters Excel "CalMag" Grower (15-5-15; 100 ppm N) (Scotts). Plants can be used for transient expression 2–4 weeks later.
2. Grow *Agrobacterium* cultures overnight at 28°C in LB broth with appropriate selection. 5–10 mL is usually sufficient to fully infiltrate several leaves. For repeat usage, make a glycerol stock (25–30% glycerol) from an overnight culture inoculated from a plate of freshly transformed bacteria and store at −70°C. As needed, streak out stocks onto LB plates with appropriate selection (50 μg/mL of kanamycin for pBIN19 and 5 μg/mL of tetracycline for pCH32). A freshly inoculated plate should be used as starting material for overnight

(O/N) cultures to allow for synchronous growth of different cultures. When setting up O/N cultures, inoculate with a generous amount of *Agrobacterium* in order to get a dense culture in 16 h.

3. After O/N growth, pellet *Agrobacterium* cultures by spinning for 15 min at 4,000 rpm in a Sorvall SH-3000 rotor. Resuspend in an equal volume of infiltration buffer (see Note 3). Leave at room temperature for 2–24 h.
4. Take an OD_{600} of each *Agrobacterium* culture to be infiltrated. The cultures should have all grown to roughly similar ODs if they were all started from fresh plates. Combine *Agrobacterium* suspensions into the desired combinations for protein co-expression. It is important to infiltrate the same amount of *Agrobacterium* for each protein to be expressed. If one co-IP calls for the expression of three proteins, and another only two; make up the difference with *Agrobacterium* carrying empty vector. Each *Agrobacterium* culture should be infiltrated at an OD = 0.1–0.2. To achieve this, simply dilute the resuspended O/N cultures into a final volume of 10 mL of infiltration buffer. For example, if three cultures have an OD = 1.0, combine 2 mL of each culture plus 4 mL of infiltration buffer which will result in a final volume of 10 mL and an overall OD = 0.6. Avoid using combinations with a final ODs greater than 1.0.
5. Infiltrate *Agrobacterium* mixtures into *N. benthamiana* leaves with a syringe. For the easiest infiltration, use a razor blade or pin to make a slight scratch in the underside of the leaf without piercing all the way through. Lightly press the syringe to the scratch and infiltrate. Either mark the infiltrated area with a marker or ensure that the entire leaf has been infiltrated (see Note 8). For each combination of co-expressed proteins, we routinely infiltrate two leaves on two different plants.

3.2. Protein Extraction

1. Harvest leaves at 1–3 days post agroinfiltration. Cut leaf material off on either side of the large middle vein with a razor blade. Weigh out 1 g of tissue for each combination and place in a prechilled mortar.
2. Extract protein in the cold room using prechilled materials. Add 2.5 mL of extraction buffer to each mortar. Grind tissue in extraction buffer for 1–2 min until it becomes a consistent slurry.
3. Pour the slurry from each sample into 2-mL Eppendorf tubes and spin at full speed in a refrigerated microcentrifuge for 2 min. Transfer supernatant to a 1.5-mL Eppendorf tube and spin for an additional 10 min (see Note 9).
4. Run 1 mL of the supernatant through a desalting column (optional; see Note 5).

3.3. Immuno-precipitation

1. For each combination of co-expressed proteins add 200–1,200 μL of extract to two separate 1.5-mL Eppendorf tubes (depending on abundance of the proteins and whether or not it has been diluted by desalting. If larger amounts are required, increase extraction volumes accordingly). One tube is used for the anti-HA immunoprecipitation and one for the anti-FLAG immunoprecipitation. Bring up the total volume to 1.4 mL with GTEN and add 10% NP-40 to bring the final concentration (v/v) to 0.15% and DTT to 1 mM. At this point, detergent is usually beneficial in order to keep protein from "sticking" to the beads and from aggregating.
2. Perform a preclearing step by adding 25 μL of agarose-conjugated IgG beads. This is to eliminate nonspecific binding and can also eliminate some proteins that become unstable and precipitate during incubation. Incubate end-over-end in a rotating microtube mixer at 4°C for 30 min. Spin 1 min, full-speed in a microcentrifuge. Add 1.3 mL of the supernatant to a new tube for the specific immunoprecipitation.
3. To this supernatant, add 25 μL of slurry of either the anti-HA or anti-FLAG affinity gel agarose beads using a pipette tip that has had the bottom ~3 mm cut off with a razor blade.
4. Incubate end-over-end for 1–16 h at 4°C.
5. Spin full-speed in a microcentrifuge for 5 s. Discard supernatant and add 1 mL of fresh IP buffer. Repeat four more times but always leave ~50 μL at the bottom of the tube to avoid removing the beads. After the last wash, spin again to remove any liquid on the sides of the tube and aspirate the remaining liquid with a 1-mL syringe with a very fine (25 Ga) needle. Point the open side of the needle to the wall of the tube to avoid aspirating beads.
6. Resuspend beads in 100–150 μL of 1× SDS–PAGE loading buffer. Use immediately or store at −70°C.

3.4. Immunoblotting

1. Boil samples for 10 min. Vortex lightly and spin in microfuge for 5 s.
2. Load 10–50 μL (or up to ~1/3rd of the total) of the sample onto two different SDS–PAGE gels (see Note 7) for immunoblotting analysis; one for anti-HA and one for anti-FLAG immunoblotting.
3. Pre-wet PVDF membrane in 100% methanol followed by equilibration in transfer buffer for 10 min.
4. Transfer proteins to PVDF membrane (see Note 7).
5. Incubate PVDF membranes in blocking buffer for 1–16 h.
6. Place membranes in a tray containing 5–10 mL of TBST. Add primary antibody HRP conjugate (see Note 10) at the

manufacturer's specified dilution. Incubate for 1 h at room temperature or O/N at 4°C.

7. Wash blots in excess TBST three times 15 min at room temperature.
8. Remove membranes, blotting off excess liquid, and transfer to a piece of Saran Wrap.
9. Perform ECL Plus™ reaction (or similar product) as per the manufacturer's instructions.
10. Wrap membranes in Saran Wrap and expose immediately (in dark room) to film. Multiple exposure times (from 1 s to 5 min) should be tested, depending on the abundance of the proteins. Develop films.

4. Notes

1. Since agroinfiltration depends only on the ability of the T-DNA to be transferred to the host, most binary vectors can be used for this purpose (1, 14). We routinely use binary vectors based on pBIN19. One of the most effective *Agrobacterium* strains in *Nicotiana* spp. is C58C1 carrying the pCH32 plasmid. The pCH32 plasmid (tetracycline selection) carries two *vir* genes (*vir-E* and *vir-G*) which increase virulence, and consequently the efficiency of T-DNA transfer and expression (15). However, most common strains of *Agrobacterium* can be used effectively in *Nicotiana* spp., including GV3101, LBA4404, and AGL1.
2. Epitope tags should be incorporated at the C terminus of CC-NB-LRR proteins as N-terminal fusions often interfere with function. With other proteins, potential interference with function by fusions at either end of the protein must be determined empirically. We routinely use single HA (YPYDVPDYA) or FLAG (DYKDDDDK) epitope tags on CC-NB-LRR protein derivatives. Many groups report difficulties detecting a single c-Myc epitope tag (EQKLISEEDLNE), and this epitope is commonly used as three to six consecutive tags. The GFP protein can also be added as a tag. GFP can add size in the case of a very small interaction motif and in some cases can stabilize proteins.
3. Acetosyringone induces virulence in *Agrobacterium* at low pH. However, when using C58C1 plus pCH32 in *N. benthamiana*, this is not strictly necessary. 10 mM $MgCl_2$ is sufficient and *Agrobacterium* can be infiltrated immediately.
4. Inclusion of DTT is very important as protein from *N. benthamiana* extracts is particularly prone to oxidation.

Proteins extracted without DTT often run as one or more higher molecular weight bands in SDS–PAGE. PVPP absorb some of the polyphenols which are abundant in *Nicotiana* spp. Although PVPP is insoluble, it is recommended that the PVPP be allowed to hydrate for several hours for maximum efficiency. Thus, one can simply make up the extraction buffer the night before. However, do not add the DTT or the protease inhibitors until just before use. Our extraction buffer functions for a number of different proteins; however, individual proteins may require extraction buffer optimization. Detergents may also be included, particularly for membrane proteins. Experiment with a number of detergents with different properties at first. For example, on the one hand, extraction with ionic detergents increases extraction of Rx, but can be denaturing. On the other hand, extraction of Rx with nonionic detergent results in a complete loss of Rx protein, possibly due to increased protease extraction. Most other proteins we have worked with, however, are fine when extracted with 0.1% NP-40 or Tween-20. Many commercial protease inhibitor cocktails are available, including Sigma plant protease inhibitor cocktail (P9599) which comes as a 100× stock solution in DMSO. If planning on performing IPs on a regular basis, however, it is considerably cheaper to make a protease inhibitor cocktail as follows:

Stocks	Preparation	(Stock)	(Working)
AEBSF	100 mg in 2.1 mL DMSO	0.02 M	1 mM
1,10-Phenanthroline	2 g in 10 mL ethanol	0.1 M	1 mM
Pepstatin A	5 mg in 5 mL methanol	1 mg/mL	1 μg/mL
E64	5 mg in 1 mL DMSO	14 mM	5 μM
Leupeptin	10 mg in 200 μL water	50 mg/mL	5 μg/mL
Bestatin	5 mg in 2.9 mL DMSO	5 mM	1 μM
PMSF	0.174 g in 10 mL isopropanol	0.1 M	1 mM

5. Passing protein extracts through a small desalting column is optional but can help reduce nonspecific modification of proteins during the IP. G25 Sephadex can be substituted for Bio-Gel P6, both of which act as size exclusion matrices that eliminate a good deal of small molecules, including harmful phenolic compounds. The difference can be seen after extended incubation of extracts: the desalted extract remains clear or greenish while the untreated extracts turn brownish. In addition, passage through a desalting column eliminates particulates that are often present even after centrifugation. Desalting columns can be used either by gravity or by spinning (similar to

a G-50 column used to clean up DNA reactions). Gravity is more effective, but spinning is sufficient for most purposes and is much more convenient when dealing with numerous samples. To make the desalting column, allow the desalting matrix to swell for at least 2 h in extraction buffer. At this point, it is not necessary to include protease inhibitors, and DTT concentration can be decreased to 1 mM. Do not include PVPP. Use about 10 mL of buffer per column. Add powder slowly to the liquid until saturation appears to be reached. If too much matrix is added, simply add more buffer. Aim to have as little liquid as possible at the top, once the slurry has settled (not more than half a centimeter). Place a screening column into a 15-mL round bottom snap cap tube. Pour slurry into the top of the column. *Spin method*: Spin in a clinical centrifuge for 1 min at 2,000 rpm to get rid of excess liquid. Transfer column to a new dry snap cap tube. There should be about 1 cm of space at the top of the tube. Load up to 25% of the column volume of protein extract (~1 mL) and spin again. Keep the flow-through and discard the rest. *Gravity method*: Allow the excess liquid to drip through until the column is settled and no more liquid is coming through. Add 1.2 mL of extract and allow it to drip through. Add 0.25 mL of buffer and allow it to drip through. Discard flow-through and transfer column to a new dry snap cap tube. Add 1.75 mL of buffer and collect what drips through. This contains the protein. This method results in dilution of the sample and/or partial loss compared to the spin method. However, this is usually not critical and gives a cleaner extract. A trial run is recommended to optimize collection with individual versions of this technique (i.e., differing columns or sample sizes). To determine what volumes to use to desalt, use a dilute solution of bromophenol blue (small molecule; is retained in the column) and dextran blue (very large molecule; is excluded from the column and remains in the void volume with the proteins). Add small amounts (0.25–0.5 mL) of buffer to the column to determine when the dextran blue elutes. Protein should elute in the same fractions as dextran blue. Fractions containing bromophenol blue (and other small molecules) should be avoided.

6. Agarose conjugated IgG is used for a nonspecific preclearing step and can be from any source and any species. Using antibodies preconjugated to agarose beads is convenient and economical if immunoprecipitating epitope-tagged proteins on a regular basis. Alternatively, nonconjugated antibodies can be used in conjunction with either protein A or protein G conjugated beads.
7. Use standard procedures (16) for SDS–PAGE and transfer of proteins to PVDF (or similar product such as nitrocellulose membrane).

8. *N. benthamiana* leaves are easy to infiltrate in their entirety under optimal conditions. Ensure that the plants are not drought stressed. Watering the plants an hour before infiltration can help. If plants are grown in a glasshouse, the best time to infiltrate is in the late afternoon or evening to avoid the effects of hot afternoon sun which can make infiltration more difficult and lead to leaf temperatures detrimental to *Agrobacterium*. Alternatively, if conditions are not optimal in the glasshouse, bring the plants to the lab a couple of hours before infiltrating. After infiltrating, leave on the bench overnight (or even for the next 1–3 days). Growth conditions can have a large effect on protein expression. Temperatures of 19–24°C and the absence of intense light are best, and the lab benchtop is often the optimal area for transient expression. Leaf position can also affect protein expression. Try to avoid the lowest leaves. These are easy to infiltrate but give lower and patchier expression. Newly emerging leaves are more venous and harder to infiltrate in large patches. The best expression and extraction can be obtained by using the two youngest fully expanded leaves. As a general rule, if the infiltration takes a lot of effort and requires more than three to six infiltration sites per leaf, conditions are nonoptimal and may yield poor results. Before initiating co-IP experiments, try experimenting with a few different aged plants and different leaf positions. It can be helpful to infiltrate with a binary vector expressing some version of mGFP and observing the plants under long-wave UV light to become familiar with which leaves give the best expression.
9. This is appropriate for soluble proteins but should be adjusted accordingly for membrane-associated proteins. An alternative approach often used is to grind in liquid nitrogen and then incubate with an extraction buffer, particularly if the extraction buffer contains detergent. This is probably fine for most proteins and protein complexes, but may make it more difficult to ensure the use of exactly the same amount of starting material for each sample.
10. This use of primary antibodies directly conjugated to HRP avoids cross-reaction with the antibodies used in the co-IP and reduces the time required for immunoblotting. Alternatively nonconjugated primary antibodies can be used, followed by incubation with an appropriate HRP-conjugated secondary antibody. However, care should be taken to use combinations of antibodies that will avoid using a secondary antibody that recognizes the antibody used for the co-IP. The latter is present in the final protein extracts and subsequently on the PVDF membranes.

References

1. Hellens, R., Mullineaux, P., and Klee, H. (2000) Technical Focus: a guide to Agrobacterium binary Ti vectors, *Trends Plant Sci* **5**, 446–451.
2. Wroblewski, T., Tomczak, A., and Michelmore, R. (2005) Optimization of Agrobacterium-mediated transient assays of gene expression in lettuce, tomato and Arabidopsis, *Plant Biotechnol J* **3**, 259–273.
3. Voinnet, O., Rivas, S., Mestre, P., and Baulcombe, D. (2003) An enhanced transient expression system in plants based on suppression of gene silencing by the p19 protein of tomato bushy stunt virus, *Plant J* **33**, 949–956.
4. McDowell, J. M., and Simon, S. A. (2008) Molecular diversity at the plant–pathogen interface, *Dev Comp Immunol* **32**, 736–744.
5. Scofield, S. R., Tobias, C. M., Rathjen, J. P., Chang, J. H., Lavelle, D. T., Michelmore, R. W., and Staskawicz, B. J. (1996) Molecular basis of gene-for-gene specificity in bacterial speck disease of tomato, *Science* **274**, 2063–2065.
6. Tang, X., Frederick, R. D., Zhou, J., Halterman, D. A., Jia, Y., and Martin, G. B. (1996) Initiation of plant disease resistance by physical interaction of AvrPto and Pto kinase, *Science* **274**, 2060–2063.
7. Bendahmane, A., Querci, M., Kanyuka, K., and Baulcombe, D. C. (2000) Agrobacterium transient expression system as a tool for the isolation of disease resistance genes: application to the Rx2 locus in potato, *Plant J* **21**, 73–81.
8. Moffett, P., Farnham, G., Peart, J., and Baulcombe, D. C. (2002) Interaction between domains of a plant NBS-LRR protein in disease resistance-related cell death, *EMBO J* **21**, 4511–4519.
9. Leister, R. T., Dahlbeck, D., Day, B., Li, Y., Chesnokova, O., and Staskawicz, B. J. (2005) Molecular genetic evidence for the role of SGT1 in the intramolecular complementation of Bs2 protein activity in *Nicotiana benthamiana*, *Plant Cell* **17**, 1268–1278.
10. Rairdan, G. J., Collier, S. M., Sacco, M. A., Baldwin, T. T., Boettrich, T., and Moffett, P. (2008) The coiled-coil and nucleotide binding domains of the potato Rx disease resistance protein function in pathogen recognition and signaling, *Plant Cell* **20**, 739–751.
11. Rairdan, G. J., and Moffett, P. (2006) Distinct domains in the ARC region of the potato resistance protein Rx mediate LRR binding and inhibition of activation, *Plant Cell* **18**, 2082–2093.
12. Sacco, M. A., Mansoor, S., and Moffett, P. (2007) A RanGAP protein physically interacts with the NB-LRR protein Rx, and is required for Rx-mediated viral resistance, *Plant J* **52**, 82–93.
13. Tameling, W. I., and Baulcombe, D. C. (2007) Physical association of the NB-LRR resistance protein Rx with a Ran GTPase-activating protein is required for extreme resistance to potato virus X, *Plant Cell* **19**, 1682–1694.
14. Komori, T., Imayama, T., Kato, N., Ishida, Y., Ueki, J., and Komari, T. (2007) Current status of binary vectors and superbinary vectors, *Plant Physiol* **145**, 1155–1160.
15. Hamilton, C. M., Frary, A., Lewis, C., and Tanksley, S. D. (1996) Stable transfer of intact high molecular weight DNA into plant chromosomes, *Proc Natl Acad Sci U S A* **93**, 9975–9979.
16. Scopes, R. K., and Smith, J. A. (2006) Analysis of Proteins, in *Current Protocols in Molecular Biology* (Ausubel, F. M., Brent, R., Kingston, R. E., Moore, D. D., Seidman, J. G., Smith, J. A., and Struhl, K., Eds.), pp 10.10.11–10.18.28, Wiley, New York.

Chapter 3

Purification of Resistance Protein Complexes Using a Biotinylated Affinity (HPB) Tag

Yiping Qi and Fumiaki Katagiri

Abstract

Plant disease resistance (R) proteins confer strong resistance against pathogens by recognizing particular pathogen effectors. Identification of proteins associated with an R protein will provide insight into the mechanism of R protein function. Many R proteins are associated with the plasma membrane (PM) and expressed at low levels. Here, we describe a method to purify such low-abundance PM R protein complexes from Arabidopsis using a biotinylated affinity tag, called the HPB tag. We have successfully applied this method to identify candidate components of the RPS2 resistance protein complex(es). This method should also be applicable to purification of other low-abundance PM protein complexes.

Key words: Arabidopsis, R protein, Plasma membrane, HPB tag, Biotinylation, Protein complexes

1. Introduction

Many plant disease resistance (R) proteins belong to the nucleotide-binding-leucine rich repeat (NB-LRR) protein class (1). There are about 125 NB-LRR *R* genes in Arabidopsis (2). Particular R proteins recognize specific pathogen effector proteins, either directly (3, 4) or indirectly (5–8). Recognition of pathogen effectors by R proteins triggers a rapid and strong defense response against the pathogen, called gene-for-gene resistance or effector-triggered immunity. This defense response can confer strong resistance against diverse plant pathogens such as viruses, bacteria, fungi, oomycetes, and insects (1, 2).

Direct recognition occurs through binding of effector molecules by R proteins. Indirect recognition may be explained by the "guard hypothesis" (1), in which R proteins "guard" the host proteins ("guardees") which are targets of pathogen effectors. A good example of a "guardee" is RIN4, which is "guarded" by

John M. McDowell (ed.), *Plant Immunity: Methods and Protocols*, Methods in Molecular Biology, vol. 712,
DOI 10.1007/978-1-61737-998-7_3, © Springer Science+Business Media, LLC 2011

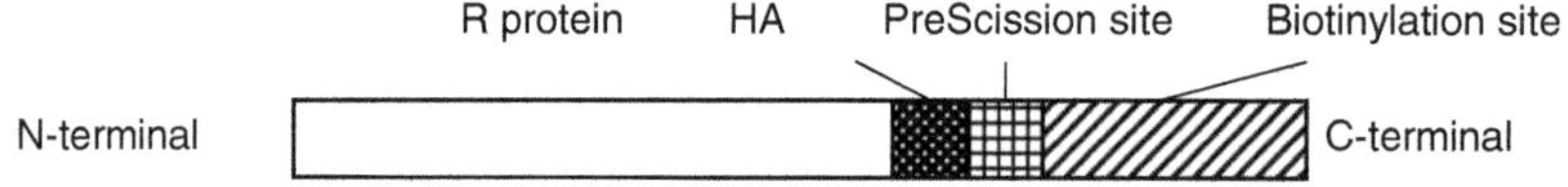

Fig. 1. Diagram of R protein fused to the HPB tag at the C-terminus.

two R proteins, RPS2 and RPM1, and targeted by three distinct bacterial effectors, AvrRpt2, AvrRpm1, and AvrB (6–8). The guard hypothesis predicts that R proteins bind and guard other plant proteins in the cell. R proteins are hypothesized to reside in large protein complexes before and after activation (1). For example, in a study using Arabidopsis leaf mesophyll protoplasts, the R protein RPS2 was co-immunoprecipitated with a plant protein of approximately 75 kD (9). Similarly, the R protein RPM1 was found to be present in native gel filtration protein fractions ranging from 440 to 1,500 kD, while the size of RPM1 by itself is about 107 kD (10).

Direct purification of plant R protein complexes and identification of the complex components can greatly advance our knowledge about the mechanism of R protein-mediated signaling. Interestingly, many known Arabidopsis R proteins are localized to the plasma membrane (PM), including RPM1 (11), RPS2 (6) and probably RPS5 (12). Also, R proteins are often expressed at low levels, probably because expression of R proteins has fitness costs for the plant (13). Thus, an efficient method is needed to purify low-abundance R protein complexes in quantities sufficient for identification of complex components by mass-spectrometry-based methods such as LC-MS/MS.

To facilitate purification of low-abundance PM R protein complexes, we have developed a method using the HPB affinity tag (14). The HPB tag consists of one copy of the HA epitope tag, a Prescission® protease recognition site, and the biotinylation site of the *Arabidopsis* 3-methylcrotonyl-CoA carboxylase (Fig. 1). The HPB tagged R proteins can be purified using a streptavidin-conjugated matrix, by taking advantage of the strong biotin–streptavidin interaction. Here we describe a detailed protocol for purifying R protein complexes using the HPB tag.

2. Materials

2.1. Construction of the Plant Transformation Vector

1. The pCR®8/GW/TOPO® TA Cloning Kit (Invitrogen).
2. Gateway® LR Clonase® II enzyme mix (Invitrogen).
3. pMDC32-HPB and/or pMDC-pRPS2-HPB vector (Genbank accessions FJ172534 and FJ172535) (see Note 1).

2.2. Preparation of R::HPB Transgenic Plants

1. Competent cells for electroporation of the *Agrobacterium tumefaciens* strain GV3101/pMP90 (see Note 2).
2. Floral dip transformation solution: 5% (w/v) sucrose, 10 mM $MgCl_2$, 0.03% (v/v) VAC-IN-STUFF (Silwet L-77) (LEHLE SEEDS, Cat # VIS-02).
3. Murashige and Skoog (MS) Basal Medium supplemented with Gamborg's vitamins (Sigma, Cat # M0404).
4. Transgenic plant selection medium: 0.8% agar plate containing 0.5× MS medium and 25 μg/mL Hygromycin B (Roche, Cat # 10843555001).

2.3. Purification of R Protein Complexes

1. Coors 60316/60317 ceramic mortars and pestles.
2. Scissors.
3. Utility funnels with a 150 mm diameter.
4. Centrifuge tubes: 2 mL microcentrifuge tubes (Fisher), 250 mL centrifuge tubes (Beckman), 70- and 15-mL ultracentrifuge tubes (Beckman).
5. Miracloth (Calbiochem).
6. Detergent stocks: 10% NP-40 (Nonidet® 40 substitute, Fluka, Cat # 74358), 10% Na-deoxycholate (Deoxycholic acid Sodium salt, Fluka, Cat # 30970); stored at room temperature.
7. Protease inhibitor stocks: 0.1 M Phenylmethylsulfonyl fluoride (PMSF, 100×, dissolved in ethanol), 1 mg/mL Leupeptin (1,000×), 1 mg/mL E64 (1,000×, dissolved in 50% ethanol), 1 mg/mL Pepstatin (1,000×, dissolved in ethanol) (all made from powder, Roche); stored at −20°C (see Note 3).
8. Cross-linker stock: 100 mM Dithiobis (succinimidylpropionate) (DSP) (Pierce, Cat # 22585) dissolved in anhydrous DMSO (Sigma); stored at −80°C (see Note 3).
9. Cross-linking quenching buffer: 1 M Tris–HCl, 330 mM sucrose, pH 7.5; stored at 4°C.
10. Dynabeads® M-280 Streptavidin (Invitrogen, Cat # 112-06D); stored at 4°C (see Note 4).
11. Mix-all™ laboratory tube mixer (Research Products International, Corp.).
12. Magnetic separation stand (Promega).
13. 1 M DTT stock (1,000×) (Promega); stored at −20°C.
14. 98% 2-mercaptoethanol (Fisher); stored at 4°C.
15. 3× SDS sample buffer without 2-mercaptoethanol: 62.5 mM Tris–HCl, 2% SDS, 10% glycerol, and 0.002% bromophenol blue; stored at room temperature.
16. Pre-stained protein markers (Bio-Rad); stored at −20°C.

17. Large protein gel set, such as Protean® II xi (Bio-Rad) or any other protein gel set which accommodates a large loading volume (more than 120 μL) for each well (see Note 5).
18. Grinding buffer: 50 mM HEPES-KOH, 10 mM EDTA, 0.6% polyvinylpolypyrrolidone, 330 mM sucrose, pH 7.5; stored at 4°C. Immediately before use, add protease inhibitor stocks and DTT stock to final concentrations of 1 mM PMSF, 1 μg/mL Leupeptin, 1 μg/mL E64, 1 μg/mL Pepstatin, and 1 mM DTT.
19. Resuspension buffer: 20 mM HEPES-KOH, 100 mM NaCl, 1 mM EDTA, 330 mM sucrose, pH 7.5; stored at 4°C. Immediately before use, add protease inhibitor stocks and DTT stock to final concentrations of 1 mM PMSF, 1 μg/mL Leupeptin, 1 μg/mL E64, 1 μg/mL Pepstatin, and 1 mM DTT.
20. RIPA (Radio Immunoprecipitation Assay) buffer 1: 50 mM Tris–HCl, 150 mM NaCl, 1 mM EDTA, pH 7.4; stored at room temperature. Immediately before use, add detergent stocks to final concentrations of 1% NP-40 and 0.5% Na-deoxycholate.
21. RIPA buffer 2: 50 mM Tris–HCl, 20 mM NaCl, 1 mM EDTA, pH 7.4; stored at room temperature. Immediately before use, add detergent stocks to final concentrations of 1% NP-40 and 0.5% Na-deoxycholate. (see Note 6).

3. Methods

3.1. Construction of the Plant Transformation Vector

1. Clone the *R* gene of interest into pCR®8/GW/TOPO® vector using the TA Cloning kit according to the manual (see Note 7).
2. Conduct an LR reaction to insert the *R* gene into the pMDC32-HPB or pMDC-pRPS2-HPB destination vector to make an R::HPB fusion construct using Gateway® LR Clonase® II enzyme mix according to the manual (see Note 8).

3.2. Preparation of R::HPB Transgenic Plants

1. Transform the *A. tumefaciens* strain GV3101/pMP90 with the transformation vector containing R::HPB using electroporation (15).
2. Transform Arabidopsis plants with the transformed *A. tumefaciens* strain using the floral dip method (16). It is best to use plants homozygous for a loss-of-function mutation in the R gene of interest, as this allows tests for proper functioning of the R::HPB construct.
3. Select transgenic R::HPB plant lines using Transgenic plant selection medium, Transfer resistant plants to soil 1–2 weeks after sowing the seeds.

4. Test the expression of the R::HPB transgene in individual T1 plants using western blots. If possible, test for the complementation of the *r* mutant phenotype using appropriate pathogen assays. Select a few good lines for further studies.
5. Grow a selected T3 homozygous R::HPB line and the negative control plant line (*r* mutant) in soil with periodic fertilization (see Note 9).

3.3. Purification of R Protein Complexes

All the procedures described in this section are performed at 4°C or on ice. The major steps of complex purification are shown in the flowchart (Fig. 2).

3.3.1. Microsomal Fractionation

1. Collect 30 g of R::HPB leaf tissue from 5- to 6-week-old healthy plants. Do the same with the control plants (see Note 10).
2. Cut leaf tissue into small pieces (~1 cm × 1 cm) with scissors and grind the tissue in Grinding buffer (about 15 g tissue/75 mL Grinding buffer/mortar) using a mortar and pestle. Grind it until most leaf pieces become invisible to the naked eye. Pool R::HPB samples. Do the same with the control samples (In total there will be four batches of grinding: 15 g × 2 of R::HPB and 15 g × 2 of control, with total 300 mL of Grinding buffer used).
3. For each sample, place a 150-mm utility funnel on a 250-mL centrifuge tube and place double-layered Miracloth on top of the funnel. Filter the ground suspension by pouring it through the funnel, collecting the homogenate in the centrifuge tube (about 150 mL for each sample).
4. Centrifuge the tubes at 18,000 × *g* for 15 min.
5. Divide the 150 mL supernatant of each sample into three 70-mL ultracentrifuge tubes (total six ultracentrifuge tubes used). Centrifuge the tubes at 100,000 × *g* for 1 h (We used a Ti45 rotor at 29,300 rpm in a Beckman Coulter Optima TLX Ultracentrifuge).

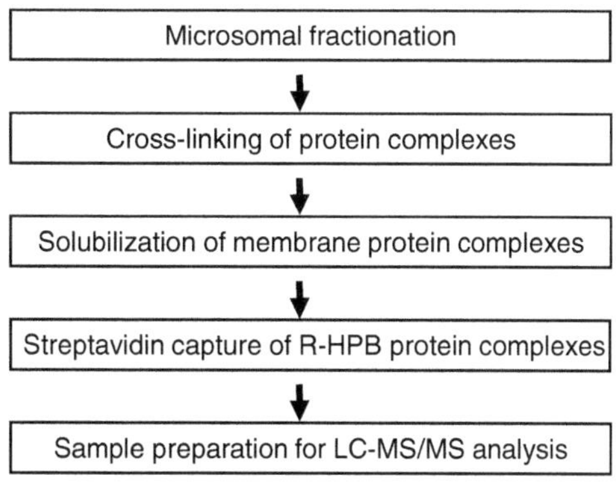

Fig. 2. Flow chart of the purification procedure.

6. Carefully discard the supernatant. Resuspend each pellet (this is the microsomal fraction) with 1 mL of resuspension buffer, to a final volume of ~1.8 mL. Transfer the microsome suspensions from the ultracentrifuge tubes to 2 mL microcentrifuge tubes (total six microcentrifuge tubes used).

3.3.2. Cross-Linking of Protein Complexes (see Note 11)

1. Add 20 μL of 100 mM DSP stock to each tube to a final concentration of 1 mM. Quickly vortex the tubes and incubate the tubes using the rotator at approximately 30 rotations per minute, for 30 min (see Note 12).
2. Add 100 μL of cross-linking quenching buffer to each tube to a final concentration of 50 mM Tris to quench the cross-linking reaction. Incubate the tubes using the rotator at approximately 30 rotations per minute for 30 min.
3. Combine all R::HPB samples (three tubes, total ~6 mL) into a 15-mL ultracentrifuge tube. Do the same with control samples.
4. Centrifuge the tubes at 100,000 × *g* for 30 min (We used an SW Ti32.1 rotor at 23,400 rpm in a Beckman Coulter Optima TLX Ultracentrifuge).
5. Discard the supernatant and keep the pellet.

3.3.3. Solubilization of Membrane Protein Complexes

1. Resuspend each pellet with resuspension buffer to a final volume of ~5.7 mL.
2. Divide the R::HPB suspension into three 2-mL tubes, with ~1.9 mL in each. Do the same with the control sample.
3. Add 100 μL of 10% Na-deoxycholate to each tube to a final concentration of 0.5%.
4. Incubate the tubes using the rotator at approximately 30 rotations per minute for 30 min.
5. Combine all R::HPB samples (three tubes, total ~6 mL) to a 15-mL ultracentrifuge tube. Do the same with control samples.
6. Centrifuge the tubes at 100,000 × *g* for 30 min (We used a SW Ti32.1 rotor at 23,400 rpm in a Beckman Coulter Optima TLX Ultracentrifuge).
7. Collect the supernatant and discard the pellet.

3.3.4. Streptavidin Capture of R::HPB Protein Complexes

1. While the samples are in the ultracentrifuge, wash 600 μL of Dynabeads® M-280 Streptavidin magnetic beads with resuspension buffer three times and resuspend them in a final volume of 600 μL.
2. Divide the supernatant of R::HPB into three 2 mL tubes, approximately 2 mL in each tube. Do the same with the control sample.

3. Add 100 μL of Dynabeads® M-280 beads to each tube. Incubate them using the rotator at approximately 30 rotations per minute, for 2 h.
4. Place each tube in the magnetic separation stand for 1 min. Then carefully remove the supernatant using a pipette.
5. Wash the beads three times with RIPA buffer 1: for each wash, gently vortex the beads in 1 mL buffer for 10 s, then place the tubes in the magnetic separation stand again for separation.
6. Wash the beads three times with RIPA buffer 2 in the same manner. At the final washing step, combine the R::HPB sample from three tubes to two tubes. Do the same with the control sample.
7. Resuspend the beads with RIPA buffer 2 to an approximate volume of 40 μL in each tube. Add 20 μL 3× SDS sample buffer supplemented with 5% (v/v) fresh 2-mercaptoethanol to a final volume of 60 μL. Mix well by pipetting. There will be a total of 120 μL for each sample (see Note 13).
8. Heat the samples at 99° C for 10 min. We did this by transferring the samples to 200 μL PCR tubes and heating them in a thermal cycler. Store the samples at –20°C, if needed.

3.4. Sample Preparation for LC-MS/MS Analysis (see Note 14)

1. Cast an 8% SDS-PAGE gel (16 cm × 16 cm × 1 cm) using a Bio-Rad Protean® II xi or other suitable protein gel set. Make the gel with a stacking gel (3% PA) and a separation gel (8% PA). Use the ten-well comb so that each well can hold more than 120 μL of protein sample (see Notes 5, 13).
2. Spin down the magnetic beads in SDS sample buffers. (see Note 15).
3. Load the R::HPB sample and the control sample into two wells in the middle of the gel (approximately 120 μL each). Load 10 μL pre-stained protein marker on each side (see Fig. 3).
4. Run the gel and carefully monitor the migration of the pre-stained protein markers.
5. Stop running the gel when all of the protein markers have moved into the separation gel. The distance of the sample front (where bromophenol blue migrates) from the boundary of the stacking gel and separation gel should be no more than 1 cm (see Fig. 3).
6. Slice the gel (about 2 mm × 10 mm, the position of each sample is shown as a black rectangle in Fig. 3) of each sample for in-gel tryptic digestion and LC-MS/MS for identification of purified proteins in both R::HPB and the control sample (see Note 16).

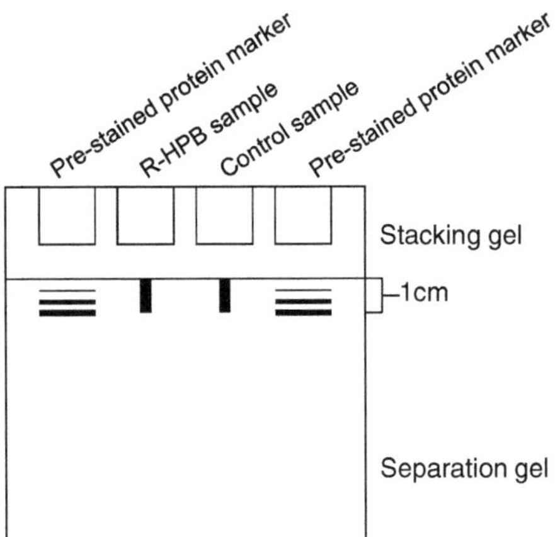

Fig. 3. Diagram of the protein gel.

4. Notes

1. The pMDC32-HPB vector is used for relatively high expression level of tagged R proteins in plant, while pMDC-pRPS2-HPB is used for a relatively low expression level. If needed, the cauliflower mosaic virus *35S* promoter on pMDC32-HPB and the *RPS2* native promoter on pMDC-pRPS2-HPB can be removed or replaced with other promoters using the unique *Kpn*I and *Hin*dIII restriction sites.
2. Other Agrobacterium transformation methods can also be used.
3. Protease inhibitors and the DSP cross-linker are toxic. They all should be handled carefully. DSP is water-sensitive and may quickly lose activity upon contact with water. The stock should be aliquoted and stored at –80°C for no more than 6 months.
4. Invitrogen also sells Dynabeads® MyOne Streptavidin, which are smaller in diameter and move more slowly than M-280 beads in a magnetic field. Although the manufacturer claims that the MyOne beads have higher capacity, we did not find an appreciable difference between these two kinds of beads. The MyOne beads seemed to have slightly more nonspecific protein binding than M-280 beads, based on silver staining of bound proteins. Other high capacity streptavidin-conjugated matrices may perform better than M-280 beads. We recommend performing a pilot experiment to compare different streptavidin-conjugated matrices.
5. A large gel set is used because of the high volume capacity of the wells. If there is a comb that allows large capacity wells

with a small gel set (like a mini-gel set), a small gel set can be used. Also, see Note 13.

6. The reason for using two RIPA buffers with different NaCl concentrations is to enhance removal of proteins nonspecifically bound to the magnetic beads because different nonspecific proteins may be removed more efficiently at different salt concentrations.
7. Other entry clones may also be used. A restriction digestion step may be needed if the entry clone has a kanamycin resistance selection marker, because the destination vectors used here also have a kanamycin resistance selection marker.
8. Screening for positive *Escherichia coli* colonies after transformation with the LR reaction mixture can be facilitated as follows: Streak each single colony to both an LB plate supplemented with spectinomycin and an LB plate with kanamycin. Pick those that only grow on a kanamycin plate for further confirmation by restriction digestion and/or sequencing.
9. Water plants about twice a week with fertilizer (Scotts, Peters Professional®) every time, at the concentration of 1 g of fertilizer per liter of water.
10. Typically, one plant will yield more than 2 g of leaf material if it is well fertilized and grown under 12-h light/12-h dark condition.
11. The cross-linking step prevents protein complexes from breaking apart in the solubilization step. However, cross-linking may reduce the specificity in identification of protein complex components since proteins in the vicinity can be cross-linked even if they do not physically form a complex. Many protein–protein interactions may survive the solubilization step. Thus, it may be a good idea to try performing a purification procedure without cross-linking.
12. A white cloud may form in the suspension when DSP is added. It is normal because the solubility of DSP is low in water.
13. A smaller elution volume (<120 μL) may be used to facilitate easy loading of one sample into one well of the protein gel.
14. This step reduces the concentration of SDS in the sample. Tryptic digestion cannot be performed in the presence of a high concentration of SDS.
15. If the samples were stored at −20°C, re-heat them at 99°C for 6 min before spinning down the beads.
16. See our publication for details about mass-spectrometry (14). A mass-spectrometry service facility can likely perform the procedure described in Subheading 3.4.

Acknowledgments

This work was supported by a grant from the National Science Foundation (Arabidopsis 2010 grant number IOB-0419648) to F.K. and a PBS Doctoral Dissertation Fellowship from the University of Minnesota to Y.Q.

References

1. Dangl, J. L., and Jones, J. D. (2001) Plant pathogens and integrated defence responses to infection. *Nature.* **411**, 826–833.
2. Jones, J. D., and Dangl, J. L. (2006) The plant immune system. *Nature.* **444**, 323–329.
3. Jia, Y., McAdams, S. A., Bryan, G. T., Hershey, H. P., and Valent, B. (2000) Direct interaction of resistance gene and avirulence gene products confers rice blast resistance. *EMBO J.* **19**, 4004–4014.
4. Deslandes, L., Olivier, J., Peeters, N., Feng, D. X., Khounlotham, M., Boucher, C., Somssich, I., Genin, S., and Marco, Y. (2003) Physical interaction between RRS1-R, a protein conferring resistance to bacterial wilt, and PopP2, a type III effector targeted to the plant nucleus. *Proc Natl Acad Sci USA.* **100**, 8024–8029.
5. Shao, F., Golstein, C., Ade, J., Stoutemyer, M., Dixon, J. E., and Innes, R. W. (2003) Cleavage of Arabidopsis PBS1 by a bacterial type III effector. *Science.* **301**, 1230–1233.
6. Axtell, M. J., and Staskawicz, B. J. (2003) Initiation of RPS2-specified disease resistance in Arabidopsis is coupled to the AvrRpt2-directed elimination of RIN4. *Cell.* **112**, 369–377.
7. Mackey, D., Holt, B. F., Wiig, A., and Dangl, J. L. (2002) RIN4 interacts with Pseudomonas syringae type III effector molecules and is required for RPM1-mediated resistance in Arabidopsis. *Cell.* **108**, 743–754.
8. Mackey, D., Belkhadir, Y., Alonso, J. M., Ecker, J. R., and Dangl, J. L. (2003) Arabidopsis RIN4 is a target of the type III virulence effector AvrRpt2 and modulates RPS2-mediated resistance. *Cell.* **112**, 379–389.
9. Leister, R. T., and Katagiri, F. (2000) A resistance gene product of the nucleotide binding site – leucine rich repeats class can form a complex with bacterial avirulence proteins in vivo. *Plant J.* **22**, 345–354.
10. Belkhadir, Y., Subramaniam, R., and Dangl, J. L. (2004) Plant disease resistance protein signaling: NBS-LRR proteins and their partners. *Curr Opin Plant Biol.* **7**, 391–399.
11. Boyes, D. C., Nam, J., and Dangl, J. L. (1998) The Arabidopsis thaliana RPM1 disease resistance gene product is a peripheral plasma membrane protein that is degraded coincident with the hypersensitive response. *Proc Natl Acad Sci USA.* **95**, 15849–15854.
12. Holt, B. F., III, Belkhadir, Y., and Dangl, J. L. (2005) Antagonistic control of disease resistance protein stability in the plant immune system. *Science.* **309**, 929–932.
13. Tian, D., Traw, M. B., Chen, J. Q., Kreitman, M., and Bergelson, J. (2003) Fitness costs of R-gene-mediated resistance in Arabidopsis thaliana. *Nature.* **423**, 74–77.
14. Qi, Y., and Katagiri, F. (2009) Purification of low-abundance Arabidopsis plasma-membrane protein complexes and identification of candidate components. *Plant J.* **57**, 932–944.
15. Mersereau, M., Pazour, G. J., and Das, A. (1990) Efficient transformation of Agrobacterium tumefaciens by electroporation. *Gene.* **90**, 149–151.
16. Clough, S. J., and Bent, A. F. (1998) Floral dip: a simplified method for Agrobacterium-mediated transformation of Arabidopsis thaliana. *Plant J.* **16**, 735–743.

Chapter 4

Biochemical Purification of Native Immune Protein Complexes

James M. Elmore and Gitta Coaker

Abstract

Protein complex purification represents a powerful approach to identify novel players in plant innate immunity. However, the identification of interacting protein partners within a natural context has been a challenge for researchers. In this chapter, we describe a method of immunoaffinity chromatography using purified, antibodies to isolate native protein complexes from wild-type tissue. We detail the antibody purification and immobilization steps in addition to the co-immunoprecipitation protocol. In addition, a method to prepare protein samples for mass spectroscopy analysis is described. This straightforward protocol has been used to isolate and identify novel components of *Arabidopsis* immunity-associated protein complexes.

Key words: Immunoprecipitation, Protein complex, Immune complex, Affinity chromatography

1. Introduction

Most biological responses require active adjustments of the cellular proteome. Emerging proteomics technologies have permitted the direct, large-scale analysis of proteins within the cell, which is a necessary component of attaining a systems understanding of plant biology. The earliest cellular responses to a given stimulus are frequently achieved through rapid alterations in protein activity and function via posttranslational regulatory modifications, changes in subcellular localization, and dynamic associations with interacting partners. Proteins are social entities; they interact with each other and work together to achieve necessary changes within each cell and the organism as a whole. However, the identification of interacting protein partners within a natural context has been a challenge for researchers. The elucidation of native protein complex composition in vivo

John M. McDowell (ed.), *Plant Immunity: Methods and Protocols*, Methods in Molecular Biology, vol. 712,
DOI 10.1007/978-1-61737-998-7_4,

and the dynamic changes that occur during active cellular signaling situations will greatly enhance our understanding of how plants respond to their environment.

In recent years, it has been demonstrated that many proteins involved in the plant innate immune response exist as members of multiprotein complexes within the cell. The molecular chaperone HSP90 (Heat Shock Protein 90 kD) (1) and co-chaperones SGT1 (Suppressor of G2 allele of *skp1*) (2) and RAR1 (Required for MLa12 Resistance) (3) are positive regulators of immune signaling. These chaperones interact with each other and various isoforms associate in different complexes with multiple disease resistance (R) proteins (1, 4–6). Moreover, specific R proteins have differential requirements of SGT1 and RAR1 for protein accumulation and activation of defense responses (6–9), indicating that distinct pools of these chaperone-R protein complexes exist within the plant cell. Additional components of immunity-associated protein complexes are now emerging. For example, a rice Rac/Rop GTPase, OsRac1, interacts with HSP90 and RAR1, but not SGT1, to regulate basal defense responses in rice (10). Most of the studies cited above utilized yeast two-hybrid library screening to identify novel interactions, and/or in vitro or in vivo overexpression and co-immunoprecipitation to explore potential interactions (1, 2, 5, 10). The ability to identify novel protein interactions in vivo under native expression levels will enhance investigations into defense signaling pathways.

Mass spectrometry-based techniques for protein identification and characterization have emerged as a powerful tool to examine complex protein samples (11). This technology represents an excellent platform to identify novel protein–protein interactions through the affinity purification of target proteins and subsequent identification of associated proteins. Many purification tools and protocols have been developed using this general strategy. Small epitope tags such as the hemagglutinin (HA) or FLAG epitopes are commonly fused to protein coding sequences and transformed into *Arabidopsis* to investigate protein function, but these tags alone may not be suitable for protein complex purification due to moderate to high amounts of contaminants resulting from washing constraints. In an accompanying chapter, Qi and colleagues describe an affinity purification strategy using a high-affinity biotinylated tag that seeks to address this issue. Additionally, various tandem affinity purification (TAP) schemes have been developed for use in plants that employ a two-step affinity purification protocol in order to isolate protein complexes at a higher purity (12–14). However, many TAP systems use relatively large protein fusion tags and multiple purification steps that could disrupt native complex formation and integrity.

Alternatively, the isolation of protein complexes can be achieved by using antibodies raised against the target protein of interest. This strategy precludes the construction of a tagged transgene and functional validation of the tagged protein in transformed plants. Moreover, the affinity purification of protein complexes in a wild-type plant background assures that native protein complex formation has been preserved in the experimental plant material. The ability to directly isolate wild-type protein complexes under natural expression levels is arguably the most biologically relevant method for complex purification.

The *Arabidopsis* protein RIN4 has been identified as an important component of plant immune responses. RIN4 is localized to the plasma membrane and associates with at least two R proteins, RPS2 and RPM1, which monitor RIN4 for modifications induced by bacterial pathogen effector proteins (15–17). RIN4 associates with both RPS2 and RPM1 and negatively regulates these R proteins in the absence of their cognate effectors. In addition, RIN4 also negatively regulates basal defense responses and *rin4* mutant plants exhibit enhanced disease resistance in the absence of RPS2 and RPM1 (18). Due to these interesting aspects of RIN4 function at two levels of the plant innate immune response, a strategy was developed in order to identify additional components of the RIN4 immunity-associated protein complex.

In this chapter, we describe a protocol for isolating native protein complexes from *Arabidopsis thaliana*. The methods that are outlined were established from our experiments with the RIN4 protein and we have successfully used this protocol to isolate additional immune complexes (19). Essentially, polyclonal antibodies are affinity purified and immobilized on protein A or G beads. These antibody beads are used to isolate native-level protein complexes containing the target protein by immunoaffinity chromatography. The entire co-immunoprecipitated sample is subjected to mass spectrometry analysis in order to identify individual complex constituents. This straightforward method represents a powerful tool for native protein complex purification and characterization, and can easily be modified to purify other native protein complexes *in planta*.

2. Materials

2.1. Immobilization of Antigen for Affinity Purification of Antibodies

1. Purified antigen (2–3 mL of 5–10 mg/mL).
2. Cyanogen Bromide-activated sepharose 4B (GE Healthcare).
3. 1 mM HCl.
4. Coupling buffer: 100 mM $NaHCO_3$, 500 mM NaCl, pH to 8.3 with NaOH.

5. Blocking buffer: 100 mM Tris, 500 mM NaCl, pH 8.0.
6. Acetate buffer: 100 mM NaOAc, 500 mM NaCl, pH to 4.0 with Glacial Acetic acid.

2.2. Affinity Purification of Antibody

1. Immobilized antigen (Subheading 3.1).
2. Crude antiserum to the protein of interest.
3. 20 mL glass econo-column (BioRad).
4. 1× Phosphate-buffered saline (PBS): 137 mM NaCl, 2.7 mM KCl, 10 mM Na_2HPO_4, 2 mM NaH_2PO_4, pH 7.4.
5. Wash buffer A: 50 mM Tris, 120 mM HCl, 0.5% NP-40, pH 8.0.
6. Wash buffer B: 50 mM Tris, 1 M LiCl, 0.5% NP-40, pH 8.0.
7. Elution buffer: 50 mM Glycine-Cl, NaCl 150 mM, pH 2.5.
8. 2 M Tris.
9. 2% NaN_3 (sodium azide) in water.

2.3. Immobilization of Antibody on Protein A-Sepharose

1. Affinity-purified antibody (Subheading 2.3).
2. Protein A sepharose (GE Healthcare).
3. Cross-linking buffer: 200 mM sodium borate ($Na_2B_2O_7$), pH 9.0.
4. Dimethyl pimelimidate (DMP) solid.
5. Wash buffer: 200 mM ethanolamine, pH 8.0.
6. Final buffer: 0.01% Merthiolate in PBS.

2.4. Arabidopsis thaliana Growth Conditions

1. Four-week-old *Arabidopsis* plants, grown vegetatively in a controlled environmental chamber at 24°C with a 10 h-light/14 h-dark photoperiod under a light intensity of 85 μE/m^2/s.

2.5. Protein Extraction and Immunoprecipitation

1. 20 mL glass econo-column (BioRad).
2. Immobilized affinity-purified antibody (Subheading 2.4).
3. 0.45 μm HPF Millex-HV high particulate filter, syringe-driven (Millipore).
4. 10 mL syringe.
5. Immunoprecipitation (IP) buffer 1 (extraction): 50 mM HEPES, 50 mM NaCl, 10 mM EDTA, 0.2% Triton X-100, 1× Complete Protease inhibitor cocktail (Roche), 0.1 mg/mL Dextran (Sigma D1037), pH7.5 (see Note 1).
6. IP buffer 2 (low salt wash): 50 mM HEPES, 50 mM NaCl, 10 mM EDTA, 0.1% Triton X-100, pH 7.5.
7. IP buffer 3 (high salt wash): 50 mM HEPES, 150 mM NaCl, 10 mM EDTA, 0.1% Triton X-100, pH 7.5.

8. Phosphate buffer: 10 mM potassium phosphate, 50 mM NaCl, pH 6.8.
9. Elution buffer (low pH): 50 mM Glycine-Cl, 150 mM NaCl, 0.1% Triton X-100, pH 2.5.
10. Neutralization buffer: 2 M Tris.
11. Strataclean Resin (Agilent).
12. 3× Laemmli buffer: 188 mM Tris, 6% SDS, 30% glycerol, 15% 2-mercaptoethanol, 0.003% bromophenol blue, pH 6.8.

2.6. Sample Analysis

1. SDS-PAGE and Western Blotting equipment and reagents.
2. SilverQuest Silver Staining Kit (Invitrogen).

2.7. In-gel Trypsin Digest for Mass Spectrometry

1. Novex Colloidal Blue Stain Kit (Invitrogen).
2. Vacuum centrifuge.
3. Ultrasonic waterbath.
4. Wash buffer: 100 mM ammonium bicarbonate, NH_4HCO_3.
5. 100% Acetonitrile.
6. Reducing buffer: 10 mM DTT in 100 mM NH_4HCO_3.
7. Alkylating buffer: 55 mM iodoacetamide, 100 mM NH_4HCO_3 (iodoacetamide is highly toxic and care should be taken to reduce exposure).
8. Digestion buffer: 13 ng/μL Sequencing grade modified trypsin (Promega) in 50 mM NH_4HCO_3.
9. 60% Acetonitrile, 1% Trifluoroacetic acid in water.
10. 0.1% Trifluoroacetic acid in water.

3. Methods

Although mass spectrometry technology breakthroughs have greatly enhanced the amount of information that can be obtained from complex protein samples, the success of this powerful tool is dependent largely on the quality of protein samples delivered into the machine. Rapid and simple tissue processing can guarantee the reliability of results across multiple experimental replicates. The ability of newer mass spectrometers to analyze increasingly complex samples allows the researcher to isolate protein complexes with minimal purification steps, which favors the preservation of intact complexes. The analysis of all precipitated proteins directly via mass spectrometry without a SDS-PAGE band excision step can facilitate the identification of low abundance interactors. If high quality mass spectrometers are not available, individual SDS-PAGE band excision or additional liquid chromatography (e.g., MuDPIT) steps prior to loading on the machine may be necessary.

Experimental replication is a critical aspect of this method. We commonly use three biological replicates for each experiment. In order to be classified as a bona fide interacting protein during data analysis, the protein should be reliably identified in every experimental replicate and never in the negative control samples. With appropriate control samples, background spectra can easily be subtracted from the test samples. As a negative control, knockout lines in your protein of interest should be used to account for nonspecific binding. If a knockout line is unavailable, wild-type tissue extracts can be subjected to protein A or G sepharose precipitation to control for nonspecific binding. If applicable, the ability to troubleshoot the protocol with a known interacting protein will facilitate the success of the experiments. After each experimental replicate, the presence of your target protein should be verified in the elution fraction prior to mass spectrometry analysis.

The methods described below have been used to reproducibly isolate native-level protein complexes *in planta*. Before purification of protein complexes, polyclonal antibodies must be affinity purified in order to ensure the maximum capture of your target protein and to reduce nonspecific binding. The general format for affinity purification includes immobilizing the antigen on sepharose beads and using this to purify antibodies from crude antisera. Affinity purification of the polyclonal antibodies with purified target protein will result in higher complex purity during co-immunoprecipitation. Purified antibodies are cross-linked to Protein A-sepharose to facilitate column chromatography steps. Co-immunoprecipitation is performed with minimal tissue processing. Sample preparation for mass spectrometry analysis is also described. Steps critical for troubleshooting each method are also highlighted.

3.1. Antigen Immobilization for Affinity Purification of Antibody

This section describes the cross-linking of purified target protein to sepharose beads which will subsequently be used to affinity purify antibodies.

1. Weigh 1.0 g CN-Br-activated sepharose 4B and swell powder for 15 min at room temperature (RT) in 15-mL conical tube with 15 mL of 1 mM HCl (see Note 2).
2. Centrifuge at 1,000 *g* for 1–2 min at RT.
3. Wash swollen gel four times with 15 mL of 1 mM HCl. Spin as described in step 2 and remove supernatant between washes.
4. Wash the gel with 2–3 mL coupling buffer, centrifuge at 1,000 *g* and remove supernatant. *Immediately* add protein (2–3 mL of 5–10 mg/mL) to be coupled. Add coupling buffer to the tube to a final volume of 10 mL (see Note 3).

5. Incubate protein and gel solution end-over-end at 4°C overnight. Remove a 20 μL aliquot, boil in 1× Laemmli buffer, and use supernatant to check coupling efficiency by running an SDS-PAGE gel (see Note 4).
6. Centrifuge sample at 1,000 *g* for 1–2 min at RT. Re-suspend in 10 mL of coupling buffer and add 1/10 bed volume of blocking buffer. Block remaining active groups for 2 h at RT.
7. Wash excess non-covalently bound protein with 5 mL coupling buffer followed by 5 mL acetate buffer. Wash alternately in coupling and acetate buffers three to four times, ending with coupling buffer. The sample should be centrifuged at 1,000 *g*, 1–2 min at RT between wash steps.
8. Store protein-sepharose conjugate at 4°C in 5 mL PBS with 0.02% Na-azide until ready for use (see Note 5).

3.2. Affinity Purification of Antibody

This step describes the purification of polyclonal antibodies from crude antisera that will be subsequently immobilized and used for co-immunoprecipitaion of the target protein.

1. Incubate 3 mL protein-sepharose conjugate "beads" from Subheading 3.1 with 40 mL crude antiserum in a 50 mL tube rotating end-over-end overnight at 4°C (see Note 6).
2. All steps below should be performed with ice-cold buffers.
3. Load serum/beads mixture into BioRad econo-column, collect flow through and save for analysis (see Note 7). Rinse tube twice with 10 mL PBS and transfer to column.
4. Wash column with 20 mL Wash buffer A.
5. Wash column with 20 mL Wash buffer B.
6. Wash column with 20 mL Wash buffer A.
7. Wash column with 20 mL PBS.
8. Add 7.5 μL of 2 M Tris buffer to 25 1.5 mL tubes (to neutralize eluted fractions).
9. Elute column twice by adding 5 mL of elution buffer each time and collecting 500 μL fractions. Mix tubes to neutralize pH.
10. Quantify the antibody in each tube via Bradford assay. Generate two pools of purified antibody: one low concentration (0.25 mg/mL) and one high concentration (>0.5 mg/mL).
11. Add 2% Na-azide to 0.02% final concentration. Antibody can be aliquoted and stored at 4°C or –20°C for long-term storage.
12. Wash column twice with 10 mL PBS. Cap the column and add 1 mL PBS with 0.02% Na-azide to protein-beads. Column can be stored at 4°C (see Note 5).

3.3. Antibody Immobilization on Protein A-Sepharose

This step describes the generation of antibody beads used during the immunoaffinity chromatography to isolate protein complexes.

1. Incubate 2 mg affinity-purified antibody with 1 mL Protein A-sepharose in 10 mL PBS buffer end-over-end for 1 h at RT (see Note 8).
2. Centrifuge 3000 × *g* for 5 min and wash beads twice with 10 mL cross-linking buffer.
3. Re-suspend beads in 10 mL cross-linking buffer and remove 100 μL of bead suspension for analysis (step 9).
4. To initiate cross-linking reaction, add DMP (solid) to final concentration of 20 mM. (see Note 9).
5. Incubate 30 min end-over-end at RT. Remove 100 μL for analysis (step 9).
6. Centrifuge at 3,000 × *g* for 5 min and discard supernatant. To stop the reaction, wash beads once with 10 mL wash buffer, centrifuge at 3,000 × *g* for 5 min and discard supernatant.
7. Add 10 mL fresh wash buffer and incubate for 2 h end-over-end at RT.
8. Centrifuge and re-suspend in 10 mL Final buffer. Antibody beads can be stored at 4°C (see Note 10).
9. Check cross-linking efficiency by boiling samples before (step 3) and after (step 5) cross-linking in 1× Laemmli buffer. (see Note 11).

3.4. Plant Growth and Tissue Collection

1. *Arabidopsis* plants are grown vegetatively in a controlled environment chamber (see Note 12).
2. Collect 5 g of *Arabidopsis* leaf tissue in foil, freeze in liquid N_2, and store at –80°C (see Note 13).

3.5. Protein Extraction and Immunoprecipitation

The quality and reproducibility of mass spectrometry data relies heavily on the quality of the protein samples that are delivered for analysis. Therefore, much care should be taken during buffer preparation and tissue processing. We have found it most useful to include the simplest number of steps during tissue processing and co-immunoprecipitation in order to ensure the most reliable and consistent results across experimental replicates.

1. All steps should be conducted in a cold room at 4°C with ice-cold buffers (see Note 14). All centrifugation steps should be conducted at 4°C.
2. Grind 5 g of leaf tissue in liquid N_2 with mortar and pestle. Re-suspend in 11 mL IP Buffer 1.
3. Centrifuge at 10,000 × *g* for 10 min at 4°C.
4. During centrifugation of cell lysate, wash antibody beads in IP buffer 1 (without protease inhibitors). Centrifuge antibody

beads at 3,000 rpm for 3 min, discard supernatant, re-suspend in IP buffer 1 to wash. Repeat (see Note 15).

5. Filter cell lysate supernatant using a 0.45 μm high particulate filter and syringe. Remove a 50 μL aliquot of input sample and save for analysis (see Note 16).
6. Add 500 μL antibody-coupled beads to filtered lysate. Incubate end-over-end for 2.5 h at 4°C (see Note 17).
7. Load material onto glass column. Pass flowthrough through column twice. Remove a 50 μL aliquot of the final flowthrough and save for analysis (see Note 16).
8. Wash column twice with 20 mL IP buffer 2 (low salt wash).
9. Wash column twice with 10 mL IP buffer 3 (high salt wash). Remove a 50 μL aliquot of the high salt wash flowthrough and save for analysis (see Note 16).
10. Wash column with 5 mL Phosphate buffer.
11. Elute bound proteins four times with 1 mL Elution buffer. Let buffer sit on column for 2 min before each elution. Collect each fraction and neutralize with 15 μL 2 M Tris.
12. Combine elution fractions and concentrate with Strataclean Resin. Add 20 μL of Strataclean resin to the pooled elution and incubate the sample end-over-end for 20 min at RT. Centrifuge at 5,000×*g* for 2 min, remove supernatant, and re-suspend resin in 40 μL of 1× Laemmli buffer (see Note 18).
13. Elute protein from Strataclean resin by boiling in 40 μL 1× Laemmli buffer. Centrifuge sample at 10,000×*g* for 1 min, transfer supernatant to a new 1.5 mL tube and store at −20°C.
14. To recharge antibody beads in the column, wash column with 30 mL (3×10 mL washes) Elution buffer. Neutralize column with 30 mL (3×10 mL) phosphate buffer. Store in PBS at 4°C (see Note 19).

3.6. Sample Assessment

1. Verify immunoprecipitation of target protein and any known interactors by running 5 μL of the sample on a 12% SDS-PAGE gel followed by western blotting.
2. Visualize co-immunoprecipitated proteins by running 5 μL of the sample on a 12% SDS-PAGE gel and stain using SilverQuest Silver Staining Kit (Invitrogen) (Fig. 1).

3.7. Sample Preparation for Mass Spectrometry

1. Run the remaining 30 μL sample on a 10% SDS-PAGE gel. Run sample about 6 mM into the separating gel.
2. Stain by colloidal coomassie (Invitrogen kit).
3. Remove protein sample smear with razor and perform an in-gel trypsin digest (Subheading 3.8) (see Note 20).

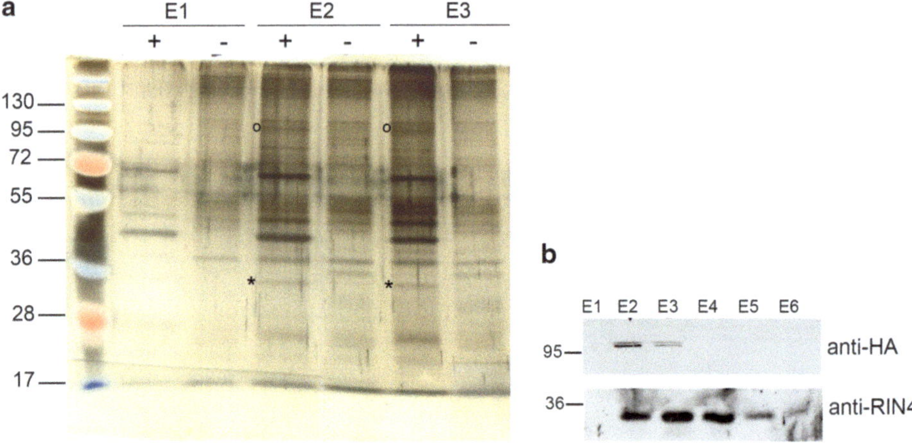

Fig. 1. Isolating the RIN4 protein complex. Affinity-purified RIN4 antiserum coupled to protein A was used to purify the RIN4 protein complex from 5 g of Arabidopsis leaf tissue. (**a**) Silver stained gel of elutions. The double mutant rps2-101c/rin4 was used as a negative control, while the transgenic line expressing npro*RPS2:HA* in the rps2-101c background was used as a positive control. *Asterisks* and *circles* indicate bands corresponding to the molecular weight of RIN4 and RPS2:HA, respectively. (**b**) Western blots of eluted fractions.

3.8. In-Gel Trypsin Digestion

1. Chop gel pieces into 1 mM fragments and transfer to a 1.5-mL tube.
2. Wash gel pieces with 100 μL Wash buffer for 5 min (see Note 21).
3. Discard buffer and add 50 μL 100% acetonitrile and dehydrate at RT for 15 min (see Note 22).
4. Remove acetonitrile and dry completely in a vacuum centrifuge for approximately 15 min (gel pieces should turn completely white).
5. Rehydrate the gel pieces with 50 μL Reducing buffer. Heat at 56°C for 30 min to reduce the sample.
6. Remove reducing buffer and add 50 μL 100% acetonitrile. Incubate for 3–5 min at RT. Repeat twice.
7. Dry in vacuum centrifuge for 15 min.
8. Add 50 μL Alkylating buffer to alkylate cysteine residues. Incubate for 20 min in dark at RT.
9. Discard supernatant and wash briefly with 50 μL Wash buffer.
10. Remove buffer and add 50 μL Wash buffer and incubate for 15 min at RT (see Note 23).
11. Remove liquid and add 50 μL 100% acetonitrile. Incubate for 15 min at RT.
12. Dry completely in vacuum centrifuge.

13. Rehydrate gel pieces in 30–50 μL Digestion buffer. Add enough digestion buffer to cover the gel pieces. Add more Digestion buffer if gel pieces absorb all the liquid.
14. Allow gel pieces to absorb digestion buffer (see Note 24).
15. Incubate at 37 C overnight.
16. Spin sample down, collect liquid into new 1.5 mL tube, and save.
17. Add 15–30 μL of 60% Acetonitrile, 1% trifluoroacetic acid in water to each gel piece.
18. Sonicate gel pieces in ultrasonic waterbath for 10 min.
19. Centrifuge tubes at 21,000 × *g* for 30 s.
20. Collect supernatant and add to solution from step 17.
21. Vacuum centrifuge solution from step 26 until *almost* dry (see Note 25).
22. Add 5–20 μL of 0.1% Trifluoroacetic acid to the tubes and sonicate tube in Branson 1200 waterbath for 5 min.
23. Centrifuge tubes at 21,000 × *g* for 30 s.
24. Store samples at 4 C if it will be analyzed within a few days or freeze at −80 C (see Note 26).

4. Notes

1. Inclusion of Dextran (MW = 400,000–500,000) in IP buffer 1 can reduce nonspecific protein–protein interactions.
2. Insoluble proteins can be coupled in coupling buffer containing 2 M urea.
3. Either recombinant protein or peptide antigen can be used. Recombinant proteins should be at least 80% pure. All steps with coupling buffer should be completed without delay because reactive groups hydrolyze at the coupling pH.
4. Coupling efficiency can be verified by boiling a small aliquot of beads in 1× Laemmli buffer and running on an SDS-PAGE gel. When compared to the input, little or no protein should be observed after coupling is completed.
5. The protein-sepharose conjugate can be stored for years at 4°C and can be repeatedly used for antibody purifications.
6. The amount of serum used depends on the titer of the antibody. For highly antigenic proteins (>10,000 titer based on ELISA results for polyclonal antibodies) 15 mL of antisera should be sufficient.

7. The flow through sera can be saved and used for western blotting experiments.
8. Depending on type of antibody IgG, either protein A- or protein G-sepharose should be used. Refer to (20). Protein A-sepharose should be used for all antibodies produced in rabbit.
9. Cross-linking must be performed at pH > 8.3 to be efficient.
10. Antibody beads can be stored at 4°C for at least 1 year and used multiple times.
11. Successful cross-linking should result in absence of heavy chain bands (55 kD) in bead supernatant after boiling.
12. We use 4-week-old *Arabidopsis* plants for most of our experiments. Exact growth conditions and tissue treatments depend on your target protein and should be determined empirically.
13. If the target protein is localized to the chloroplast or mitochondria, purification of intact organelles prior to complex purification will reduce the number of contaminating proteins.
14. For protein extraction and immunoprecipitation, use freshly sterilized buffers made specifically for MS analysis that are free of common contaminants such as keratin. Buffers must be absolutely free of contamination in order to obtain robust mass spectrometry data.
15. We include Dextran protein in IP buffer 1 during complex binding but not during wash and elution steps. The addition of Dextran acts to decrease nonspecific protein binding. If nonspecific interactions are a problem, incubate antibody beads with IP Buffer 1 containing Dextran for 15 min to preblock beads.
16. Save aliquots of the cell lysate input, flow through, and high salt wash for troubleshooting purposes. Boil aliquots in 1× Laemmli and subject to SDS-PAGE and western blotting to visualize target protein and any known interactors at each step of the process. This information can be used to optimize immunoprecipitation conditions for the target protein complex. Depending on the target protein, it may be necessary to troubleshoot ideal IP buffer conditions. Solubility of target proteins can be achieved by increasing detergent concentrations as well as including different detergents. Nonspecific protein binding can be decreased by increasing the salt concentration. We have never purified interacting proteins when the salt concentration was increased above 300 mM.
17. It is ideal to use the shortest incubation time possible during this step to reduce nonspecific binding.

18. It is possible to precipitate proteins with a traditional Trichloroacetic Acid protocol. We have found that using Strataclean resin is more reproducible and results in higher yields than traditional precipitation protocols.
19. The low ph (Elution buffer) column washes must be completed rapidly to avoid denaturing the antibodies. The column can be reused multiple times (we use up to six times). We separate the columns and use one column for a specific genotype.
20. Alternatively, the sample can be sent to a mass spectrometry facility for sample preparation.
21. All solutions should be made fresh each time. Ensure that solutions are sterilized and free of common contaminants such as keratin.
22. The gel piece will turn white upon dehydration. Repeat if gel piece is not completely dehydrated.
23. If the gel pieces are not completely destained, wash with 1:1 acetonitrile: Wash buffer solution at 37°C for 30 min.
24. Incubate samples at 37°C for 15–25 min to facilitate absorption of the digestion buffer. If all the digestion buffer is absorbed by the gel pieces after the 15–25 min at 37°C, then add 10–30 μL of 0.5× Wash buffer (enough to cover samples).
25. Do not dry out the samples completely or for an extended amount of time.
26. Freezing the samples can result in sample loss.

Acknowledgments

We thank Brett Phinney at UC Davis Genome Center Proteomics Core Facility for providing the in-gel trypsin digestion protocol.

References

1. Takahashi, A., Casais, C., Ichimura, K., and Shirasu, K. (2003) HSP90 interacts with RAR1 and SGT1 and is essential for RPS2-mediated disease resistance in Arabidopsis, *Proc Natl Acad Sci USA* **100**, 11777–11782.
2. Azevedo, C., Sadanandom, A., Kitagawa, K., Freialdenhoven, A., Shirasu, K., and Schulze-Lefert, P. (2002) The RAR1 interactor SGT1, an essential component of R gene-triggered disease resistance, *Science* **295**, 2073–2076.
3. Muskett, P. R., Kahn, K., Austin, M. J., Moisan, L. J., Sadanandom, A., Shirasu, K., Jones, J. D. G., and Parker, J. E. (2002) Arabidopsis RAR1 exerts rate-limiting control of R gene-mediated defenses against multiple pathogens, *Plant Cell* **14**, 979–992.
4. Hubert, D. A., Tornero, P., Belkhadir, Y., Krishna, P., Takahashi, A., Shirasu, K., and Dangl, J. L. (2003) Cytosolic HSP90 associates with and modulates the Arabidopsis

RPM1 disease resistance protein, *EMBO J* **22**, 5679–5689.
5. Liu, Y. L., Burch-Smith, T., Schiff, M., Feng, S. H., and Dinesh-Kumar, S. P. (2004) Molecular chaperone Hsp90 associates with resistance protein N and its signaling proteins SGT1 and Rar1 to modulate an innate immune response in plants, *J Biol Chem* **279**, 2101–2108.
6. Shirasu, K., and Schulze-Lefert, P. (2003) Complex formation, promiscuity and multi-functionality: protein interactions in disease-resistance pathways, *Trends Plant Sci* **8**, 252–258.
7. Austin, M. J., Muskett, P., Kahn, K., Feys, B. J., Jones, J. D. G., and Parker, J. E. (2002) Regulatory role of SGT1 in early R gene-mediated plant defenses, *Science* **295**, 2077–2080.
8. Bieri, S., Mauch, S., Shen, Q. H., Peart, J., Devoto, A., Casais, C., Ceron, F., Schulze, S., Steinbiss, H. H., Shirasu, K., and Schulze-Lefert, P. (2004) RAR1 positively controls steady state levels of barley MLA resistance proteins and enables sufficient MLA6 accumulation for effective resistance, *Plant Cell* **16**, 3480–3495.
9. Azevedo, C., Betsuyaku, S., Peart, J., Takahashi, A., Noel, L., Sadanandom, A., Casais, C., Parker, J., and Shirasu, K. (2006) Role of SGT1 in resistance protein accumulation in plant immunity, *EMBO J* **25**, 2007–2016.
10. Thao, N. P., Chen, L., Nakashima, A., Hara, S. I., Umemura, K., Takahashi, A., Shirasu, K., Kawasaki, T., and Shimamoto, K. (2007) RAR1 and HSP90 form a complex with Rac/Rop GTPase and function in innate-immune responses in rice, *Plant Cell* **19**, 4035–4045.
11. Aebersold, R., and Mann, M. (2003) Mass spectrometry-based proteomics, *Nature* **422**, 198–207.
12. Rohila, J. S., Chen, M., Cerny, R., and Fromm, M. E. (2004) Improved tandem affinity purification tag and methods for isolation of protein heterocomplexes from plants, *Plant J* **38**, 172–181.
13. Rubio, V., Shen, Y. P., Saijo, Y., Liu, Y. L., Gusmaroli, G., Dinesh-Kumar, S. P., and Deng, X. W. (2005) An alternative tandem affinity purification strategy applied to Arabidopsis protein complex isolation, *Plant J* **41**, 767–778.
14. Van Leene, J., Witters, E., Inze, D., and De Jaeger, G. (2008) Boosting tandem affinity purification of plant protein complexes, *Trends Plant Sci* **13**, 517–520.
15. Mackey, D., Holt, B. F., Wiig, A., and Dangl, J. L. (2002) RIN4 interacts with Pseudomonas syringae type III effector molecules and is required for RPM1-mediated resistance in Arabidopsis, *Cell* **108**, 743–754.
16. Mackey, D., Belkhadir, Y., Alonso, J. M., Ecker, J. R., and Dangl, J. L. (2003) Arabidopsis RIN4 is a target of the type III virulence effector AvrRpt2 and modulates RPS2-mediated resistance, *Cell* **112**, 379–389.
17. Axtell, M. J., and Staskawicz, B. J. (2003) Initiation of RPS2-specified disease resistance in Arabidopsis is coupled to the AvrRpt2-directed elimination of RIN4, *Cell* **112**, 369–377.
18. Kim, M. G., da Cunha, L., McFall, A. J., Belkhadir, Y., DebRoy, S., Dangl, J. L., and Mackey, D. (2005) Two Pseudomonas syringae type III effectors inhibit RIN4-regulated basal defense in Arabidopsis, *Cell* **121**, 749–759.
19. Liu J, Elmore JM, Fuglsang AT, Palmgren MG, Staskawicz BJ, Coaker G. (2009) RIN4 functions with plasma membrane H+-ATPases to regulate stomatal apertures during pathogen attack, *PLoS Biol* 7(6), e1000139.
20. Harlow, E., and Lane, D. (1988) Antibodies: A Laboratory Manual, *Cold Spring Harbor Laboratory Press.*

Chapter 5

Chromatin Immunoprecipitation to Identify Global Targets of WRKY Transcription Factor Family Members Involved in Plant Immunity

Mario Roccaro and Imre E. Somssich

Abstract

The completion of the alfalfa, *Arabidopsis*, papaya, poplar, and rice genome sequences along with ongoing sequencing projects of various crop species, offers an excellent opportunity to study gene expression at the whole genome level and to unravel the complexity of gene networks underlying the reprogramming of plant defense toward pathogen challenge. Gene expression in eukaryotic cells is mainly controlled by regulatory elements that recruit transcription factors (TFs) to modulate transcriptional outputs. Therefore, methods allowing the identification of all cognate TF binding sites (TFBS) within the regulatory regions of target genes on a genome-wide basis are the next obvious step to elucidate the plant defense transcriptome. Chromatin immunoprecipitation (ChIP) is one such powerful technique for analyzing functional *cis*-regulatory DNA elements. The ChIP assay allows the identification of specific regulatory DNA regions associated with *trans*-acting regulatory factors in vivo. ChIP assays can provide spatial and temporal snapshots of the regulatory components involved in reprogramming host gene expression upon pathogen ingress. Moreover, the use of ChIP-enriched DNA for hybridization to tiling microarrays (ChIP-chip) or for direct sequencing (ChIP-Seq) by means of massively parallel sequencing technology has expanded this methodology to address global changes in gene expression.

Key words: Antibody, Arabidopsis, *cis*-acting DNA elements, Cross-linked DNA–protein complex, Defense transcriptome, Epitope-tag, Formaldehyde (X-ChIP), Gene expression, *Golovynomices orontii*

1. Introduction

Chromatin immunoprecipitation (ChIP) technology provides a vital tool to investigate dynamic changes occurring on chromatin in vivo, for example, in response to various exogenous or endogenous signals or during development (1). This method has become widely used in yeast and animal studies and more recently also in plant research (2). In plants, it is now well established that

John M. McDowell (ed.), *Plant Immunity: Methods and Protocols*, Methods in Molecular Biology, vol. 712,
DOI 10.1007/978-1-61737-998-7_5,

massive changes in host gene expression occur following pathogen attack with a subsequent major redistribution and recruitment of a large number of specific transcription factors (TFs) to their cognate binding sites. Pathogen-triggered host transcriptional reprogramming can occur very rapidly and in a highly dynamic fashion, with diverse sets of TFs triggering distinct immune responses during the different stages of the plant–pathogen interaction (3). Our group is interested in determining the global contribution of the WRKY family of TFs in modulating the Arabidopsis defense transcriptome (4). Thus, ChIP is an essential part of such investigations. ChIP often employs the cell wall/membrane diffusible and dipolar chemical agent formaldehyde (HCHO), which is able to react with amino- and imino-groups of amino acids and of DNA. The cross-link reaction occurs within minutes upon addition of formaldehyde, and it is readily reversible by heat or in the presence of low pH and available protons (5). Formaldehyde produces protein–protein, protein–DNA, and protein–RNA complexes. It also can be used to study proteins that are not directly making contact to the DNA but that are in close proximity to genuine DNA-interacting proteins. However, one should note that formaldehyde is not adequate to detect all interactions, and thus alternative cross-linking procedures should also be considered (6). Once the biological material has been cross-linked and the chromatin extracted in a soluble form, the protein–DNA complexes can be immunoprecipitated by specific antibodies or using an antibody for an epitope-tagged version of the factor under investigation. In a final step, the immunoprecipitated protein–DNA complexes are de-cross-linked and the ChIP DNA fragments can be used in qPCR experiments to validate candidate target genes, hybridized to DNA tiling arrays (7), or cloned in bulk and subjected to second generation sequencing to enable genome-wide discovery of target genes for DNA-associated proteins (8).

2. Materials

2.1. Extraction of Soluble Cross-Linked Chromatin

1. 37% Formaldehyde (molecular biology grade) stabilized with 10% methanol. Formaldehyde is a volatile toxic compound; take all precautions to avoid inhalation or direct contact.
2. 1.25 M glycine: 18.76 g glycine, in 200 mL of water. Sterilize by filtration and store at room temperature.
3. 10× Phosphate-Buffer Saline (10× PBS): 80 g NaCl, 2 g KCl, 1.44 g Na_2HPO_4, 2.4 g KH_2PO_4, 1 L Milli-Q water. Adjust the pH of the buffer to 7.5 to obtain 10× PBS stock solution.
4. 2% Triton X-100 pure in water.

5. 2% Tween 20 in water.
6. Protease inhibitor cocktail tablets (ROCHE, Complete-Mini EDTA-free).
7. Phosphatase inhibitor cocktail-1, (Sigma-Aldrich). Required when the DNA binding factor under investigation may undergo phosphorylation to exert its DNA binding ability.
8. 100 mM Phenylmethanesulfonyl fluorid (PMSF): 0.174 g PMSF (Sigma-Aldrich) in 10 mL absolute ethanol. Aliquot into 1 mL microcentrifuge tubes and store at –20°C. (Half-life of 1 mM PMSF upon addition to the Extraction Buffers is about 30 min, therefore add just prior to use).
9. 4′,6-Diamidino-2-phenylindole dihydrochloride (DAPI) (Sigma-Aldrich). Dissolve the powder in water to a concentration of 5 mg/mL. Protect from light. The working concentration is 2.5 μL/mL. DAPI is a cell permeable fluorescent dye that binds to the minor groove of DNA. It is used to visualize nuclei and the release of the cross-linked chromatin after nuclear lysis.
10. Fixing Solution 250 mL: 25 mL 10× PBS-buffer, 2.5 mL 2% Triton X-100, 2.5 mL 2% Tween 20, 213.25 mL cold Milli-Q quality water; keep cold until use, 6.75 mL 37% formaldehyde (added just before fixation of the plant material).
11. Nuclei Extraction Buffer 1 (NEB1): 0.5 M sucrose, 10 mM Tris–HCl pH 8.0, 10 mM MgCl, 5 mM β-mercaptoethanol (β-ME), 0.1 mM PMSF, Protease inhibitor two tablets to 100 mL of NEB1 (ROCHE, Complete-Mini EDTA-free), Phosphatase inhibitor cocktail-1 (suggested dilution 1:1,000).
12. Nuclei Extraction Buffer 2 (NEB2): 0.25 M sucrose, 10 mM Tris–HCl pH 8.0, 10 mM MgCl, 1% Triton X-100, 5 mM β-ME, 0.1 mM PMSF, Protease inhibitor one tablet to 10 mL of NEB2, Phosphatase inhibitor cocktail-1 (suggested dilution 1:100).
13. Nuclei Extraction Buffer 3 (NEB3): 1.7 M sucrose, 10 mM Tris–HCl pH 8.0, 0.15% Triton X-100, 2 mM MgCl, 5 mM β-ME, 0.1 mM PMSF, Protease inhibitor one tablet to 10 mL of NEB3, Phosphatase inhibitor cocktail-1 (suggested dilution 1:100).
14. Nuclei Lysis Buffer (NLB): 50 mM Tris–HCl ph 8.0, 10 mM EDTA, 0.5% SDS, 0.1 mM PMSF, Protease inhibitor one tablet to 10 mL of NLB, Phosphatase inhibitor cocktail-1 (1:100 dilution).

2.2. Chromatin Immunoprecipitation

1. BSA, Albumin Bovine Serum Fraction V (Sigma-Aldrich, cat. no. A4503).
2. ChIP Dilution Buffer: 1% Triton X-100, 0.1% SDS, 0.1% sodium deoxycholate (Sigma-Aldrich), 140 mM NaCl,

10 mM sodium-pyrophosphate (Sigma-Aldrich), Protease inhibitor one tablet to 10 mL of NLB.

3. High salt wash solution: 500 mM NaCl, 0.1% SDS, 1% Triton X-100, 2 mM EDTA, 20 mM Tris–HCl pH 8.0.
4. Low salt wash solution: 150 mM NaCl, 0.1% SDS, 1% Triton X-100, 2 mM EDTA, 20 mM Tris–HCl pH 8.0.
5. LiCl wash solution: 250 mM LiCl, 1% Nonidet P-40, 0.1% sodium deoxycholate.
6. Proteinase K, (Invitrogen).
7. RNase, DNase free (ROCHE).
8. Dynabeads® M-280 Sheep anti-Rabbit IgG, (INVITROGEN).
9. Anti-HA antibody – ChIP Grade. Rabbit polyclonal to HA tag (ABCAM, cat. no. ab9110).
10. Ethanol.
11. Phenol/chloroform/isoamyl alcohol (25:24:1).
12. 3 M sodium acetate, pH 5.2.
13. Jiffy pots (Alwaysgrows).
14. *Golovynomices orontii.* Virulent *Arabidopsis* powdery mildew pathogen, *G. orontii*, was maintained on Col-0 WT *Arabidopsis* plants at 20°C, 16 h light/8 h darkness, 80% relative humidity.
15. Bioruptor™ sonication instrument (DIAGENODE, Liege, Belgium).

2.3. Detection of the HA-Tag

1. 10% SDS-polyacrylamide gel 10% (9).
2. 1× Tris-saline (TBS): 8 g NaCl, 20 mL of 1 M Tris–HCl pH 7.6. Dilute to 1,000 mL with distilled water. Add 1 mL of Tween 20 to obtain a TBS buffer with 0.1% Tween 20 (TBS-T) to use as diluent of the antibody and as wash buffer.
3. Western blotting (according to manufacturer instruction for ECL western blotting, GE Healthcare).
4. Anti-HA antibody (for Western). IgG1 Rat monoclonal antibody clone (suggested dilution 1:2,500 in 5% dry milk), (ROCHE, cat. no. 1867423001).
5. Secondary Anti-Rat IgG (whole molecule) Peroxidase Conjugated (suggested dilution 1:5,000 in 5% dry milk) (Sigma-Aldrich, cat. no. A9037).

3. Method

In our laboratory, we use ChIP to identify direct targets of WRKY TFs in transcriptional reprogramming host responses induced by the powdery mildew *G. orontii* during early stages of fungal establishment. Arabidopsis *wrky* mutant lines functionally complemented with the respective WRKY TF carrying a C-terminal hemagglutinin (HA) tag are used to perform ChIP assays. As a prerequisite, western blots on ChIP material are performed to check for the specificity of the antibody and to demonstrate accessibility of the epitope-tag within the WRKY proteins. If the tagged protein is clearly detectable (see Subheading 3.7), ChIP assays are performed to isolate DNA fragments representing the candidate target sequences.

For ChIP to be successful, it is important that sufficient amounts of the cross-linked chromatin are released and brought into solution. Former protocols included a CsCl purification step of the cross-linked chromatin (5). This time-consuming step has been omitted because it was shown that sonication with a high-detergent buffer is sufficient to solubilize cross-linked chromatin and that the quality of this chromatin is suitable for ChIP assays (2). However, CsCl gradients cannot be avoided when yields of cross-linked plant chromatin are very low or when high background from non-cross-linked DNA contamination is present.

Although precaution is needed to avoid protein degradation, the cross-linked biological material is relatively stable and can be stored at –80°C for several months without significantly reducing its quality. Below we outline all steps performed to obtain ChIP DNA fragments starting from pathogen infected plant material to the immunoprecipitated DNA.

3.1. Infection of Plant Leaf Material with *Golovynomices orontii*

1. *Arabidopsis* seeds were germinated in Jiffy pots (Alwaysgrows).
2. Three seeds were sown per single Jiffy pot. All of the plants were maintained at short-day conditions (16 h light/8 h darkness) at 20°C, in sterilized closed cabinets (Schneider-chambers). Five- to six-week-old plants were used for all experiments.
3. Subsequently, ten Jiffy pots were transferred to a small tray and the plants were subjected to pathogen challenge.
4. Plants were inoculated with *G. orontii* conidiospores by touching the upper leaf surface with previously inoculated heavily infected leaf material (12 day post inoculation).

3.2. Formaldehyde Cross-linking of Protein–DNA Complex *In Vivo*

1. Prepare 250 mL of fixing solution with 1% formaldehyde (add formaldehyde just prior to harvesting the plant material, work under a hood) in a ½ L beaker. This volume of fixing solution is sufficient for the plant material collected from ten Jiffy pots.
2. Harvest the *Arabidopsis* plants at appropriate time points postinoculation by cutting them at the base of the rosette leaves with a scissor.

3. Cut plant material into small stripes and allow the leaf stripes to fall directly into the fixative solution. Cutting facilitates more efficient penetration of the material by formaldehyde and enables rapid protein–DNA cross-linking. If wound response is a concern, cross-linking of the plant material without cutting is also possible.
4. Place the beaker inside to a vacuum chamber and apply vacuum 2–3 times for 5 min each within the 30-min fixation time-period (see Note 1) until the leaf material becomes translucent (see Note 2).
5. Add Glycine to the final concentration of 0.125 M to quench excess formaldehyde, and apply vacuum for another 5 min.
6. Remove the fixing solution and dry material on paper towels. Fast-freeze cross-linked leaf material in liquid nitrogen and store at −80°C.

3.3. Extraction of Soluble Cross-Linked Chromatin

1. Cool on ice all plastic-/glassware to be used in the subsequent steps described below.
2. Grind about 2 g of the cross-linked plant material in the presence of liquid nitrogen to a fine powder using a mortar and pestle.
3. Collect the fine powder into a 50 mL Falcon tube with a small spoon cooled in liquid nitrogen and add 25 mL ice cold NEB1.
4. (Optional) After cross-linking, the plant material often has a leathery appearance due to a matrix created by protein–protein cross-linkages. Homogenization with a Polytron (KINEMATICA) may facilitate the dispersion and breakage of the cross-linked plant material resulting in a better disruption of the cells and efficient release of the nuclei. The plant material kept on ice is homogenized by setting the power knob to 22, for two times 30 s allowing an interval of a few minutes to prevent heating of the material.
5. Dilute the disrupted plant material (= sample) by adding additional ice-cold NEB1 solution to 50 mL. Filter the sample through a series of nylon meshes (100, 65, and 40 μm) using a filter holder with a receiver (Nalgene) attached to a vacuum system and with the sample receiver on ice. Add the sample in small portions and replace filters if they become clogged.
6. Split the sample in two Corex centrifuge tubes and spin at 3,000 × *g*, at 4°C for 15 min using a swing-out rotor. Put aside 1 mL of the supernatant for western blot analysis and

store it until use at −20°C. Carefully discard the remaining supernatant. The pellet should appear green consisting of several layers of plant extracts and organelles mainly composed of chloroplasts, starch grains and nuclei, as depicted in Fig. 1.

7. Add 2 mL of ice-cold NEB2 solution to the pellet. Resuspend pellet by pipetting gently up and down in the NEB2 solution using a 5 mL plastic tip or a trimmed 1 mL tip. Complete resuspension of the pellet at this stage is difficult and small clumps of nuclei embedded on green material will still be visible.
8. Centrifuge the sample at 10,000 × *g* at 4°C for 5 min, using a swing-out rotor.
9. Put aside 250 μL of this second supernatant for western blot analysis and store it until use at −20°C. Decant the remaining supernatant and resuspend/wash the pellet in 2 mL of NEB2 solution.
10. Repeat the centrifugation step and the washes with NEB2 until most of the pellet appears white (mainly nuclei and starch grains depleted of chloroplasts; Fig. 1).
11. After the final centrifugation step, resuspend the nuclei in 500 μL of ice-cold NEB3 solution (see Note 3).
12. Prepare at least two 1.5 mL microcentrifuge tubes per sample (two subsamples) containing 300 μL of ice-cold NEB3 solution. Slowly overlay this with 300 μL of the resuspended

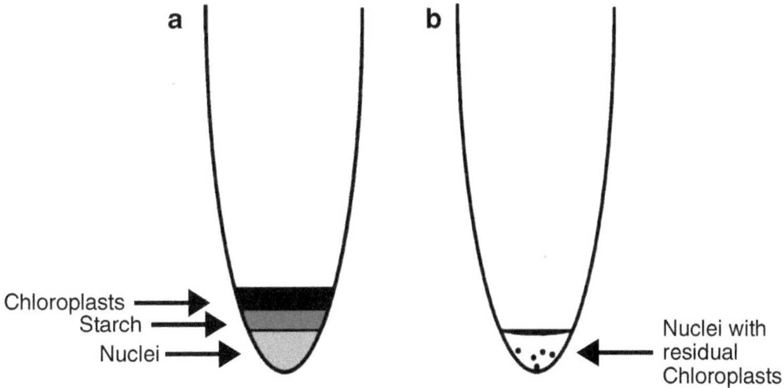

Fig. 1. Appearance of the pelletted plant material from the first and final centrifugation steps to isolate the nuclei. (**a**) After the first centrifugation, the pelletted plant material consists of three layers as indicated here by the three different shades of gray. (**b**) After final centrifugation, the pellet appears white with patches of green material, here depicted in black, containing residual chloroplasts here depicted as black dots.

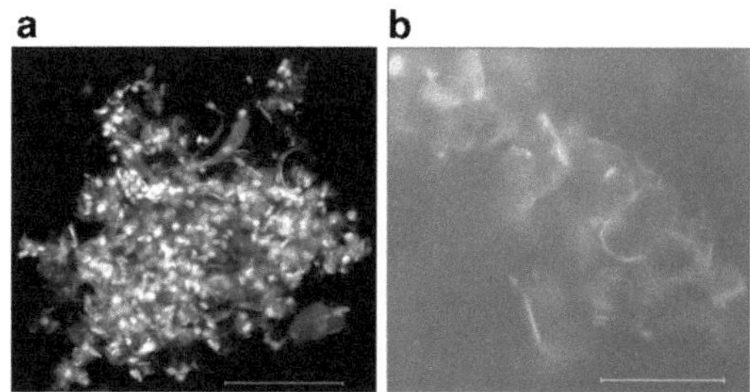

Fig. 2. DAPI staining of extracted nuclei before and after sonication. (**a**) Arabidopsis nuclei just prior to sonication. In this black and white figure, the stained nuclei appear as bright white spots within the cells. (**b**) The same nuclei sample after sonication. Mostly "ghost" cells void of nuclei are visible. Under UV light microscopy a blue-light coloration of the background indicates successful release of chromatin into the solution. Bars in (**a**) and (**b**) are 100 and 50 μm, respectively.

pellet and centrifuge at 16,000 × *g* at 4°C for 30 min (see Note 4).

13. Carefully remove the supernatant and resuspend the pellet containing the nuclei in 250 μL of NLB. Transfer 3 μL of the resuspended sample in a microcentrifuge tube containing 5 μL of water and 2 μL of DAPI solution. Place the 10 μL sample on a glass slide and cover with the cover slip. Visualize the presence and quality of the nuclei under a UV microscope (Fig. 2).
14. Put aside 25 μL of each resuspended subsample for western analysis and store them at −20°C until use. At this stage, the remaining nuclei sample can be also stored at −80°C until further processing.

3.4. Lysis and Sonication of the Cross-Linked Chromatin

1. Chromatin release from the nuclei and subsequent fragmentation is achieved by sonication of the sample. We use a Bioruptor™ sonication instrument (see Note 5).
2. The subsamples are loaded into the appropriate holder and immersed in the sonication chamber containing ice-cold water. Each sub-sample is then sonicated for a total time of 8 min with 30 s continuous pulses and 90 s interruption periods, with instrumentation settings at medium power (see Note 6).
3. Nuclei breakage and the release of the cross-linked chromatin can be visualized by pipetting 20 μL of sonicated nuclei on a glass slide and adding 4 μL of DAPI solution (see Note 7). The DAPI stain reveals increased blue background coloration visualized under a UV microscope, while the plant cells appear as broken "ghosts" without nuclei (Fig. 2).
4. Centrifuge the subsamples at 12,000 × *g* for 5 min at 4°C to remove cellular debris. Recover and combine the supernatants

of the sub-samples in a fresh ice-cold 1.5 mL tube to reconstitute the original sample (about 500 μL).

5. Vortex briefly and aliquot 25 μL of the reconstituted sample in a fresh 1.5 mL tube, then add 225 μL of ChIP dilution buffer containing 0.5% SDS. Add 2 μg of DNase-free RNase, and 2 μg of Proteinase K (see Note 8). Incubate the sample at 55°C for 3–4 h followed by incubation over-night at 65°C to reverse the formaldehyde cross-link. This aliquot serves as input material for the quantification of the extracted chromatin and in a dilution series, for qPCR. Store the remaining sample at −80°C until further use.
6. Following reverse cross-linking of the input material, extract the DNA by addition of 250 μL of phenol/chloroform/isoamyl alcohol, followed by one chloroform extraction (1:1) step and precipitation of the DNA with ethanol in the presence of 0.3 M sodium acetate (pH 5.2). Wash the pellet with 80% ethanol, air dry, and resuspend pellet in 100 μL of Tris–HCl pH 8.0. Measure the concentration of the DNA in the sample (see Note 9).
7. Prepare a 0.7% agarose gel. Load 20 μL of the input material per lane on the agarose gel. Following electrophoresis visualize the size distribution of DNA fragmentation under UV light (Fig. 2). Fragment distribution should be in the range of 3.0 and 0.3 kb.

3.5. Chromatin Immunoprecipitation with Magnetic Beads

1. Aliquot 125 μL of magnetic beads per each sample in ice-cold tubes (see Note 10). This amount of beads is sufficient to bind 3 μg of anti-HA ChIP-grade antibody. 3 μg of anti-HA are required for about 25 μg of cross-linked chromatin (see Note 11).
2. Place the tube on a magnet-rack to separate the beads from the solution. Discard the supernatant. Wash the beads by adding 250 μL of 1× PBS buffer containing 0.1% BSA. Vortex until the beads are resuspended and spin down by brief centrifugation. Repeat the washing step two more times.
3. After removal of the supernatant from the last wash, resuspend the beads in ChIP dilution buffer with 0.1% BSA and 3 μg of anti-HA, ChIP-grade. Incubate the beads with the antibody by slow tilt rotation mixing for 6 h or even overnight at 4°C.
4. After the incubation step, wash the coupled antibody beads once with 250 μL of ChIP dilution buffer. Remove the supernatant and keep the conjugated antibody-beads on ice.
5. In parallel, thaw stored sample from Subheading 3.4, step 5. Extract a volume corresponding to 25 μg of sonicated cross-linked chromatin, as calculated from the input, and dilute

fivefold with ChIP dilution buffer to reduce to 0.1% SDS. If the volume of the sonicated sample after addition of the ChIP buffer is greater than 1.5 mL, divide the diluted sample evenly in two 1.5 mL tubes. Similarly, divide evenly the conjugated antibody-beads into two 1.5 mL tubes. Add the sonicated sample to the conjugated antibody-beads and incubate for a minimum of 6 h or over-night at 4°C on a slow tilt rotator.

3.6. Wash and Recovery of the ChIP DNA Fragments

1. Following incubation, place the sample on the magnet rack and remove the supernatant. (Save the supernatant for western blot analysis).
2. Wash the chromatin-bound coupled antibody beads twice with 500 μL of low salt wash buffer. The first wash involves a short vortexing followed by a brief centrifugation and subsequent removal of the supernatant. The second wash is performed for 5 min at 4°C rocking gently.
3. Repeat the washing step as described above sequentially with high salt wash buffer, then with LiCl wash buffer and finally with TE buffer.
4. Resuspend the beads in 250 μL of TE buffer containing 0.5% SDS. Add 2 μg of RNase, DNase-free, and 2 μg of proteinase K (see Note 8). Incubate the sample at 55°C for 3–4 h followed by an incubation overnight at 65°C to reverse the formaldehyde cross-link.
5. After reverse cross-linking of the sample, extract the DNA by addition of 250 μL of phenol/chloroform/isoamyl alcohol, followed by one chloroform extraction (1:1) step and precipitation of the DNA with ethanol in the presence of 0.3 M sodium acetate (pH 5.2). Wash the pellet with 80% ethanol, air dry, and resuspend pellet in 50 μL of Tris–HCl pH 8.0 (see Note 9).
6. Determine the DNA concentration of the sample using, for instance, a NanoDrop spectrophotometer (Thermo Scientific) or a quartz cuvette suited for small volumes (<100 μL). Normally, a total of 50–200 ng of ChIP DNA fragments can be immunoprecipitated using this method.

3.7. Detection of HA Epitope-Tagged Protein

1. Reverse cross-link all of the samples that were put aside for the detection of the tagged protein. The supernatant of the NEB1 sample can be concentrated to about 250 μL using an Amicon concentrator column (Millipore) with a nominal molecular weight limit (NMWL) smaller than the protein under investigation.
2. Prepare a 1.5 mm thick 10% SDS-PAGE gel and load 25 μL per lane of each individual sample. Include a positive and a negative reference control sample plus a protein size marker.

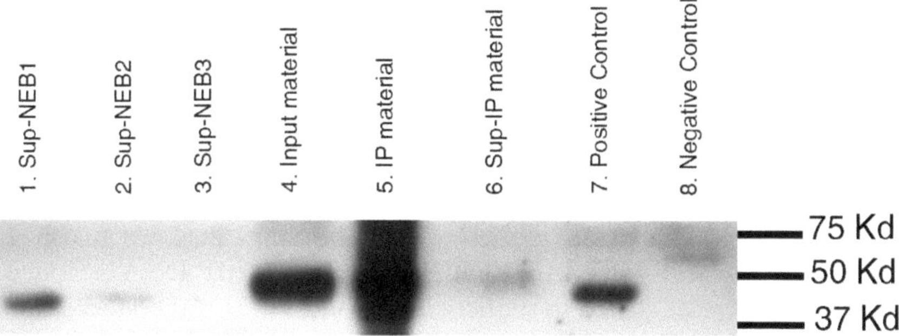

Fig. 3. Western blot performed to detect a hemagglutinin (HA)-tagged WRKY40 transcription factor during the extraction of soluble cross-linked chromatin (*lane 1–4*) and after ChIP to verify the accessibility of the HA-tagged protein by the HA-antibody in formaldehyde cross-linked plant material. Cross-linking was reversed in the samples shown in *lane 1–6* prior to loading on the SDS-PAGE gel. The positive control is a total leaf protein extract aliquot derived from a *wrky40* mutant line complemented with a HA-tagged WRKY40 protein. The negative control is a total leaf protein extract aliquot derived from the *wrky40* mutant line. Sup-NEB1, 2, 3: supernatants postcentrifugation of the three extraction buffers used for nuclei isolation.

3. Blot the SDS-PAGE gel on a PVDF transfer membrane, followed by a blocking step with 5% milk powder in 1× Tris-saline with 0.1% Tween 20 (TBS-T) buffer for overnight at 4°C under gentle agitation.
4. Wash the membrane three times with 1× TBS-T buffer for 10–15 min per wash. Add the primary antibody (clone 3F10) diluted in 5% milk powder in 1× TBS-T buffer. Incubate under gentle agitation for 1 h at room temperature.
5. Remove the primary antibody and wash the membrane as in step 4. Add the secondary antibody diluted in 5% milk power in 1× TBS-T buffer. Incubate under gentle agitation for 1 h at room temperature.
6. Remove the secondary antibody and wash the membrane as above.
7. Develop the chemiluminescence reaction to detect the tagged protein (Fig. 3).

3.8. Use of the ChIP DNA for Target Identification

The ChIP DNA obtained in this protocol can subsequently be used to:

1. Verify potential candidate target loci. For this, primers are designed to amplify the genomic locus under investigation as well as for control loci. The amount of specific genomic regions in control and ChIP DNA samples are determined individually by quantitative PCR (qPCR). The fold enrichment of the specific locus relative to the control loci provides quantitative information about the relative association level of a given protein with different genomic regions (10). A detailed protocol is given in Aparicio et al. (11).

2. Identify genome-wide location DNA binding sites using genome tiling microarrays (ChIP-chip). For this, the ChIP DNA material is often fluorescently labeled using Cy5-dCTP and Cy3-dCTP and labeled probes hybridized to DNA microarrays. Detailed protocols for this procedure have been published recently (7, 12, 13).
3. Define global protein binding sites by massively parallel sequencing (ChIP-Seq). Total ChIP DNA is cloned and subjected to sequencing using 454, SOLiD or Solexa technology (14). Deep sequencing allows for the detection of protein binding positions with relatively high accuracy (8, 15).

4. Notes

1. The time of fixation can vary depending on the plant species and biological material used and needs to be empirically optimized for each ChIP experiment.
2. The translucent nature of the leaf material is a good indicator of proper fixation. A patchy appearance of the leaf material with translucent and nontranslucent areas may hint toward a malfunction of the vacuum system resulting in insufficient cross-linking.
3. It is important at this stage that the pellet containing the nuclei has been efficiently resuspended. Resuspension can be difficult and pipetting is insufficient, so be patient! Apply a little pressure to small aggregates and gently pressing them against the inner wall of the centrifugation tube helps to better disrupt them.
4. The dense nuclei pellet while residual disrupted chloroplasts and other cellular debris remain on the sucrose cushion.
5. Other sonication devices can also be used; however, it is important that sonication conditions be optimized for efficient breakage of the nuclei as well as for sufficient fragmentation of the cross-linked chromatin.
6. The duration of sonication depends on the time of cross-linkage and needs to be empirically determined and the size of the DNA fragments verified by agarose gel electrophoresis (Fig. 4). The sonication conditions were optimized for 30 min formaldehyde cross-linkage. Longer periods of cross-linking may require longer sonication times and/or higher sonication settings.
7. In most cases, the increased blue coloration of the background is a good indication of chromatin release in the buffer solution. If this coloration is not observed additional

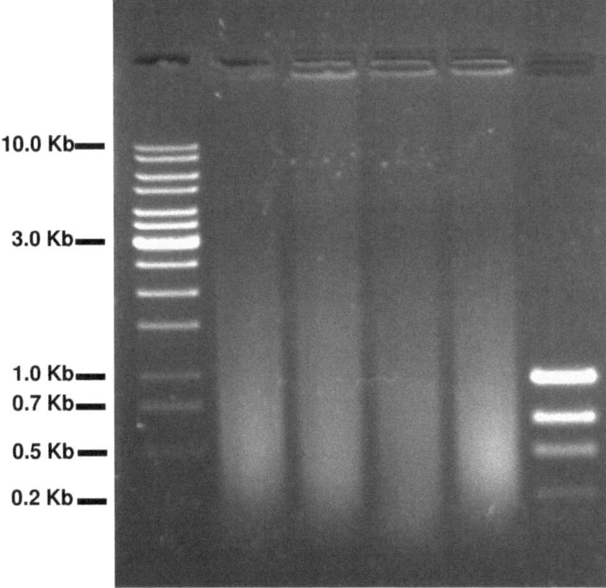

Fig. 4. Agarose gel electrophoresis of four independent formaldehyde cross-linked chromatin samples sonicated in parallel using the Bioruptor™. The four samples show similar and regular DNA size fragmentation (in the range between 3 and 0.3 Kb) indicating proper duration of cross-linkage. An extensive or insufficient cross-linkage would result in lower or higher molecular weight DNA fragmentation, respectively.

supplementary cycles of sonication can be administered. However, prolonged sonication shears the DNA to very small size fragments that are not suitable for ChIP (verify size of DNA fragments by agarose gel electrophoresis).

8. Omit the proteinase K treatment in the material used to detect the tagged protein on a western blot.
9. Omit this step if the sample will serve for the detection of the tagged protein on western blots.
10. It is advisable to also prepare an aliquot of magnetic beads to perform a mock ChIP reaction. In this case, the magnetic beads are not coupled to a specific antibody but used in the ChIP step together with the same amount of cross-linked chromatin. Alternatively, this aliquot of magnetic beads can be coupled with the specific antibody and used in the ChIP step together with cross-linked chromatin extracted from a plant line that does not contain the epitope-tagged version of the protein under investigation. This resulting ChIP material serves as a control sample for the downstream applications (see Subheading 4) to evaluate background levels of the ChIP target DNA-fragments.
11. The manufacturer advises that the final bead concentration should be $1–2 \times 10^7$ beads/mL for efficient target isolation.

References

1. Massie, C.E., and Mills, I.G. (2008). ChIPping away at gene regulation. EMBO Rep. **9**, 337–343.
2. Gendrel, A.V., Lippman, Z., Martienssen, R., and Colot, V. (2005). Profiling histone modification patterns in plants using genomic tiling microarrays. Nat. Methods **2**, 213–218.
3. Eulgem, T. (2005). Regulation of the *Arabidopsis* defense transcriptome. Trends Plant Sci. **10**, 71–78.
4. Eulgem, T., and Somssich, I.E. (2007). Networks of WRKY transcription factors in defense signaling. Curr. Opin. Plant Biol. **10**, 366–371.
5. Orlando, V., Strutt, H., and Paro, R. (1997). Analysis of chromatin structure by *in vivo* formaldehyde cross-linking. Methods **11**, 205–214.
6. Kurdistani, S.K., and Grunstein, M. (2003). In vivo protein-protein and protein-DNA crosslinking for genomewide binding microarray. Methods **31**, 90–95.
7. Kim, T.H., Barrera, L.O., and Ren, B. (2007). ChIP-chip for genome-wide analysis of protein binding in mammalian cells. Curr. Protoc. Mol. Biol. **79**, 21.13.21–21.13.22.
8. Kharchenko, P.V., Tolstorukov, M.Y., and Park, P.J. (2008). Design and analysis of ChIP-seq experiments for DNA-binding proteins. Nat. Biotechnol. **26**, 1351–1359.
9. Sambrook, J., Fritsch, E.F., and Maniatis, T. (1989). Molecular cloning: a laboratory manual. Cold Spring Laboratory Press, New York.
10. Haring, M., Offermann, S., Danker, T., Horst, I., Peterhansel, C., and Stam, M. (2007). Chromatin immunoprecipitation: optimization, quantitative analysis and data normalization. Plant Methods **3**, 11.
11. Aparicio, O., Geisberg, J.V., Sekinger, E., Yang, A., Moqtaderi, Z., and Struhl, K. (2005). Chromatin immunoprecipitation for determining the association of proteins with specific genomic sequences *in vivo*. Curr. Protoc. Mol. Biol. 69, 21.23.21–21.23.33.
12. Lee, T.I., Johnstone, S.E., and Young, R.A. (2006). Chromatin immunoprecipitation and microarray-based analysis of protein location. Nat. Protoc. **1**, 729–748.
13. Sandmann, T., Jakobsen, J.S., and Furlong, E.E.M. (2006). ChIP-on-chip protocol for genome-wide analysis of transcription factor binding to *Drosophila melanogaster* embryos. Nat. Protoc. **1**, 2839–2855.
14. Shendure, J., and Ji, H. (2008). Next-generation DNA sequencing. Nat. Biotechnol. **26**, 1135–1140.
15. Valouev, A., Johnson, D.S., Sundquist, A., Medina, C., Anton, E., Batzoglou, S., Myers, R.M., and Sidow, A. (2008). Genome-wide analysis of transcription factor binding sites based on ChIP-Seq data. Nat. Methods **5**, 829–834.

Chapter 6

Dose–Response to and Systemic Movement of Dexamethasone in the GVG-Inducible Transgene System in Arabidopsis

Xueqing Geng and David Mackey

Abstract

Construction of transgenic plants is central to modern plant molecular genetics. Inducible systems permit spatial and temporal control of transgene expression. One commonly used inducible system relies on the use of dexamethasone to activate an endogenously expressed hybrid transcription factor, which positively regulates the expression of the gene of interest (Aoyama and Chua, Plant J 11:605–612, 1997). We have developed Arabidopsis plants using this inducible system to drive expression of a bacterial type III effector protein. The effector, AvrRpm1, elicits either strong cell death or weak cell death and chlorosis depending on the genetic background of the plant. Using these reagents, we examine several properties of the inducible system in Arabidopsis, including the timing of induction, the ability to tune the level of transgene expression by altering the concentration of applied dexamethasone, and the movement of dexamethasone within the plant.

Key words: Dexamethasone, Inducible expression, Dex inducible, GVG, Transgenic plant, Transgene

1. Introduction

Construction of transgenic plants is central to modern plant molecular genetics. Inducible systems permit spatial and temporal control of transgene expression. In 1997, Aoyama and Chua described a dexamethasone (Dex) inducible system for transgenic plants (1). This system is contained on a single binary vector for Agrobacterium-mediated plant transformation. The transferred DNA (T-DNA) portion of the vector contains two key elements. The first is a gene fusion encoding a protein composed of the DNA-binding domain of yeast GAL4 (2), the potent transcriptional activation domain of herpes simplex virus VP16 (3),

John M. McDowell (ed.), *Plant Immunity: Methods and Protocols*, Methods in Molecular Biology, vol. 712,
DOI 10.1007/978-1-61737-998-7_6, © Springer Science+Business Media, LLC 2011

and the hormone-binding domain of the rat glucocorticoid receptor (GR) (4). This hybrid transcription factor, termed GVG, is constitutively expressed from the highly active cauliflower mosaic virus 35S promoter. The GR domain renders the transactivating capacity of this protein responsive to exogenously applied glucocorticoids, including Dex. The GAL4 DNA-binding domain targets the transcription factor specifically to GAL4 binding sites. Such sites are not features of plant promoters, but are a key feature of the second element of the inducible system – the inducible promoter. Adjacent to GVG, within the T-DNA, is a minimal promoter from CaMV 35S flanked by six repeated GAL4 binding sites, which thus render expression of the introduced gene inducible by the GVG protein, following its activation by Dex.

Since its introduction, the GVG system has proven widely useful for plant biologists. One research area to which the system has been applied is the study of effector proteins from bacterial and other plant pathogenic microbes. These effectors, such as type III secreted effectors from Gram-negative bacteria, are pathogen-derived proteins that are active following delivery into the inside of plant cells. Thus, expression of these effectors from plant transgenes places them in the appropriate location for their biological activity. The GVG-inducible system permits in planta expression of effectors that would be lethal or deleterious if constitutively expressed. Also, the system permits the controlled and synchronous induction of an effector thus allowing timed expression of the effector within the context of an experiment and/or a chronological description of the events that follow effector expression. Finally, and perhaps most importantly, this approach permits the study of an individual effector in isolation, apart from the medley of effectors, elicitors, and other factors typically associated with a plant–pathogen interaction.

AvrRpm1 is a type III secreted effector from *Pseudomonas syringae*. The consequences of AvrRpm1 expression on defense signaling in Arabidopsis have been widely studied using the GVG system (5–14). AvrRpm1 elicits defense responses when detected by either of two resistance (R) proteins, RPM1 or RPS2 (9). The induction of AvrRpm1 in wild-type (Col-0) plants (Dex:AvrRpm1-HA Col-0) induces rapid and strong cell death reminiscent of the HR induced by bacteria expressing AvrRpm1. This strong cell death is dependent on RPM1 since it is not observed in *rpm1* mutant plants. However, these Dex:AvrRpm1-HA *rpm1* plants do demonstrate slower and weaker symptoms, including cell death and chlorosis. These weaker symptoms are RPS2-dependent since they are absent in plants also lacking *rps2* (Dex:AvrRpm1-HA *rpm1 rps2*). We have employed the strong and weak AvrRpm1-induced responses observed in Dex:AvrRpm1-HA Col-0 and Dex:AvrRpm1-HA *rpm1* plants, respectively, to characterize the GVG system in Arabidopsis. Below we describe several key parameters for successful utilization of the Dex-GVG system.

2. Materials

1. Dexamethasone (Sigma) suspended at 20 mM in 100% ethanol and stored at –20°C.
2. Silwet L-77 (Lehle Seeds).
3. Paint sprayer (Preval Power Unit, available from Bodi Company, http://www.bodico.com).
4. Lint-free wipes, cotton swabs, aluminum foil, and toothpicks.
5. Arabidopsis plants – vegetatively grown in 8-h light at 23°C and 16-h dark at 16°C.

3. Methods

3.1. Dose–Response of Transgene Induction as Assessed by Transgene Protein Accumulation and Plant Cell Death

In the original description of the GVG system from Aoyama and Chua (1), the dose–response was measured in transgenic tobacco plants. We tested whether a similar dose–response is observed in Arabidopsis. Also, the work from Aoyama and Chua (1) was done via root feeding of Dex. In research applications, Dex is commonly applied by spraying on aerial tissues rather than through root feeding. Thus, we tested the dose–response and timing of response following spray application of Dex. Using Dex:AvrRpm1-HA Col-0 and Dex:AvrRpm1-HA *rpm1* plants, we monitored two distinct readouts in Arabidopsis – transgene protein accumulation (Fig. 1a) and symptoms induced in the plants (Fig. 1b).

3.1.1. Procedure

1. Dilute the Dex stock solution to the desired final concentration in distilled water with 0.005% Silwet L-77 (see Note 1).
2. Use the Preval sprayer to apply the different solutions to plants and then keep humidity high (clear dome on) for 2–3 h (see Notes 2–4).
3. Collect samples at different times after Dex treatment for anti-HA Western blotting to detect AvrRpm1-HA (Fig. 1a) (see Notes 5–7).
4. Observe and photograph plants at different times following Dex treatment (Fig. 1b).

The dose–response to Dex in Arabidopsis plants to which Dex was spray applied closely parallels that published for luciferase induction in tobacco plants to which Dex was applied to roots (1). Differing degrees of transgene induction were observed with concentrations of Dex ranging from 0.2 to 20 μM (Fig. 1). Above 20 μM Dex, the response appears to be saturated – application of 200 μM Dex did not induce any stronger AvrRpm1-HA accumulation or plant symptoms than did 20 μM. Below 0.2 μM Dex,

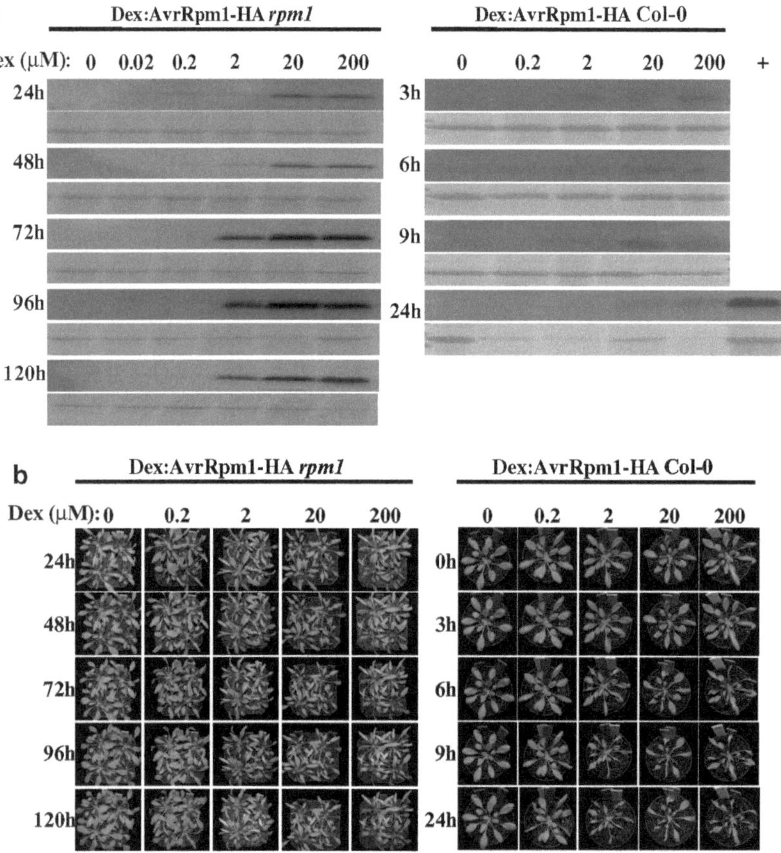

Fig. 1. Responsiveness of the GVG system in Arabidopsis to spray application of dexamethasone (Dex) to aerial tissues. Dex:AvrRpm1-HA Col-0 and Dex:AvrRpm1-HA *rpm1* plants (4–5 weeks old) were sprayed with the indicated concentrations of Dex. At indicated times after spraying (in hours, h), (**a**) accumulation of AvrRpm1-HA or (**b**) plant phenotypes were monitored by western blot or photograph, respectively. In the anti-HA western blots, + is a positive control extract expressing AvrRpm1-HA and ponceau S staining in the *lower panels* indicates the amount of protein that was loaded.

responses were not detectable; 0.2 μM Dex induced barely detectable protein expression and plant phenotypes, neither of which were detectable with 0.02 μM Dex. The weak detection of AvrRpm1-HA protein accumulation in Dex:AvrRpm1 Col-0 is probably a consequence of the *RPM1*-dependent cell death.

The timing of transgene induction in Dex-sprayed Arabidopsis closely parallels that in root-exposed tobacco. Accumulation of AvrRpm1-HA protein in Dex:AvrRpm1-HA *rpm1* was apparent by 24 h and reached a maximum by 96 h after spraying with 20 μM Dex (Fig. 1a). Cell death following spraying of Dex:AvrRpm1 Col-0 with 20 μM Dex was apparent by 3 h and all tissue was fully desiccated by 9 h (Fig. 1b). Aoyama and Chua observed that the inducible luciferase mRNA in tobacco was diminished by 48 h and undetectable by 72 h after removing Dex from the roots (1). However, levels of AvrRpm1-HA protein in Dex:AvrRpm1-HA *rpm1* increase until 96 h and high levels persist beyond 144 h (data not shown) following a single spray application of 20 μM Dex. The significance of the longer duration of transgene expression in Arabidopsis is unclear without knowing the half-lives of the luciferase mRNA and AvrRpm1-HA mRNA and protein. Nonetheless, these results as well as the systemic response experiments described in the next section indicate that Dex may be long-lived in Arabidopsis.

3.2. Systemic Movement of Dex Within Arabidopsis Plants

In many research applications, it is desirable to induce transgene expression in one part of a plant and assess the consequent effects in another part of the plant. The utility of Dex induction for such approaches depends on the mobility of Dex itself within the plant. In the original description from Aoyama and Chua (1), it was shown in tobacco that transfer of seedlings onto agar resulted in transgene expression in aerial tissues after 48 h, but that application of Dex to the lateral half of a tobacco leaf did not induce luciferase expression in the other half of the same leaf. Here we examined Dex movement in Arabidopsis from roots to aerial tissues, from lower leaves to upper leaves and vice versa, and from one portion of a leaf to another portion of the same leaf. Based on the timing and sensitivity of the plants to different doses of spray applied Dex (from Subheading 3.1), we make estimates about the timing and efficiency of Dex movement within Arabidopsis.

3.2.1. Procedure

1. Dilute the Dex stock solution to the desired concentration in distilled water with 0.005% Silwet L-77.
2. (a) Gently pick up 2-week-old seedlings from soil, wash roots in distilled water, and place the roots on lint-free wipes saturated with the desired Dex solutions (see Note 8)

   or

   (b) Use cotton swabs to paint the desired Dex solutions onto lower leaves, upper leaves, or distal, basal, or lateral halves of individual leaves (see Notes 9 and 10).
3. Observe and photograph plants at different times following Dex treatment.

The root to shoot movement of Dex was detectable at the concentration of 0.2 μM and higher, similar to that observed for spray application. The most severe cell death was observed following root feeding of Dex:AvrRpm1-HA Col-0 with 20 or 200 μM Dex – plant tissues were starting to collapse by 5 h and were completely collapsed by 8 h (data not shown). Comparing the first apparent symptoms to those when Dex was spray applied indicates a delay of approximately 2 h for movement of Dex from roots into aerial tissues. Application of 200 μM Dex did not show any more severe or rapid collapse than 20 μM Dex, the response to 2 μM Dex was intermediate, exhibiting less severe collapse at 8 and 24 h than plants fed 20 μM Dex, and the response to 0.2 μM Dex was weak but apparent by 24 h (data not shown and Fig. 2a). The similarity in the dose–response between spray application and root feeding indicates that the concentration of Dex applied to Arabidopsis roots leads to an accumulation in aerial tissue similar to spray application of the same concentration.

The movement of Dex from treated leaves to other leaves was also tested. Using Dex:AvrRpm1-HA Col-0 and Dex:AvRpm1 *rpm1* plants to monitor transgene expression, we tested for movement of Dex from lower to upper leaves and vice versa. Using 200 μM Dex, we treated either the oldest five true leaves (true leaves number 1–5) or the next oldest five leaves (true leaves number 6–10) by painting with Dex (Fig. 2b). In all cases, we observed symptoms on the painted leaves that were comparable in timing and intensity to symptoms after spray application of Dex. We also observed that movement of Dex from painted leaves to the remainder of the plant is more efficient from upper leaves than from lower leaves. Symptoms on upper leaves were mild to undetectable 6 days after painting lower leaves of Dex:AvrRpm1-HA in *rpm1*. Contrasting this observation, treatment of upper leaves led to more efficient symptoms throughout the rest of the plants. For example, symptoms in non-painted leaves of Dex:AvrRpm1 *rpm1* were apparent by 4 days with 200 μM Dex

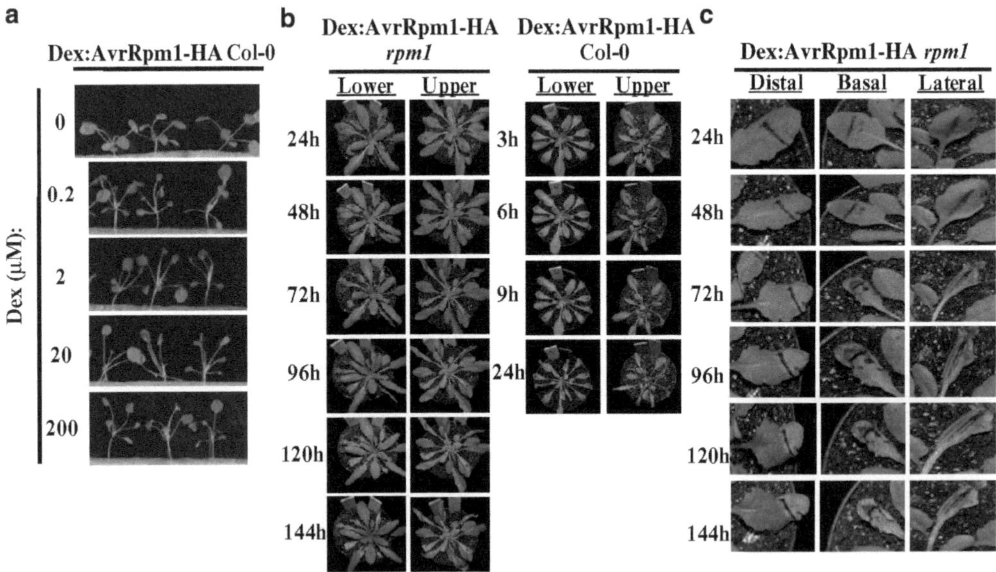

Fig. 2. Movement of Dex within Arabidopsis plants. Symptoms observed in Dex:AvrRpm1-HA Col-0 and Dex:AvrRpm1-HA *rpm1* plants were used to monitor movement of Dex (**a**) from the roots to the shoots of 2-week-old seedlings, (**b**) from the upper and lower leaves to other leaves, and (**c**) from one part of a leaf to another part of a leaf. In (**a**), pictures are from 24 h after seedling transfer. In (**b**), older or younger leaves treated with 200 μM Dex are indicted with *marker spots*. In (**c**), 2 μM Dex was pointed onto the halves of leaves with the *marker dot*.

or by 5 days with 20 μM Dex (Fig. 2b and data not shown). Direct spray application of 2 μM Dex to Dex:AvRpm1 *rpm1* plants induces symptoms by 72 h (Fig. 1b). Based on these observations, we estimate that when Dex is applied to true leaves 6–10, it arrives in other leaves of the plant after 1 day at ~1% of the applied concentration and after 2 days at ~10% of the applied concentration. Parallel experiments with Dex:AvRpm1 Col-0 showed that cell death is apparent by 24 h whether younger or older leaves are painted with Dex, but that the death of untreated leaves is stronger and more rapid following painting of younger leaves (Fig. 2b).

The movement of Dex between different portions of the same leaf was examined by painting Dex onto the distal, basal, or lateral halves of leaves and observing the timing and intensity of symptoms in the remainder of those leaves (Fig. 2c). We observed that Dex movement was efficient from the basal to the distal portion of leaves and, in contrast to the observation in tobacco (1), that Dex also moved efficiently between lateral halves of Arabidopsis leaves. In Dex:AvrRpm1 *rpm1* leaves painted with 2 μM Dex on lateral or basal halves of leaves, symptoms were apparent throughout the leaves by 72 h and the leaves were uniformly collapsed by 120 h. Contrastingly, Dex:AvrRpm1 *rpm1* leaves painted on their distal half showed symptoms on only the painted half of the leaf, indicating that the movement of Dex from distal to basal portions of Arabidopsis leaves is inefficient. Parallel experiments in leaves of Dex:AvRpm1 Col-0 demonstrated similar results, although cell death was more readily apparent in the basal half of leaves painted on the distal half (data not shown).

4. Notes

1. Dex in ethanol is stable in the freezer, but dilutions in water are made fresh for each use.
2. Dexamethasone is an active hormone in humans, so spray in a fume hood and take care to avoid bodily contact with the solution.
3. For spray treatments with different concentrations of Dex, it is helpful to grow plants in separate pots. This allows plants to be kept away from contaminating mist of other Dex concentrations.
4. In addition to transgenic plants, the GVG system can also be used in transient assays. The system is delivered into plant leaves with Agrobacteria, and gene expression is subsequently induced by spraying the infiltrated tissue with Dex. Optimal gene expression is obtained if Dex spray follows 48 h after Agro-mediated delivery of the GVG system (data not shown).

This works very efficiently in *Nicotiana benthamiana*, and also works less efficiently in Arabidopsis. The *efr* mutant (15) permits more efficient agro-transient than wild-type Arabidopsis, but the efficiency is still less than that observed in *N. benthamiana* (data not shown).

5. Silencing of the GVG system is one of the biggest problems with stably transgenic Dex inducible plants. Often, an individual transgenic line will express the transgene in the T1 or T2 generation only to silence in some or all plants of subsequent generations. One key factor that triggers silencing is the presence of other CaMV 35S promoters. We have observed that lines using the 35S promoter to drive constitutive expression of another transgene or T-DNA mutants from the SALK insertion collection (which have a fragment of the 35S promoter driving expression of the KanR gene) cause efficient silencing of the GVG system (data not shown). The nature of silencing is unknown, but the correlation of silencing with the presence of the 35S promoter indicates that post-transcriptional silencing may be the cause.
6. Another factor that we postulate affects silencing is the efficiency of expression of the GVG fusion protein. When establishing lines in a background where transgene expression is lethal (e.g., Dex:AvrRpm1-HA Col-0), we have obtained lines that infrequently silence (data not shown). We have used one Dex:AvrRpm1-HA Col-0 line to cross the GVG system into numerous genetic backgrounds and the system has maintained its low silencing phenotype. On the contrary, establishing Dex:AvrRpm1-HA lines in an *rpm1* mutant background (where it is not lethal and higher constitutive expression is thus tolerated), we obtained lines that frequently silenced (data not shown). We have observed similar results with other transgenes able to cause lethality. Thus, to minimize subsequent silencing, we recommend obtaining initial transgenic lines in a genetic background in which the transgene is potentially lethal. In cases where lethality is not associated with the transgene, we predict that testing initial transgenic lines for those with low levels of GVG RNA or protein will identify those with a low proclivity to silencing.
7. When working with GVG transgenic lines, it is important to consider silencing. Silencing is heritable, so when moving the transgene into new genetic backgrounds via crosses or when propagating a transgenic line, it is wise to collect seed from multiple plants. Silencing must also be considered in the design of experiments. It is important to know the frequency with which individuals silence in a particular line. Making this determination can be trivial if the inducible gene causes a macroscopic phenotype (as for Dex:AvrRpm1-HA).

Otherwise, protein- or RNA-based screening may be required. The frequency of silencing and the nature of an individual experiment will determine whether it is necessary to monitor expression in individual plants. For an open-ended experiment such as a forward genetic screen, this frequency will dictate the necessary rigor of screening steps to eliminate silenced plants.

8. Take care to avoid touching any aerial tissues to the Dex solution.
9. Use aluminum foil and/or toothpicks to isolate treated leaves in order to be certain that the Dex solution does not contaminate other leaves by physical contact.
10. Do not apply Dex solution so heavily that liquid runs down the petiole into the rosette center.

Acknowledgments

This work was supported by NSF (MCB – 0718882) and the Ohio Agricultural Research and Development Center (OARDC) of The Ohio State University.

References

1. Aoyama, T., and N.-H. Chua, A glucocorticoid-mediated transcriptional induction system in transgenic plants. Plant J, 1997. **11**: 605–12.
2. Keegan, L., G. Gill, and M. Ptashne, Separation of DNA binding from the transcription-activating function of a eukaryotic regulatory protein. Science, 1986. **231**: 699–704.
3. Triezenberg, S.J., R.C. Kingsbury, and S.L. McKnight, Functional dissection of VP16, the trans-activator of herpes simplex virus immediate early gene expression. Genes Dev, 1988. **2**: 718–29.
4. Rusconi, S., and K.R. Yamamoto, Functional dissection of the hormone and DNA binding activities of the glucocorticoid receptor. EMBO J, 1987. **6**: 1309–15.
5. Al-Daoude, A., M. de Torres Zabala, J.H. Ko, and M. Grant, RIN13 is a positive regulator of the plant disease resistance protein RPM1. Plant Cell, 2005. **17**: 1016–28.
6. Andersson, M.X., M. Hamberg, O. Kourtchenko, A. Brunnstrom, K.L. McPhail, W.H. Gerwick, C. Gobel, I. Feussner, and M. Ellerstrom, Oxylipin profiling of the hypersensitive response in *Arabidopsis thaliana*. Formation of a novel oxo-phytodienoic acid-containing galactolipid, arabidopside E. J Biol Chem, 2006. **281**: 31528–37.
7. Andersson, M.X., O. Kourtchenko, J.L. Dangl, D. Mackey, and M. Ellerstrom, Phospholipase-dependent signalling during the AvrRpm1- and AvrRpt2-induced disease resistance responses in *Arabidopsis thaliana*. Plant J, 2006. **47**: 947–59.
8. Kim, M.G., L. da Cunha, A.J. McFall, Y. Belkhadir, S. DebRoy, J.L. Dangl, and D. Mackey, Two *Pseudomonas syringae* type III effectors inhibit RIN4-regulated basal defense in Arabidopsis. Cell, 2005. **121**: 749–59.
9. Kim, M.G., X. Geng, S.Y. Lee, and D. Mackey, The *Pseudomonas syringae* type III effector AvrRpm1 induces significant defenses by activating the Arabidopsis nucleotide-binding leucine-rich repeat protein RPS2. Plant J, 2009. **57**: 645–53.
10. Kourtchenko, O., M.X. Andersson, M. Hamberg, A. Brunnstrom, C. Gobel, K.L. McPhail, W.H. Gerwick, I. Feussner, and M. Ellerstrom, Oxo-phytodienoic acid-containing galactolipids in Arabidopsis: jasmonate signaling dependence. Plant Physiol, 2007. **145**: 1658–69.

11. Mackey, D., Y. Belkhadir, J.M. Alonso, J.R. Ecker, and J.L. Dangl, Arabidopsis RIN4 is a target of the type III virulence effector AvrRpt2 and modulates RPS2-mediated resistance. Cell, 2003. **112**: 379–89.
12. Mackey, D., B.F. Holt III, A. Wiig, and J.L. Dangl, RIN4 interacts with *Pseudomonas syringae* type III effector molecules and is required for RPM1-mediated disease resistance in Arabidopsis. Cell, 2002. **108**: 743–54.
13. Nimchuk, Z., E. Marois, S. Kjemtrup, R.T. Leister, F. Katagiri, and J.L. Dangl, Eukaryotic fatty acylation drives plasma membrane targeting and enhances function of several type III effector proteins from *Pseudomonas syringae*. Cell, 2000. **101**: 353–63.
14. Widjaja, I., K. Naumann, U. Roth, N. Wolf, D. Mackey, J.L. Dangl, D. Scheel, and J. Lee, Combining subproteome enrichment and Rubisco depletion enables identification of low abundance proteins differentially regulated during plant defense. Proteomics, 2009. **9**: 138–47.
15. Zipfel, C., G. Kunze, D. Chinchilla, A. Caniard, J.D. Jones, T. Boller, and G. Felix, Perception of the bacterial PAMP EF-Tu by the receptor EFR restricts Agrobacterium-mediated transformation. Cell, 2006. **125**: 749–60.

Chapter 7

Quantifying Alternatively Spliced mRNA via Capillary Electrophoresis

Xue-Cheng Zhang and Walter Gassmann

Abstract

Alternative splicing (AS) significantly contributes to transcriptome and proteome complexity. Transcriptome-wide studies concluded that approximately 22% of Arabidopsis and rice genes are subject to AS. Despite increasing recognition of AS in plants, little is known about the function of individual products of AS. In our studies of the Arabidopsis *RPS4* resistance gene, which requires AS transcripts for function, the need to quantify AS transcripts became apparent. Because *RPS4* expression levels are very low and the pattern of *RPS4* splicing is complex, existing mRNA quantification methods were not adequate. We therefore developed a new method based on reverse transcription (RT) PCR amplification of all transcript variants with a common set of primers and separation of the PCR products by size via capillary electrophoresis. Products were quantified by analysis of several PCR cycles per sample using the signal quantification procedures developed for microsatellite genotyping on capillary sequencing machines. With this method, we were able to measure differential regulation of individual *RPS4* alternative transcripts specifically during effector-triggered immunity. This method is especially suitable for quantification of alternative transcripts of low-expressed genes exhibiting complex splicing patterns.

Key words: Alternative splicing, Arabidopsis, Transcript quantification, Capillary array, GeneMapper, Disease resistance gene, RPS4

1. Introduction

Alternative splicing (AS) is a ubiquitous feature of eukaryotic organisms, although the proportion of genes undergoing alternative splicing varies from species to species. AS affects approximately 22% of Arabidopsis and rice genes (1–3), but large variations exist among plant species (4). While the incidence of AS is much higher in animals and most recently was concluded to be near-universal for human genes (5–7), AS in plants is most likely still underestimated because of limited transcript databases,

John M. McDowell (ed.), *Plant Immunity: Methods and Protocols*, Methods in Molecular Biology, vol. 712, DOI 10.1007/978-1-61737-998-7_7, © Springer Science+Business Media, LLC 2011

and an overrepresentation of low-expressed stress-induced genes among known AS genes (8). The transcript variants produced by alternative splicing display a wide range of transcriptome complexity and encode diverse protein isoforms. Conceptually, deletion of protein domains in products of AS transcripts can counteract full-length isoforms (such as with receptor isoforms) or lead to unregulated protein activity (such as when autoinhibitory domains are deleted), and the production of AS transcripts can be static or dynamic.

Despite the increasing importance of AS, very little is known about the function of individual AS transcripts in plants. One exception is the requirement for AS for effector-triggered immunity mediated by members of the Toll/interleukin-1 receptor (TIR) – nucleotide binding site (NBS) – leucine-rich repeat (LRR) class of resistance genes to enable effector-triggered immunity (9, 10). While the AS mechanisms vary, these alternative transcripts generally encode truncated TIR or TIR–NBS proteins. A first indication that these alternative transcripts are crucial for function came from the observation that intronless cDNAs of the tobacco *N* gene were nonfunctional (11). Similarly, intronless Arabidopsis *RPS4* cDNAs, despite being expressed, failed to complement an *rps4* mutant line (12). However, combining full-length and truncated cDNAs that mimic the prevalent *RPS4* transcripts provided resistance, demonstrating directly that a combination of *RPS4* transcripts is required for function (12).

AS in *RPS4* is limited to a region consisting of two introns that can be retained flanking an exon. This exon contains a cryptic intron that can be spliced out in AS transcripts. Theoretically, *RPS4* can therefore generate eight transcripts if all three AS events occur independently of each other (12). With the goal to determine whether *RPS4* AS is temporally regulated after stimulus perception and is fine-tuned to optimally regulate the plant innate immune response, we considered established mRNA quantification techniques for suitability of *RPS4* AS quantification. RNA gel blots do not have the resolution to distinguish between AS transcripts with retained introns, and are often not sensitive enough to reliably detect low-expressed genes such as *RPS4* (13). Quantitative real-time PCR (qRT-PCR), a technique commonly used to measure gene expression levels, was found to be unfeasible because AT-rich sequences within introns and in adjacent exonic sequences precluded the design of transcript-specific primer pairs. Other techniques for quantifying RNA abundance, such as RNase protection and primer extension, also suffer from a lack of suitable probes or primers that would unequivocally identify specific transcripts when AS patterns are complex.

Here, we describe the method that we consequently developed to measure *RPS4* alternative transcript abundance. This method couples signal sensitivity with unbiased detection of transcripts combining individual AS events. It is based on amplification of reverse-transcribed cDNA with a single set of PCR primers, one of which is fluorescently labeled. Products of individual transcripts are then identified based on size using capillary electrophoresis. Quantification is achieved by using GeneMapper features developed for genotyping on capillary sequencing machines, and by running several PCR cycles per sample, followed by linear regression analysis of log-transformed fluorescence values. AS transcript levels are expressed as a percentage of regular transcript levels in each sample. Absolute changes in transcript levels can be calculated by combining AS quantification with quantification of total mRNA levels of the gene of interest by qRT-PCR (14).

The method described here is ideal for sensitive AS quantification of individual genes with low expression and complex AS patterns. It relies on well-established protocols developed for other applications, and therefore is not technically challenging. Higher throughput could be achieved by automating the quantification and analysis steps, and by improving the dynamic range of the capillary polymer.

2. Materials

2.1. RNA Extraction, Reverse Transcription, and PCR Amplification

1. Tri Reagent (Sigma, St. Louis, MO).
2. Turbo DNase (Ambion, Austin, TX).
3. Poly(A) Purist Mag mRNA purification kit (Ambion, Austin, TX) (see Note 1).
4. SuperScriptII reverse transcription kit (Invitrogen, Carlsbad, CA).
5. Random hexamer primers (see Note 1).
6. 6-Carboxy-flourescein (6-FAM)-labeled primer and reverse primer (see Note 2).
7. *Taq* polymerase, unlabeled dNTPs, and thermal cycler.

2.2. Capillary Electrophoresis and Data Analysis

1. ABI PRISM® 3100 Genetic analyzer (Applied Biosystems, Foster City, CA).
2. POP-4™ performance optimized polymer: 4% poly(dimethylacrylamide) (PDMA) in 100 mM TAPS (pH 8.0), 8 M urea, and 5% 2-pyrrolidinone. Preformulated and low viscosity (Applied Biosystems, Foster City, CA).
3. Internal size standard: GeneScan™ 500 ROX™ dyed with 6-carboxy-X-rhodamine (Applied Biosystems, Foster City, CA).

The sizes of the single-stranded labeled fragments are: 35, 50, 75, 100, 139, 150, 160, 200, 250, 300, 340, 350, 400, 450, 490, and 500 nucleotides (see Note 3).

4. Capillary data analysis software: GeneMapper v3.5 (Applied Biosystems, Foster City, CA).

3. Methods

This AS transcript quantification procedure is conceptually similar to qRT-PCR, except that multiple PCR products separated by size are quantified in a single sample. Because capillary sequencers do not have a comparable dynamic signal range as qRT-PCR machines, an important step in this procedure is to develop transcript amplification parameters that result in quantifiable (detectable but nonsaturated) signals on the capillary sequencer. These parameters include the amount of mRNA for reverse transcription, the amount of cDNA for PCR amplification, and the PCR cycles that are picked for analysis, and will depend on the expression level of the gene of interest. The parameters listed below were optimized for *RPS4*, a gene of very low expression levels. Total *RPS4* mRNA levels in uninduced tissue correspond to approximately one-tenth of the mRNA levels of At2g28390, a *SAND* family qRT-PCR reference gene (14, 15). Nevertheless, the dynamic range of sequencing capillaries is sufficient to quantify regular and AS transcripts that differ in expression levels by 25- to 50-fold.

3.1. mRNA Extraction, Reverse Transcription, and PCR Amplification

1. Extract total RNA from plant tissue (equivalent to six mature rosette leaves) with Tri Reagent and treat with Turbo DNase. From 5 μg total RNA, isolate mRNA using Purist Mag beads.
2. Reverse transcription is performed using SuperScriptII with 20 ng of mRNA as template and random hexamer primers. The final volume of reverse transcription reactions is 50 μl.
3. Set up PCR master mixes of 180 μl total volume per sample with a gene-specific primer pair that will capture all transcript variants and 9 μl first-strand cDNA per master mix as template. One of the primers must be fluorescently labeled (see Note 2). Divide each sample into eight aliquots of 20 μl. Run a PCR for 26 cycles on a thermal cycler and remove one aliquot at the end of cycles 19, 20, 21, 22, 23, 24, 25, and 26 (see Note 4).
4. Run 5 μl of PCR product from three to four aliquots (e.g., from cycles 19, 21, 23, and 25) on an agarose gel to check the quality and amount of PCR products. Bands that are visible but not saturated on an ethidium bromide-stained agarose

gel will in general correspond to well-resolved fluorescence signals on the capillary genetic analyzer.

3.2. Capillary Running Procedure

1. Mix 1 μl fluorescently labeled PCR products with 24 μl deionized formamide and 0.5 μl internal size standard and transfer 10 μl to a 96-well plate compatible with the capillary genetic analyzer (see Notes 5 and 6).
2. Denature samples in a thermal cycler at 95°C for 5 min and snap cool to 4°C on an ice bath before loading the plate onto the genetic analyzer.
3. Run samples by capillary electrophoresis with the following run module settings:

   Oven_temperature: 60°C, range between 18 and 65°C

   PreRun_voltage: 15 kV, range between 0 and 15 kV

   Pre_Run_Time: 180 s, range between 1 and 1,000 s

   Injection_voltage: 1 kV, range between 1 and 15 kV

   Injection_Time: 10 s, range from 1 to 600 s

   Voltage_Number_of_Steps: 10 nk, range from 1 to 100 nk

   Voltage_Step_Interval: 60 s, range between 1 and 60 s

   Data_Delay_Time: 1, range between 1 and 3,600 s

   Run_Voltage: 15 kV, range between 0 and 15 kV

   Run_Time: 1,500 s, range between 300 and 14,000 s

3.3. Data Collection and Transcript Abundance Analysis

1. Fluorescence intensity is analyzed with GeneMapper v3.5 using the microsatellite module and a detection threshold set at 50. Peak and area values are exported to an Excel spreadsheet. An example of 6-FAM-labeled *RPS4* regular transcript (TV3) and one of the transcript variants (TV4) is shown in Fig. 1.
2. Extract and tabulate the peak and area values of all *RPS4* transcripts (Table 1).
3. Transform the fluorescence values on a $\log_{10}$ scale and plot these values as a function of the PCR cycle numbers as a "scattered dotted line" in Excel. Most data points should fall onto a linear curve representing constant amplification efficiency, although later PCR cycles may show a reduction in amplification efficiency (16). After manually removing these later data points, fit a linear trend line and display the regression equation and the correlation coefficient (r^2). An example of such a plot is shown in Fig. 2.
4. When PCR amplification efficiency is constant, the amount of cDNA produced after n cycles of amplification (cDNAn) is a function of the initial cDNA concentration (cDNA_0),

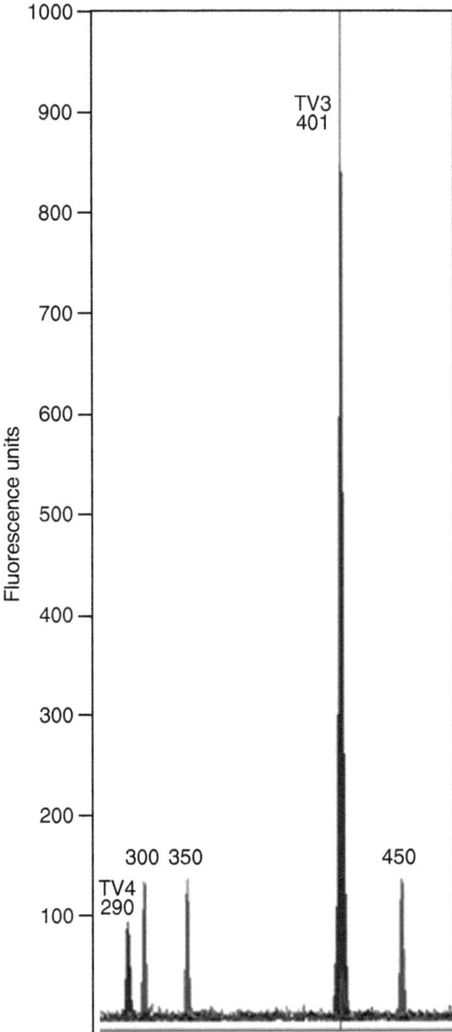

Fig. 1. Capillary electrophoresis of fluorescently labeled *RPS4* transcript products separated by size along the *x*-axis. Two *RPS4* transcripts, TV3 and TV4, and three internal size standards are shown. Numbers in peaks indicate DNA size in bp. The *y*-axis indicates the fluorescence signal peak height.

the cycle number, and the amplification efficiency constant R according to the equation (16):

$$\mathrm{cDNA}_n = \mathrm{cDNA}_0\ (1 + R)^n$$

After log transformation, this equation becomes:

$$\log(\mathrm{cDNA}_n) = n \log(1 + R) + \log(\mathrm{cDNA}_0)$$

At constant amplification efficiency, $\log(1 + R)$ is the slope of the linearized plot, and $\log(\mathrm{cDNA}_0)$ is the y-axis intercept. From this value, the initial amount of cDNA is calculated (see Note 7).

Table 1
Fluorescence signals of *RPS4* transcript products measured in consecutive PCR cycles[a]

	TV4[b] (288/288[c])		TV3[d] (402/401)		TV2A[b] (512/512)	
Cycles	Peak	Area	Peak	Area	Peak	Area
19	7	83	411	4,379	19	244
20	10	99	842	9,049	40	520
21	27	264	1,596	18,081	73	1,011
22	42	412	2,959	32,889	130	1,695
23	85	788	5,612	60,136	252	3,083
24	125	1,204	7,589	88,393	308	4,069
25	206	2,341	7,613	115,617	380	5,356
26	298	3,438	7,629	123,664	373	5,135

[a] Profile of *RPS4* transcripts from one RNA sample isolated from Columbia wild-type plants at 4 h post inoculation with *Pseudomonas syringae* pathovar tomato strain DC3000. The area fluorescence units shown in this table are plotted in Fig. 2

[b] TV4 and TV2A are the two most abundant alternative transcripts

[c] PCR fragment size via size calling/predicted fragment size

[d] TV3 is the regular, most abundant transcript with the longest open reading frame

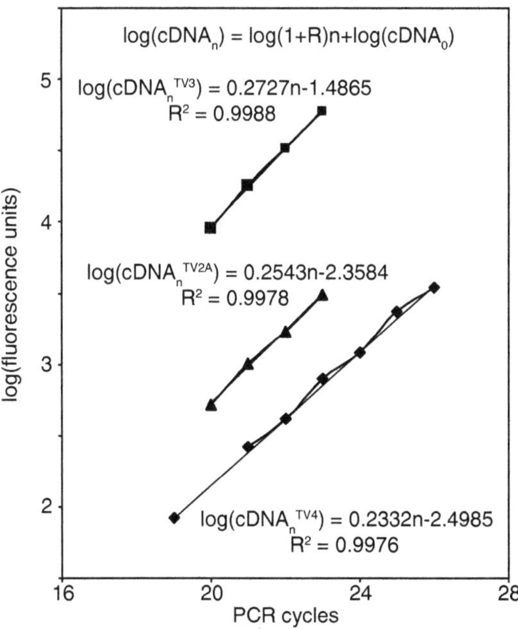

Fig. 2. Quantification of *RPS4* transcripts. PCR amplification was quantified over multiple cycles by measuring the fluorescence units of 6-FAM-labeled products in a capillary genetic analyzer. Fluorescence units were plotted on a logarithmic scale against the PCR cycle number. The linear portion of plots was fitted to a trend line by regression analysis. The resulting regression equations and correlation coefficients for each transcript are shown in the plot.

4. Notes

1. Isolation of mature polyadenylated mRNA improves cDNA quality and minimizes signal from heteronuclear RNA. Random hexamers are preferred for reverse transcription compared to oligo-dT primers because the latter lead to biased amplification of mRNA 3′-ends. The RNeasy plant mini kit (Qiagen, Valencia, CA) for total RNA isolation is an alternative to Tri Reagent.
2. The ABI PRISM® 3100 Genetic Analyzer supports multiple dyes for detection. Among others, dye choices include 5-FAM, TAMARA, ROX, and JOE. Be sure to place the fluorescence group at the 5′-end of a primer. With one primer labeled, the fluorescence signal of a given PCR product is directly proportional to the number of DNA molecules, regardless of the size of the fragment.
3. The choice of size standard will depend on the range of expected PCR product sizes. Size standards should be as close as possible to the products of most interest to ensure accurate size calling. The GeneScan™ 500 ROX™ size standard ranging from 35 to 500 nucleotides allowed accurate identification of *RPS4* transcript products ranging from 288 to 601 nucleotides.
4. Run these PCR aliquots on the same PCR cycler and remove tubes immediately after the extension step of a given cycle, before denaturation for the next cycle occurs. When the last PCR cycle is completed, place all aliquots back in the thermal cycler and extend for 5 min.
5. Use high quality formamide (<100 µS/cm). Single-stranded DNA samples will remain denatured for at least 3 days in the presence of high quality formamide.
6. The relative ratio of PCR products and internal size standards is critical for accurate size calling. Unbalanced ratios often cause a shift in size calling. Optimization experiments are strongly recommended.
7. The initial cDNA amount is calculated as a percentage of the regular transcript. To calculate absolute changes in transcript levels, determine the total amount of mRNA of your gene of interest by qRT-PCR with primers flanking a constitutively spliced intron using the same cDNA samples that were used for AS quantification.

Acknowledgment

We thank Dr. Joe Forrester and Ellen Krueger of the MU DNA Core for assistance in developing the transcript quantification procedure, Dr. Henry Nguyen, Sean Blake, and Dr. Xiaolei Wu for use of the capillary genetic analyzer and technical assistance. This work was supported by a Monsanto Senior Graduate Student Fellowship (to X.-C.Z.), by the US Department of Agriculture/ National Research Initiative (grant no. 2002–35319–12639 to W.G.), and in part by the Missouri Agricultural Experiment Station (project no. MO–PSSL0603).

References

1. Barbazuk, W. B., Fu, Y., and McGinnis, K. M. (2008) Genome-wide analyses of alternative splicing in plants: opportunities and challenges. *Genome Res.* **18**, 1381–92.
2. Campbell, M. A., Haas, B. J., Hamilton, J. P., Mount, S. M., and Buell, C. R. (2006) Comprehensive analysis of alternative splicing in rice and comparative analyses with Arabidopsis. *BMC Genomics* 7, 327.
3. Wang, B. B., and Brendel, V. (2006) Genomewide comparative analysis of alternative splicing in plants. *Proc. Natl. Acad. Sci. U. S. A.* **103**, 7175–80.
4. Ner-Gaon, H., and Fluhr, R. (2006) Whole-genome microarray in Arabidopsis facilitates global analysis of retained introns. *DNA Res.* **13**, 111–21.
5. Brett, D., Pospisil, H., Valcarel, J., Reich, J., and Bork, P. (2002) Alternative splicing and genome complexity. *Nat. Genet.* **30**, 29–30.
6. Johnson, J. M., Castle, J., Garrett-Engele, P., Kan, Z., Loerch, P. M., Armour, C. D., Santos, R., Schadt, E. E., Stoughton, R., and Shoemaker, D. D. (2003) Genome-wide survey of human alternative pre-mRNA splicing with exon junction microarrays. *Science* **302**, 2141–4.
7. Wang, E. T., Sandberg, R., Luo, S., Khrebtukova, I., Zhang, L., Mayr, C., Kingsmore, S. F., Schroth, G. P., and Burge, C. B. (2008) Alternative isoform regulation in human tissue transcriptomes. *Nature* **456**, 470–6.
8. Ner-Gaon, H., Halachmi, R., Savaldi-Goldstein, S., Rubin, E., Ophir, R., and Fluhr, R. (2004) Intron retention is a major phenomenon in alternative splicing in *Arabidopsis*. *Plant J.* **39**, 877–85.
9. Gassmann, W. (2008) Alternative splicing in plant defense. *Curr. Top. Microbiol. Immunol.* **326**, 219–33.
10. Jordan, T., Schornack, S., and Lahaye, T. (2002) Alternative splicing of transcripts encoding Toll-like plant resistance proteins – what's the functional relevance to innate immunity? *Trends Plant Sci.* 7, 392–98.
11. Dinesh-Kumar, S. P., and Baker, B. J. (2000) Alternatively spliced *N* resistance gene transcripts: their possible role in tobacco mosaic virus resistance. *Proc. Natl. Acad. Sci. U. S. A.* **97**, 1908–13.
12. Zhang, X. C., and Gassmann, W. (2003) *RPS4*-mediated disease resistance requires the combined presence of *RPS4* transcripts with full-length and truncated open reading frames. *Plant Cell* **15**, 2333–42.
13. Gassmann, W., Hinsch, M. E., and Staskawicz, B. J. (1999) The *Arabidopsis RPS4* bacterial-resistance gene is a member of the TIR-NBS-LRR family of disease-resistance genes. *Plant J.* **20**, 265–77.
14. Zhang, X.-C., and Gassmann, W. (2007) Alternative splicing and mRNA levels of the disease resistance gene *RPS4* are induced during defense responses. *Plant Physiol.* **145**, 1577–87.
15. Czechowski, T., Stitt, M., Altmann, T., Udvardi, M. K., and Scheible, W. R. (2005) Genome-wide identification and testing of superior reference genes for transcript normalization in Arabidopsis. *Plant Physiol.* **139**, 5–17.
16. Golde, T. E., Estus, S., Usiak, M., Younkin, L. H., and Younkin, S. G. (1990) Expression of beta amyloid protein precursor mRNAs: recognition of a novel alternatively spliced form and quantitation in Alzheimer's disease using PCR. *Neuron* **4**, 253–67.

Chapter 8

Constructing Haustorium-Specific cDNA Libraries from Rust Fungi

Ann-Maree Catanzariti, Rohit Mago, Jeff Ellis, and Peter Dodds

Abstract

The haustorium is a distinguishing feature of biotrophic plant pathogens. Several highly diverged pathogen classes have independently evolved haustoria, suggesting that they represent an effective adaptation for growing within living plant tissue. Despite their clear importance in biotrophy, they have been difficult to study due to the close association of biotrophic pathogens with their host and the inability to produce haustoria in vitro. These drawbacks have been circumvented in the study of rust fungi by the development of a haustoria isolation technique. The strong binding of the lectin concanavalin A (ConA) to rust haustoria allows these structures to be purified from infected plant tissue by affinity chromatography on a ConA–Sepharose macrobead column. The isolation process results in substantial yields of intact haustoria that retain their cytoplasmic contents, making them amenable to experimentation. The construction of cDNA libraries from isolated rust haustoria and their subsequent sequence analysis have provided significant insight into haustoria function at a molecular level, revealing important roles in nutrient acquisition and the delivery of pathogenicity effector proteins. The generation of a rust haustorium-specific cDNA library is described in this chapter.

Key words: Haustoria, Rust fungi, Obligate plant pathogens, Flax rust, Wheat stem rust

1. Introduction

The haustorium, which is produced only during infection of the host plant, is a distinguishing feature of biotrophic fungal plant pathogens. This specialised structure is formed after the host cell wall is breached, and expands within the plant cell while invaginating the host plasma membrane, and thus remains outside the host cytoplasm creating a unique host–pathogen interface. The ultrastructure of this interface has been extensively examined and consists of distinct regions showing significant differentiation (1). These include the haustorial neck which spans the host cell wall,

John M. McDowell (ed.), *Plant Immunity: Methods and Protocols*, Methods in Molecular Biology, vol. 712, DOI 10.1007/978-1-61737-998-7_8, © Springer Science+Business Media, LLC 2011

and the haustorial body, which is encased by the extrahaustorial membrane derived from the invaginated host plasma membrane, and between the haustoria and this membrane, a region known as the extrahaustorial matrix (2).

Haustorial cells provide the closest contact between the fungus and the host and are therefore a likely site for molecular signalling, particularly for the establishment and maintenance of biotrophy (3, 4). However, early research in the molecular basis of biotrophy and gene-for-gene recognition was constrained by the growth of haustoria within host tissue making them relatively inaccessible to experimentation, and the inability to produce these structures in vitro. Although there is one account of rust haustoria forming in the absence of a living cell, they appeared immature and not completely differentiated, indicating that additional plant signals or nutrients are required to complete haustorial development (5, 6). The ability to isolate fully developed haustorial cells from infected host tissue has been a major factor in allowing molecular, genetic, and biochemical studies of the host–fungus interaction in a way that was not previously achievable in planta or in vitro. Recent results stemming from this technique have identified amino acid and sugar transporters in the haustorial membrane, and thus confirmed the role of haustoria in nutrient uptake (7–9). It has also now been shown that rust fungi deliver pathogenicity effector proteins into host cells from their haustoria, and these include avirulence proteins that are recognised by host-resistance proteins (10–12).

A robust haustorial isolation method was developed by Hahn and Mendgen (13), and involves affinity chromatography with the lectin concanavalin A (ConA). This method has proven to be very successful in isolating high yields of haustoria from various rust species (order Pucciniales), including broad bean rust (*Uroymces fabae*), cowpea rust (*Uroymces vignae*), wheat leaf rust (*Puccinia triticina*), and maize rust (*Puccinia sorghi*), and has subsequently been used to isolate haustoria from flax rust (*Melampsora lini*) (12, 14) and wheat stem rust (*Puccinia graminis* f. sp. *tritici*; Rohit Mago unpublished data). Electron microscopy showed isolated rust haustoria retain their cytoplasmic contents due to a blockage of the neck space by electron dense material, but lose the extrahaustorial membrane, while the extrahaustorial matrix is only partly removed (13). ConA-affinity for rust haustoria is thought to be due the presence of surface-exposed α-linked D-mannose within the haustorial cell wall and haustorial matrix. However, a non-specific component also contributes to binding as haustoria cannot be released from a ConA column by elution with methyl α-D-mannopyranoside (13).

ConA-affinity chromatography has not been successful for the isolation of haustoria from powdery mildew fungi including *Sphaerotheca fuliginea*, *Erysiphe pisi*, and *E. graminis* f. sp. *hordei*, due to their weak binding to the ConA lectin (13). However, haustoria from this class of fungi have been successfully isolated by density gradient centrifugation (15–17). In this case, the haustoria are isolated as whole haustorial complexes, enclosed by the extrahaustorial membrane, and thus the differences between mildews and rust haustoria in their affinity for ConA may be due to differences in exposed carbohydrates.

This chapter describes the isolation of haustoria by ConA-affinity chromatography that has proved successful in our studies on flax and wheat stem rust for the generation of haustorium-specific cDNA libraries.

2. Materials

2.1. Coupling ConA to Cyanogen Bromide-Activated Sepharose Macrobeads

1. One litre vacuum flask with sintered glass funnel (4 cm diameter).
2. Whatman filter paper, Grade No. 1, (4.25 cm diameter).
3. Rotary wheel.
4. Cyanogen bromide (CNBr)-activated Sepharose 6MB (see Note 1) (GE Healthcare, cat # 17-0820-01).
5. 1 mM HCl.
6. ConA (Amersham Biosciences) dissolved in coupling buffer just before use.
7. Coupling buffer: 0.1 M $NaHCO_3$, 0.5 M NaCl, pH 8.3.
8. Blocking buffer: coupling buffer containing 0.5 M glycine.
9. Acid wash buffer: 0.5 M NaCl, 0.1 M sodium acetate pH 4.
10. Alkaline wash buffer: 0.5 M NaCl, 0.1 M Tris–HCl pH 8.
11. Storage buffer: 0.15 M NaCl, 10 mM Tris–HCl pH 7.2, 1 mM $CaCl_2$, 1 mM $MnCl_2$, 0.02% (w/v) NaN_3.

2.2. Haustoria Isolation

1. Infected plant material (see Note 2).
2. Glass Econo-Columns, 1.5 × 10 cm, 18 ml (Bio-Rad).
3. Waring blender.
4. Homogenisation buffer: 0.3 M sorbitol, 20 mM MOPS pH 7.2, 0.1% (w/v) BSA, 0.2% (w/v) PEG 6000, 0.2% (v/v) β mercaptoethanol (added fresh): chilled to 4°C before use.
5. Two nylon meshes, one with a pore size of 100 μm, and the other with 20, 15, or 11 μm.
6. Suspension buffer: 0.3 M sorbitol, 10 mM MOPS pH 7.2, 0.2% BSA, 1 mM KCl, 1 mM $MgCl_2$, 1 mM $CaCl_2$.

2.3. Generation of Haustorium-Specific cDNA Library

1. RNeasy Plant mini kit (QIAGEN).
2. PolyATract mRNA isolation system (Promega) or equivalent.
3. SMART cDNA library kit (BD Biosciences) or equivalent.
4. Lambda packaging extract (Epicentre Biotechnologies).

3. Methods

3.1. Coupling ConA to CNBr-Activated Sepharose Macrobeads

The amount of ConA–Sepharose required will depend on the amount of tissue to be processed. As a guide, aim to make at least 15–20 ml, keeping in mind that small amounts of Sepharose are lost during the coupling and bead regeneration process. All washes described below are carried out using a funnel with a sintered glass filter under vacuum filtration. Placing a circular filter paper on top of the glass filter will assist with flow speed and the transfer of beads between steps.

1. Suspend the CNBr-activated Sepharose 6MB in 1 mM HCl. The gel will swell immediately (1 g equals about 3.5 ml of swollen gel; see Note 3) and should then be washed with 1 mM HCl, using about 200 ml/g of the freeze-dried powder.
2. Dissolve ConA (5–10 mg/ml of swollen gel) in coupling buffer (use 5 ml per gram of CNBr sepharose powder). Add the swollen Sepharose and mix end-over-end on a rotary wheel for 2 h at room temperature (see Note 4).
3. After incubation, wash the gel once with coupling buffer (~400 ml) to remove any excess ligand.
4. Block any remaining active groups by adding blocking buffer (~40 ml) and mix end-over-end for 2 h at room temperature.
5. Wash the ConA–Sepharose with three cycles of alternating pH. Use the acid wash buffer followed by the alkaline wash buffer for each cycle, with ~10 ml of buffer per ml of gel for each wash. Resuspend the beads in the wash buffer briefly and then remove the buffer by vacuum filtration.
6. Equilibrate the ConA–Sepharose with storage buffer so that the volume above the gel is equal to the gel bed volume and store at 4°C.

3.2. Isolation of Haustoria from Infected Tissue

Several steps will require empirical adjustments according to the fungal pathogen being used, in particular the timing of tissue collection (see Note 2), the amount of infected tissue required (see Note 5) and the pore size of the nylon mesh used to filter the homogenised tissue (see Note 6). After weighing the infected tissue, all subsequent steps should be carried out at 4°C.

1. Harvest 30 g of heavily infected tissue before sporulation; 5–10 days post-inoculation, depending on the advancement of the infection (see Note 2).
2. Using a Waring blender, homogenise the infected plant material in 180 ml cold homogenisation buffer at maximum speed for 30 s.
3. Filter the homogenate through a 100-μm nylon mesh by gravity flow to remove larger cell debris, and then through a 20-, 15-, or 11-μm nylon mesh (see Note 6).
4. Centrifuge the filtrate at 6,900 × *g* for 5 min and resuspend the pellet in 6 ml of suspension buffer.
5. Add a 5-ml bed volume of ConA–Sepharose to three columns (see Note 7) and equilibrate with suspension buffer.
6. Load 1 ml of the resuspended pellet onto each column. Allow the suspension to migrate into the gel, then incubate for 15 min without flow, and then repeat with the remaining suspension.
7. Wash each column by carefully layering 10–15 ml of suspension buffer on top of the ConA–Sepharose, taking care not to disturb the gel, and then allow the buffer to flow through until it runs clear (it will initially be green due to the chloroplasts washing off the column).
8. Release the bound haustoria by adding 5 ml of suspension buffer to the column and agitating the beads by pipetting up and down using a wide bore pipette; a 1-ml pipette tip with the end (2–3 mm) cut off works well.
9. Immediately after the beads settle, collect the supernatant containing the haustoria on top of the beads.
10. Repeat steps 8 and 9 and pool the eluted haustoria samples.
11. Centrifuge the suspension at 14,700 × *g* for 5 min. Resuspend in 1 ml of suspension buffer and transfer to a microcentrifuge tube, retain an aliquot for microscopic analysis. The haustoria can be seen as intact structures among chloroplasts and some hyphal fragments (see Note 8). Figure 1 shows isolated flax rust haustoria.
12. Re-pellet the haustorial suspension, remove the supernatant, and weigh sample before freezing in liquid nitrogen, then store at –80°C until RNA extraction.

3.3. Regeneration of Column

To regenerate the ConA–Sepharose beads, remove from the column (see Note 9) and wash as given in step 5 of Subheading 3.1. The gel can then be equilibrated with storage buffer and stored at 4°C.

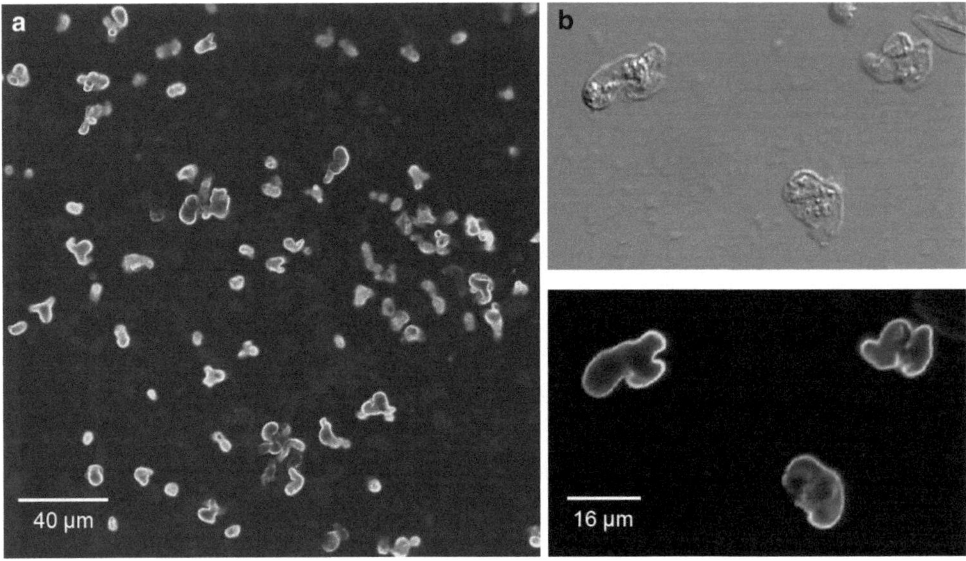

Fig. 1. Isolated flax rust haustoria labelled with monoclonal antibody ML1, which specifically binds to a surface carbohydrate epitope present in the haustorial cell wall of *Melampsora lini* (14). Images were collected on a Leica TCS SP2 confocal microscopy. (**a**) Fluorescence image showing immunolabelled haustoria in green. (**b**) Higher magnification showing haustoria as intact structures under bright-field (*top*) and fluorescence (*bottom*).

3.4. Generation of Haustorium-Specific cDNA Library

3.4.1. Isolation of mRNA

1. Isolate total RNA from frozen haustoria samples using the RNeasy Plant Mini Kit from QIAGEN, following the protocol for "plant cells and tissues and filamentous fungi."
2. Grind the frozen haustoria sample (no more than 100 mg per column; 30 g of infected flax tissue will give a haustorial pellet of approximately 80 mg) thoroughly with a mortar and pestle in liquid nitrogen.
3. Add the ground powder to 450 μl of Buffer RLT and proceed with the manufacturer's protocol.
4. The RNA yield can be increased by using two aliquots of RNase-free water when eluting from the column. All processed samples can then be pooled and the volume reduced to ~100 μl in a vacuum centrifuge.
5. Optional (see Note 10): treat the RNA with DNase and cleanup the reaction using the "RNA cleanup and concentration" protocol. As a guide, 30 g of infected flax leaf tissue will yield ~5 μg of total RNA, and similar yields were obtained for stem rust.
6. From the total RNA, isolate mRNA using polyATract mRNA isolation kit (Promega).

3.4.2. Generation of cDNA Library

1. Using 0.025–0.5 μg of haustorial mRNA (see Note 10), generate cDNA with the SMART cDNA Library kit (BD

Biosciences) following the cDNA synthesis by long distance (LD) PCR protocol.

2. Package the library into lambda phage using the MaxPlax lambda packaging extract (Epicentre Biotechnologies), then titrate and amplify following the SMART cDNA library kit protocol.

   Lambda clones can be converted to plasmids in *Escherichia coli* strain BM25.8 using the protocol given in the SMART cDNA library kit manual. Colonies can be picked using blue/white selection on X-Gal plates and grown overnight in 96- or 384-well microtiter plates. Cultures can then be used directly for PCR and subsequent sequencing to characterise the insert, and then frozen as glycerol stocks.

4. Notes

1. It is important to use 6MB beads which are designed specifically for cell affinity chromatography and have a large bead size (250–350 μm) to allow minimal trapping of non-specific cells when passing mixed cell suspensions through the column.
2. Harvest tissue as late as possible to increase the number of haustoria, but should be at least 1 day before pustule eruption; 5–6 days for flax rust, 8–10 days for stem rust (for the latter, collection should be as soon as the first signs of infection are visible, i.e. the appearance of yellowing spots).
3. Significantly less swelling of the freeze-dried powder (than the stated 1 g to 3.5 ml) is an indication that the Sepharose macrobeads have expired and should not be used.
4. The ConA coupling step can also be performed overnight at 4°C, and while other gentle mixing methods can be employed, do not use magnetic stirring as this will damage the beads.
5. For flax rust, RNA yields were ~5 μg of total RNA from 30 g of infected plant material. Thus multiple isolations were required to generate sufficient material for library construction. For stem rust, we obtained similar RNA yields but scaled up each isolation to ~80 g of infected material.
6. The size exclusion provided by the mesh pore size is an important step for pure yields of haustoria as large numbers of contaminating cell types can become trapped within the column of Sepharose beads (note, the Bio-Rad Econo-columns have a 28-μm sintered filter). Rust fungi display a

diverse range of haustorial size and morphology, so to choose the pore size of the second nylon mesh, perform a small-scale isolation and microscopic analysis with each of the three nylon meshes to determine which is optimal for purity and yield. Alternatively, collect haustoria using the 20-μm mesh and measure the size of haustoria and any intact plant cell sizes that are present by microscopy using an eyepiece graticule. As an example, isolation of flax rust haustoria using the 20-μm mesh will give a higher yield of haustorial cells than the 11-μm mesh (number of haustoria per 30 g tissue: 2.9×10^6 versus 4.5×10^5), but the 11-μm mesh is required to exclude intact plant cells and thus will result in the greatest purity (purity as the percentage of haustorial cells among total number of intact cells: 49% versus >99%). For *P. graminis* haustoria 15-μm mesh was used. In this case no plant cell contamination was observed with the 20-μm mesh, but some rust spores were not excluded by this mesh size.

7. The use of three columns each containing a 5-ml bed volume of ConA–Sepharose, rather than one containing 15 ml, is more convenient during the elution of the bound haustoria due to the bead agitation that is required. Using two columns of 7.5 ml was also found to be relatively convenient.
8. Retain an aliquot of the eluted haustoria for microscopic examination to determine the integrity of the haustorial cells and the purity of the sample. The number of isolated haustoria can be calculated using a hemocytometer (see Note 5 above). Other cell types observed within the haustoria suspension include some hyphal fragments, which unlike the isolated haustoria have sheared ends and are devoid of cellular contents, and chloroplasts. However, the contribution of total RNA from contaminating chloroplasts was investigated and found to be relatively insignificant (18).
9. The column can be used several times before needing regenerating. To remove the gel from the column, add elution buffer and invert several times, then pour out into the funnel, with sintered glass filter and filter paper, for washing with pH buffers.
10. Although it is stated that a cDNA library can be constructed using total RNA with this kit, we highly recommend the use of mRNA as we observed non-specific priming of the modified oligo(dT) primer (CDS III/3′ PCR Primer) on ribosomal RNA or contaminating DNA during first-strand synthesis. If using total RNA for library construction, a DNase treatment is also recommended after RNA isolation.

References

1. Harder, D. E. (1989) Rust haustoria – past, present, future. *Can. J. Plant Pathol.* **11**, 91–99.
2. Voegele, R. T. and Mendgen, K. (2003) Rust haustoria: nutrient uptake and beyond. *New Phytol.* **159**, 93–100.
3. Heath, M. C. (1997) Signalling between pathogenic rust fungi and resistant or susceptible host plants. *Ann. Bot.* **80**, 713–720.
4. Panstruga, R. (2003) Establishing compatibility between plants and obligate biotrophic pathogens. *Curr. Opin. Plant Biol.* **6**, 320–326.
5. Heath, M. C. (1989) *In vitro* formation of haustoria of the cowpea rust fungus, *Uromyces vignae*, in the absence of a living plant cell. I. Light microscopy. *Physiol. Mol. Plant Pathol.* **35**, 357–366.
6. Heath, M. C. (1990) *In vitro* formation of haustoria of the cowpea rust fungus *Uromyces vignae* in the absence of a living plant cell. II. Electron microscopy. *Can. J. Bot.* **68**, 278–287.
7. Voegele, R. T., Struck, C., Hahn, M. and Mendgen, K. (2001) The role of haustoria in sugar supply during infection of broad bean by the rust fungus *Uromyces fabae*. *Proc. Natl Acad. Sci. U. S. A.* **98**, 8133–8138.
8. Struck, C., Ernst, M. and Hahn, M. (2002) Characterization of a developmentally regulated amino acid transporter (AAT1p) of the rust fungus *Uromyces fabae*. *Mol. Plant Pathol.* **3**, 23–30.
9. Struck, C., Mueller, E., Martin, H. and Lohaus, G. (2004) The *Uroymces fabae UfAAT3* gene encodes a general amino acid permease that prefers uptake of *in planta* scare amino acids. *Mol. Plant Pathol.* **5**, 183–189.
10. Dodds, P. N., Lawrence, G. J., Catanzariti, A., Ayliffe, M. A. and Ellis, J. G. (2004) The *Melampsora lini AvrL567* avirulence genes are expressed in haustoria and their products are recognized inside plant cells. *Plant Cell* **16**, 755–768.
11. Kemen, E., Kemen, A. C., Rafiqi, M., Hempel, U., Mendgen, K., Hahn, M. and Voegele, R. T. (2005) Identification of a protein from rust fungi transferred from haustoria into infected plant cells. *Mol. Plant Microbe Interact.* **18**, 1130–1139.
12. Catanzariti, A., Dodds, P. N., Lawrence, G. J., Ayliffe, M. A. and Ellis, J. G. (2006) Haustorially expressed secreted proteins from flax rust are highly enriched for avirulence elicitors. *Plant Cell* **18**, 243–256.
13. Hahn, M. and Mendgen, K. (1992) Isolation by ConA binding of haustoria from different rust fungi and comparison of their surface qualities. *Protoplasma* **170**, 95–103.
14. Murdoch, L. J., Kobayashi, I. and Hardham, A. R. (1998) Production and characterisation of monoclonal antibodies to cell wall components of the flax rust fungus. *Eur. J. Plant Pathol.* **104**, 331–346.
15. Dekhuijzen, H. M. (1966) The isolation of haustoria from cucumber leaves infected with powdery mildew. *Neth. J. Plant Pathol.* **72**, 1–11.
16. Gil, F. and Gay, J. L. (1977) Ultrastructural and physiological properties of the host interfacial components of haustoria of *Erysiphe pisi in vivo* and *in vitro*. *Physiol. Plant Pathol.* **10**, 1–12.
17. Manners, J. M. and Gay, J. L. (1977) The morphology of haustorial complexes isolated from apple, barley, beet and vine. *Physiol. Plant Pathol.* **11**, 261–266.
18. Hahn, M. and Mendgen, K. (1997) Characterization of in planta-induced rust genes isolated from a haustorium-specific cDNA library. *Mol. Plant Microbe Interact.* **10**, 427–437.

Chapter 9

Microaspiration of Esophageal Gland Cells and cDNA Library Construction for Identifying Parasitism Genes of Plant-Parasitic Nematodes

Richard S. Hussey, Guozhong Huang, and Rex Allen

Abstract

Identifying parasitism genes encoding proteins secreted from a plant-parasitic nematode's esophageal gland cells and injected through its stylet into plant tissue is the key to understanding the molecular basis of nematode parasitism of plants. Parasitism genes have been cloned by directly microaspirating the cytoplasm from the esophageal gland cells of different parasitic stages of cyst or root-knot nematodes to provide mRNA to create a gland cell-specific cDNA library by long-distance reverse-transcriptase polymerase chain reaction. cDNA clones are sequenced and deduced protein sequences with a signal peptide for secretion are identified for high-throughput in situ hybridization to confirm gland-specific expression.

Key words: Secretory proteins, Transcriptome analysis, Effector proteins, Plant parasitism

1. Introduction

Sedentary endoparasitic nematodes have evolved sophisticated parasitic relationships with their host plants to obtain nutrients that are necessary to support their development and reproduction. These biotrophic parasites feed from the cytoplasm of living root cells they have modified into elaborate, discrete feeding cells (1). Three enlarged secretory gland cells in the nematode's esophagus are the principal sources of secretions used by the sedentary parasites to modify root cells by modulating complex changes in cell gene expression, physiology, morphology, and function (2, 3). The cloning and identification of parasitism genes encoding the esophageal gland proteins secreted from the nematode stylet into host tissue is significantly increasing our understanding of the molecular mechanisms of nematode parasitism of plants.

John M. McDowell (ed.), *Plant Immunity: Methods and Protocols*, Methods in Molecular Biology, vol. 712,
DOI 10.1007/978-1-61737-998-7_9, © Springer Science+Business Media, LLC 2011

The most successful approach employed to clone parasitism genes directly analyzes the transcriptome of the esophageal gland cells by microaspiration of the cytoplasm of the gland cells to use in cDNA library construction. The gland-specific cDNA libraries were sequenced and predicted proteins containing signal peptides for secretion were identified. Expression of the cloned parasitism genes in the esophageal glands was confirmed by in situ mRNA hybridization, leading to the identification of more than 60 parasitism genes for the soybean cyst nematode, *Heterodera glycines* (4–6) and over 50 for the root-knot nematode, *Meloidogyne incognita* (7, 8).

2. Materials

2.1. Nematode Preparation

1. While root-knot nematodes are used for the methods described herein, this strategy for identifying parasitism genes has been used for cyst and can be used for other plant-parasitic nematodes as well. Propagate root-knot nematodes in greenhouse on susceptible tomato or eggplant by inoculating 7-weeks old plants with 40,000–50,000 root-knot nematode eggs (9). Pre-parasitic second-stage juveniles were collected by hatching eggs on sieves with 25-μm openings suspended over deionized water in plastic bowls.
2. Sieves (20 cm diameter) used to collect parasitic stages of root-knot nematodes. Recommended sizes: #60 (250-μm opening), #80 (180 μm), #100 (150 μm), #500 (25 μm).
3. Magnesium sulfate ($MgSO_4 \cdot 7H_2O$) 25% (w/v) dissolved in water. Store at room temperature.
4. Ultra pure low-melting-point agarose (25 mL) 3% dissolved in water by heating in a microwave. Store in a dri-bath at 40°C before use.

2.2. Gland Aspiration

1. Aluminosilicate capillary tubes (o.d. 1 mm, i.d. 0.68 mm, 10 cm length; Sutter Instruments, Novato, CA).
2. mRNA extraction buffer: 100 mM Tris–HCl, pH 7.5, 500 mM LiCl, 10 mM EDTA, pH 8.0; 1% LiDS, 5 mM dithiothreitol. Store at 4°C.
3. Microinjection station consisting of: Nikon Eclipse TE300 inverted microscope equipped with a Plan 4×/0.10 objective and two Nomarski differential-interference contrast objectives (Plan Fluor 10×/0.30 DIC and Plan Fluor ELWD 40×/0.60 DIC); Sony color video camera (Model DXC-107A) and Trinitron color video monitor (Model PVM-1353MD) (Sony Electronics, Montvale, NJ); micromanipulator (Model 5171),

CellTram oil microinjector-aspirator with Universal capillary holder (Model 5176), and microloader (#930-00-100-7) (Eppendorf, Hamburg, Germany). Microinjection station sets on a Micro-G vibration isolation table (Technical Manufacturing Corporation, Peabody, MA).

4. Automated horizontal micropipette puller (Model P-97) and micropipette beveler (Model BV-10) (Sutter Instruments).
5. Aquarium air pump (#SA-X2, Penn-Plax, Inc., Hauppauge, NY).

2.3. Isolation of mRNA

1. Dynabeads Oligo(dT)$_{25}$ in phosphate-buffered saline (PBS) pH 7.4 containing 0.02% NaN_3 (5 μg/μL) (Invitrogen).
2. Washing buffer I: 10 mM Tris–HCl, pH 7.5, 0.15 M LiCl, 1 mM EDTA, 0.1% LiDS. Store at 4°C.
3. Washing buffer II: 10 mM Tris–HCl, pH 7.5, 0.15 M LiCl, 1 mM EDTA. Store at 4°C.

2.4. cDNA Amplification

1. 3′ SMART CDS primer II A (12 μM): 5′-AAGCAGTGGTATCAACGCAGAGTACT$_{(30)}$ N$_{-1}$N-3′ (N=A, C, G, or T; N$_{-1}$=A, G, or C) (Clontech, Palo Alto, CA). Store at −80°C.
2. SMART II A Oligo (12 μM): 5′-AAGCAGTGGTATCAACGCAGAGTACGCGGG-3′ (Clontech). Store at −80°C.
3. First-strand buffer (5×): 250 mM Tris–HCl, pH 8.3, 375 mM KCl, 30 mM $MgCl_2$. Store at −20°C.
4. 5′ PCR primer II A (12 μM): 5′-AAGCAGTGGTATCAACGCAGAGT-3′. Store at −20°C.
5. dNTP mix (50×): 10 mM of each dNTP. Store at −20°C.
6. MMLV reverse transcriptase (200 U/μL) (Clontech). Store at −20°C.
7. NucleoSpin extract II kit (Clontech). Store at room temperature.
8. RNase inhibitor (20 U/μL) (Clontech). Store at −20°C.
9. Advantage cDNA polymerase mix: 50× Advantage cDNA polymerase mix, 10× cDNA PCR reaction buffer (Clontech). Store at −20°C.
10. BenchTop 1 kb DNA ladder (Promega, Madison, WI). Store at −20°C.
11. TAE electrophoresis buffer (50×): 24.2% (w/v) Tris base, 5.71% (v/v) glacial acetic acid, 3.72% (w/v) $Na_2EDTA \cdot 2H_2O$. Store at room temperature.
12. RadPrime DNA-labeling system (Invitrogen). Store at −20°C.
13. QIAquick PCR purification kit (QIAGEN, Valencia, CA). Store at room temperature.

2.5. Library Construction

1. Rapid ligation buffer (2×): 60 mM Tris–HCl, pH7.8, 20 mM $MgCl_2$, 20 mM DTT, 2 mM ATP, 10% polyethylene glycol (PEG). Store at –20°C.
2. T4 DNA ligase (3 Weiss units/μL): (Promega). Store at –20°C.
3. pGEM-T Easy vector (50 ng/μL) (Promega). Store at –20°C.
4. *Escherichia coli* XL10-GOLD ultracompetent cells (Stratagene, La Jolla, CA). Store at –80°C.
5. IPTG (Promega) is dissolved at 0.1 M in water, filter-sterilized and stored at 4°C.
6. X-Gal (5-bromo-4-chloro-3-indolyl-β-D-galactoside) (Promega) is dissolved at 50 mg/mL in N,N′-dimethyl-formamide (DMF), covered with aluminum foil, and stored at –20°C.
7. Luria-Bertani (LB) medium: 1% (w/v) bacto-peptone, 0.5% (w/v) bacto-yeast extract, 0.5% NaCl, pH 7.0. Store at room temperature.
8. Ampicillin (Sigma, St. Louis, MO) is dissolved at 100 mg/mL in water, and filter-sterilized and stored at –20°C.
9. QIAprep Spin Miniprep kit: (QIAGEN). Store at room temperature.
10. *Eco*RI (20 U/μL) (New England Biolabs, Ipswich, MA). Store at –20°C.
11. MICROTEST III Tissue Culture plates from Becton Dickinson, Franklin Lakes, NJ.

2.6. Sequencing

1. Hybond-XL nylon membrane from Amersham Parmacia Biotech, Piscataway, NJ.
2. Denaturing solution: 1.5 M NaCl, 0.5 M NaOH. Made fresh.
3. Neutralizing solution: 1.5 M NaCl, 0.5 M Tris–HCl, pH 7.2, 1 mM EDTA. Store at room temperature.
4. Intestinal cDNA probes are generated from purified LD-PCR products from the intestinal cells of parasitic nematodes and labeled from 20 ng of the PCR products with [α^{32}P]dCTP in one labeling reaction using an RadPrime DNA-labeling system (Invitrogen).
5. Millipore Montage Plasmid $Miniprep_{96}$ kit from Millipore-Fisher Scientific, Billerica, MA.
6. $MultiScreen_{96}$ PLASMID plates (Millipore-Fisher Scientific).
7. BigDye Terminator v3.1 cycle sequencing kit (Applied Biosystems, Foster City, CA).
8. Sequencing primer: 5′-GGTATCAACGCAGAGTACGCG-3′.

2.7. In Situ Hybridization

1. PBS buffer: 137 mM NaCl, 2.7 mM KCl, 4.3 mM Na_2HPO_4, 1.47 mM KH_2PO_4, pH 7.4. Store at room temperature.

2. Formalin, neutral, buffered in 10% (w/v) in phosphate buffer, pH 7.4 from Electron Microscopy Sciences, Hatfield, PA. Store at room temperature.
3. Proteinase K, recombinant, PCR grade (1 mg/mL) (Roche, Indianapolis, IN). Store at 4°C.
4. SSC buffer (20×): 3 M NaCl, 0.3 M sodium citrate, pH 7.0. Store at room temperature.
5. Denhardt's (100×): 2% (w/v) Ficoll 400, 2% (w/v) polyvinlpyrrolidone (PVP), 2% (w/v) bovine serum albumin (Fraction V). Store at −20°C.
6. Digoxigenin (DIG)-hybridization buffer (see Note 1): 50% deionized formamide, 4× SSC (diluted from 20× SSC), 1× Denhardt's (diluted from 100× Denhardt's), 1 mM EDTA, 0.02% (w/v) fish sperm DNA, 0.015% (w/v) yeast tRNA. Store at −20°C.
7. PCR DIG-labeling mix: 2 mM dATP, dCTP, dGTP each, 1.9 mM dTTP, 0.1 mM DIG-11-dUTP (Roche). Store at −20°C.
8. DIG-labeled cDNA probes (see Note 2) are synthesized through two subsequent PCRs. First, gene-specific forward and reverse primers for each gene are used to amplify the cDNA fragments in the plasmids by regular PCR amplification. Second, the PCR products purified by QIAquick PCR purification kit are used to synthesize DIG-labeled sense and antisense cDNA probes by asymmetric PCR amplification with the forward or reverse primer. The PCR reactions are performed in a 20 μL reaction mixture with PCR DIG-labeling mix (Roche).
9. PCR DIG-labeled DNA probes are denatured at 100°C for 5 min in a thermal cycler and rapidly cooled on ice water before use.
10. Maleic acid buffer: 0.1 M maleic acid, 0.15 M NaCl, pH 7.5. Store at room temperature.
11. Blocking solution is diluted from 10× blocking stock solution (Roche) with 1× maleic acid buffer. Store at 4°C.
12. Detection buffer: 0.1 M Tris–HCl, 0.1 M NaCl, 50 mM MgCl2, pH9.5. Store at 4°C.
13. Antibody solution is freshly prepared by diluting 1:1,000 anti-DIG-alkaline phosphatase (AP) Fab fragment (Roche) in 1× blocking solution.
14. Staining solution (1 mL) is prepared freshly by adding 11.25 μL of 4-nitro blue tetrazolium chloride (NBT) stock (50 mg/mL) and 8.75 μL of 5-bromo-4-chloro-3-indolyl-phosphate (BCIP) stock (100 mg/mL) to 980 μL of 1× detection buffer.

3. Methods

Since the three esophageal gland cells of Tylenchid and Aphelenchid nematodes are the source of the secretions that enable these nematodes to infect and parasitize plants, the key to understanding the molecular basis of parasitism lies in the identification and functional analysis of effector proteins encoded by the parasitism gene expressed in the gland cells. In addition, changes in the esophageal gland cells during the parasitic cycle indicate various roles for the gland effector proteins during different stages of nematode parasitism of plants (1). Thus it is critical to obtain a profile of the parasitism genes expressed in the esophageal gland cells throughout the parasitic cycle.

Targeting nematode esophageal gland cells to identify the effector proteins encoded by parasitism genes requires that cDNA library construction utilizes technology that generates high yields of quality full-length, double-stranded cDNA from small amounts of RNA. Full-length cDNAs with complete 5′ ends that contain signal peptide-coding regions for secretion are identified by SignalP analysis of the predicted open reading frames. Parasitism genes are identified from the cDNAs encoding signal peptides by confirming their expression in the esophageal gland cells by in situ mRNA hybridization (5–8, 10).

3.1. Nematode Preparation

1. Microwave 25 mL of a 3% solution of low-melting-point agarose and transfer 3 mL to several 14 mL polypropylene round-bottomed tubes. Hold tubes in a dri-bath at 40°C. Prepare the agarose solution the previous day so the temperature can stabilize.
2. Harvest plants to collect a mixture of parasitic stages ranging from second-stage juveniles to young adults (see Note 3).
3. Place roots into 1-L blender container in 500 mL water and blend approximately 8 s at high speed; repeat blending two to three times.
4. Pour water and roots through five stacked sieves (20 cm diameter) – #60 (250-μm opening), #80 (180 μm), #100 (150 μm), and #500 (25 μm) on the bottom.
5. Rinse stacked sieves thoroughly with running tap water using a short garden hose attached to a faucet and spray with hose to rinse nematodes through sieves.
6. Remove top sieve (#60) and save roots. In a stepwise fashion, spray water over each sieve and discard the debris, until the nematodes are collected on the bottom #500 sieve.
7. Collect nematodes and some root pieces in a 250-mL beaker.

8. Using roots recovered from the top sieve (#60) repeat blending three more times, collect nematodes as in step 4, and add nematodes collected on the #500 sieve to the beaker.
9. After root blending is completed reduce the water in the beaker to 50 mL by letting the collected nematodes settle to the bottom and carefully pouring off the excess water. Transfer the suspended nematodes to two glass 250-mL centrifuge bottles (Corning 1260).
10. Slowly inject approximately 150 mL magnesium sulfate solution with a 60-mL syringe fitted with a 10.2-cm-long 14-gauge pipetting needle into the bottom of each centrifuge bottle under the nematodes. The nematodes and root pieces should become layered on top of the magnesium sulfate solution.
11. After centrifuging the nematode suspensions at 1,400 × *g* for 10 min (e.g, 2,500 rpm in IEC Model K centrifuge with swinging bucket rotor, #277), gently remove the bottles from the centrifuge and carefully collect the nematodes floating on top of the magnesium sulfate with the syringe. Avoid removing root pieces.
12. Add nematodes to a small #500 sieve (8 cm diameter) and rinse with tap water.
13. Rinse nematodes with approximately 20–30 mL of distilled water into a small beaker.
14. Pour nematodes into petri dish and under a dissecting microscope pick out 100–200 parasitic stages ranging from second-stage juveniles to young adults and collect in a small dish for microaspiration. The remaining parasitic nematodes are fixed in phosphate-buffered 10% formalin for in situ mRNA hybridization.
15. Wash nematodes three times with sterile water and place 50 nematodes in a drop of water on a glass slide.
16. Excess water is removed and the nematodes are immobilized by covering with 3% low-melting-point agarose (prepared in step 1) using a Pasteur pipette. Quickly align nematodes with a fine needle perpendicular to the long axis of the slide before the agarose solidifies (see Note 4).
17. After nematodes are aligned, overlay the entire slide with the agarose to provide a smooth surface. Place the slide in a covered petri dish with moist filter paper to keep the agarose from drying out.

3.2. Microaspiration

1. Needles used for microaspiration of the esophageal glands are made by pulling aluminoscilicate glass tubes on the horizontal micropipette puller with a setting that gives a 3–4 mm taper and a tip opening of less than 1 µm diameter.

2. The tip of the needle is beveled at a 45° angle with the micropipette beveler (see Note 5).
3. Using the Eppendorf microloader, back fill the needle with 10 μL of the mRNA extraction buffer to protect the RNA from RNase activity. Insert the filled needle into the capillary holder attached to the CellTram unit.
4. Place the slide with the nematodes on the microscope stage under the 4× objective and while viewing nematodes on the Sony monitor identify living nematodes by looking for muscular contraction of the metacorpus. Only living nematodes should be selected for microaspiration of gland contents. Insert the buffer-filled needle tip in the agarose and carefully move the tip toward the immobilized nematode while maintaining a slight positive flow pressure in the micropipette to keep agarose from plugging the needle tip (see Note 6). Use the 10× objective for positioning the needle here.
5. Microposition the needle tip next to the cuticle directly opposite the esophageal gland cells in the nematode while observing under the 40× objective. As the needle tip punctures the cuticle and enters an esophageal gland cell, negative pressure is applied to the micropipette with the CellTram unit to aspirate the contents of the gland cells directly into the buffer within the needle. As the needle enters the gland cells the contents of the gland cells can be observed moving into the needle (Fig. 1).
6. The needle content (approximately 0.5 pL of cytoplasm of the nematode esophageal gland cells mixed with the buffer) is then emptied into a separate 0.5 mL microfuge tube by applying positive pressure and the very tip of the needle is broken off in the tube as well (see Note 7). Each tube is immediately placed in a –12°C freezer and then transferred to a –80°C freezer until further processed.
7. Intestional region of nematodes were also microaspirated as described above for use in preparing cDNA probes to remove cDNAs from common housekeeping and structural genes in the esophageal gland libraries (see below).

3.3. Isolation of mRNA

1. Transfer the desired volume of Dynabeads Oligo$(dT)_{25}$ from the suspended stock tube to a RNase-free 1.5 mL microcentrifuge tube (see Note 8) and place the tube on a Dynal MPC magnet. After 30 s remove the supernatant with a micropipette without disturbing the beads.
2. Remove the tube from the magnet and wash the Dynabeads Oligo $(dT)_{25}$ once by re-suspending in an equivalent volume of mRNA extraction buffer. When nematode gland or intestine sample is ready for combination with the Dynabeads Oligo $(dT)_{25}$, place the tube on the magnet and remove the supernatant after 30 s.

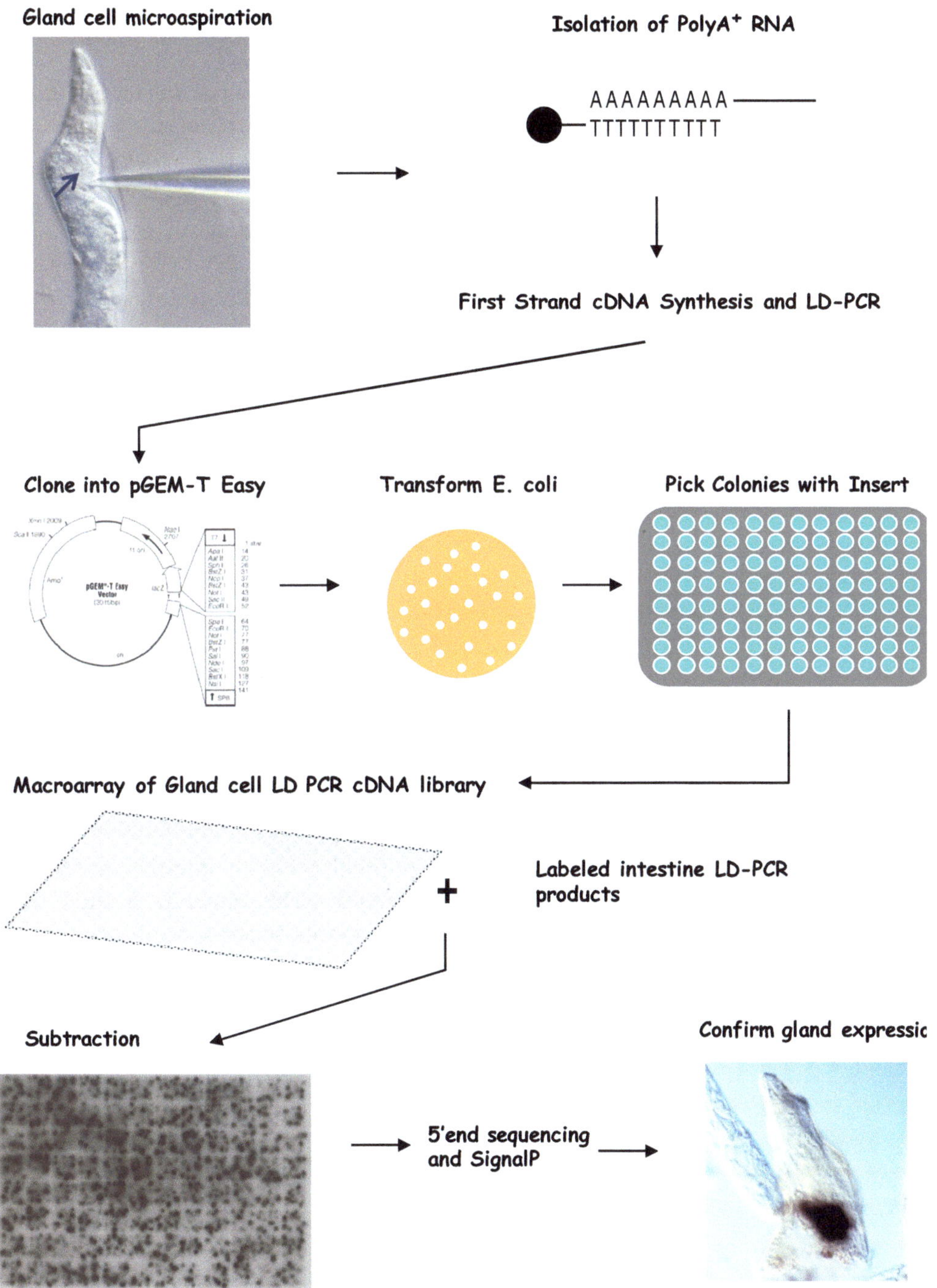

Fig. 1. Schematic presentation of the method used for construction of the gland-specific cDNA libraries. The content of esophageal gland cells (*arrow*) from the different parasitic stages of nematodes immobilized in agarose was aspirated by using a glass micropipette under inverted microscope on a microinjection station. Root-knot nematode gland-cell cDNA library clones (6,144 clones) were macroarrayed onto a nylon membrane. The macroarrayed gland-cell cDNA library was hybridized with cDNA probes labeled from purified LD-PCR products generated from the intestinal cells of ten parasitic root-knot nematodes. Over 1,000 root-knot nematode cDNA clones (*dark spots*) were subtracted. The nonhybridizing root-knot nematode gland-cell-specific cDNA library clones were picked for sequencing. In situ mRNA hybridization was used to confirm parasitism gene expression in the esophageal gland cells of root-knot nematodes. The dark staining signal indicates hybridization of the digoxigenin-labeled antisense-specific cDNA probes of a parasitism gene clone to transcripts expressed exclusively within the dorsal esophageal gland cell of root-knot nematode. Reprinted from (2) with permission from CAB International.

3. Transfer microaspirated contents of gland cells (or intestine regions) to a RNase-free 1.5 mL microcentrifuge tube.
4. Combine the gland or intestine material with the pretreated Dynabeads Oligo $(dT)_{25}$ in 0.5–1 mL of 1× mRNA extract buffer, and mix and shake the tube gently on a rotator for 15 min at room temperature to achieve hybridization of the polyA-tail of the mRNA with the Oligo$(dT)_{25}$ on the beads.
5. Place the tube on the Dynal MPC for 2 min and remove the supernatant.
6. Wash the Dynabeads/mRNA complex thoroughly twice with 0.5 mL of washing buffer I and twice with 0.5 mL of washing buffer II at room temperature using the Dynal MPC. Transfer the suspension to a new RNase-free 0.5 mL tube between the two last washing steps.
7. Wash the Dynabeads/mRNA complex once with 50 μL of 1× first-strand buffer and remove the supernatant (see Note 9).
8. Add 50 μL of DEPC-treated water to the Dynabeads/mRNA complex and incubate at 70°C for 2 min. Place the tube immediately on the Dynal MPC and transfer the supernatant containing the mRNA to a new 0.5 mL RNase-free tube. Place the mRNA immediately on ice or store it at –80°C (see Note 10).

3.4. cDNA Amplification

1. Combine 50 μL of purified gland or intestine mRNA with 7 μL of 12 μM 3′ SMART CDS primer II A and 7 μL of 12 μM SMART II A Oligo in a sterile 0.5 mL reaction tube (see Note 11).
2. Mix contents and spin the tube briefly in a microcentrifuge.
3. Incubate the tube at 65°C in a thermal cycler for 2 min and then reduce the temperature to 42°C.
4. Add 42 μL of the mixture containing 20 μL of 5× first-strand buffer, 2 μL of 100 mM DTT, 10 μL of 10 mM dNTP, 5 μL of RNase inhibitor (20 U/μL), and 5 μL of MMLV reverse transcriptase (200 U/μL) to each reaction tube.
5. Gently pipet up and down to mix, and then spin the tubes briefly in a microcentrifuge.
6. Incubate the tubes at 42°C for 90 min in an air incubator.
7. Place the tubes on ice to terminate first-strand cDNA synthesis.
8. Purify gland-cell or intestine first-strand cDNA from unincorporated nucleotides and small (<100 bp) cDNA fragments by using Clontech's NucleoSpin extract II kit according to the manufacturer's instructions and in the last step elute the first-strand cDNA in 85 μL of sterile distilled H_2O.
9. Prepare 100 μL of LD-PCR reactions each containing 5 μL of gland-cell or intestine first-strand cDNA, 79 μL of H_2O,

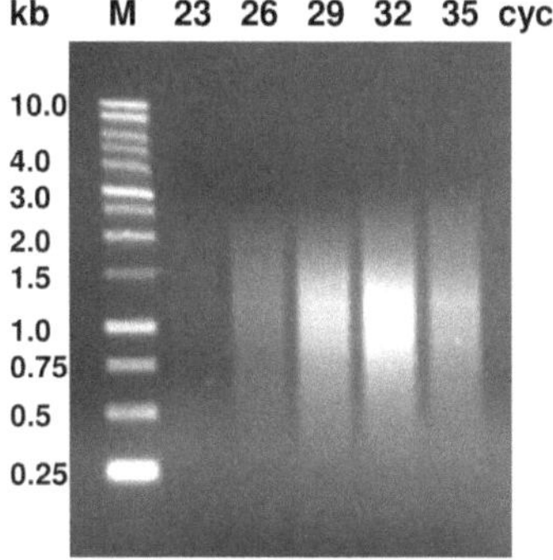

Fig. 2. Analysis for optimizing LD-PCR parameters. The contents of gland cells of 25 parasitic root-knot nematodes were used to generate the first-strand cDNA (80 μL). 5 μL of the gland-cell first-strand cDNA was used for LD-PCR amplification. A range of PCR cycles were performed (23, 26, 29, 32, and 35) as indicated. 15 μL of each LD-PCR product was electrophoresed on a 1.0% agarose/EtBr gel in 1× TAE buffer, along with 500 ng of BenchTop 1 kb DNA ladder (M). The LD-PCR products showed a standard normal pattern, ranging from 0.3–3.0 kb in size. The optimal number of cycles determined in this experiment was 26 for large-scale LD-PCR amplification for cDNA library construction.

10 μL of 10× cDNA PCR reaction buffer, 2 μL of 10 μM 50× dNTP, 2 μL of 12 μM 5′ PCR primer II A and 2 μL of 50× Advantage cDNA polymerase mix.

10. Perform LD-PCR with hot-start at 94°C (1 min), followed by 35 cycles at 94°C (15 s), 65°C (30 s), 72°C (6 min) to determine the optimal number of cycles for gland or intestine first-strand cDNA (see Note 12), by pausing the program every three cycles at the range of 20–35 cycles, transferring 15 μL from each PCR reaction tube to a clean 0.5 mL microcentrifuge tube, and then storing the tube on ice for agarose/EtBr gel analysis. An example result is shown in Fig. 2.
11. Run LD-PCR reactions containing the components as described above for cDNA amplification for gland-cell cDNA library construction using the optimal PCR cycling profiles with hot-start at 94°C for 1 min, followed by 26 cycles of 94°C for 15 s, 65°C for 30 s, 72°C for 4 min, and a final step of 72°C for 10 min. Intestine LD-PCR products will also be amplified from the first-strand cDNA, purified with QIAquick PCR purification kit, and then labeled with [α^{32}P]dCTP using a RadPrime DNA-labeling system to generate intestine cDNA probes for subtraction.

3.5. Library Construction

1. Purify gland-cell LD-PCR products from unincorporated nucleotides (primers) and small (<100 bp) PCR products (see Note 13) by using QIAquick PCR purification kit, and elute the LD-PCR products in 50 μL deionized, sterile (d.s.) H_2O (see Note 14). An overview of library construction is presented in Fig. 1.

2. Determine the concentration of the purified gland-cell LD-PCR products by UV spectrophotometry.
3. Set up 10 µL of ligation reactions in 0.5-mL sterile tubes each containing 5 µL of 2× rapid ligation buffer, 1 µL of T_4 DNA ligase (3 Weiss units/µL), 1 µL of pGEM-T Easy vector (50 ng/µL) and 1–3 µL of purified gland-cell LD-PCR products at a mass ratio 3:1 or 1:1 with the vector. Incubate the reactions overnight at 4°C to generate the maximum number of transformants to improve the quality of gland-cell cDNA library.
4. Add 10 µL of d.s. H_2O to each reaction tube, mix well, and incubate the tubes at 65°C for 10 min to stop the ligation reactions.
5. Transform 2–4 µL of the diluted ligation products into 50 µL of *E. coli* XL10-GOLD ultracompetent cells in sterile 1.5 mL microcentrifuge tubes by heat-shock at 42°C in a water bath for 50–60 s.
6. Assay the efficiency of transformation by plating 1:10 and 1:100 dilutions on LB agar plates with ampicillin/X-Gal/IPTG for blue-white selection and incubate the plates overnight at 37°C. The efficiency of transformation should be approximately 10^9 clones/µg of vector and over 98% colonies should be white (recombinants).
7. Randomly pick 24 white colonies to check the insert size range in the pGEM-T Easy vector, by using QIAprep Spin Miniprep kit to isolate the plasmid DNA followed by *Eco*RI digestion. An example result is shown in Fig. 3.

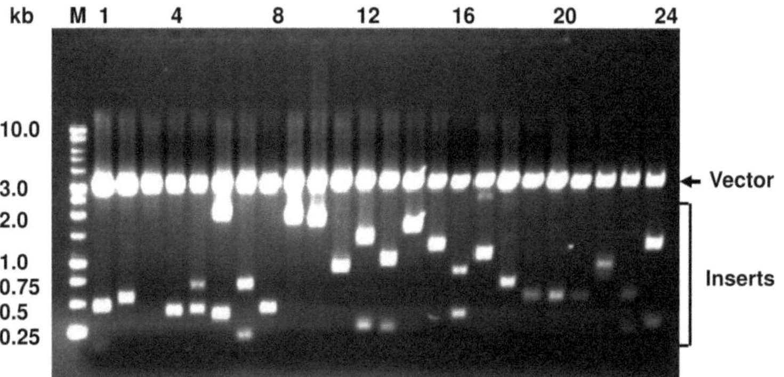

Fig. 3. Determination of inserts of gland-cell cDNA library clones in the cloning vector. Two recognition sites for the restriction enzyme *Eco*RI flanked in the cloning site of the pGEM-T Easy vector provide the enzyme digestion for the release of the insert. 2 µL of each plasmid DNA sample of 24 root-knot nematode gland-cell cDNA library clones in the pGEM-T Easy vector was digested with 0.5 µL of *Eco*RI (20 U/µL) in 20 µL of digestion reactions containing 1× *Eco*RI digestion buffer and 1× BSA at 37°C for 1 h. After digestion, 15 µL of each *Eco*RI-digested DNA sample was electrophoresed on a 1.2% agarose/EtBr gel in 1× TAE buffer, along with 500 ng of BenchTop 1 kb DNA ladder (M). The 24 clones contained inserts ranging from 0.4–2.4 kb in size, excluding one clone with no insert (*lane 3*). The arrow indicates the linearized pGEM-T Easy vector at 3 kb.

8. Perform large-scale *E. coli* transformations by heat-shock aiming for approximately 250–300 colonies per plate based on the titer calculated in step 6. Plate the transformed bacteria on 60–80 large LB agar plates (150 × 15 mm) with ampicillin/X-Gal/IPTG overnight at 37°C. Randomly hand pick 6,144 white colonies to 64 96-well MICROTEST III Tissue Culture plates containing 150 μL of 10% glycerol LB medium with ampicillin, and incubate the plates by shaking at 250 rpm on an orbital shaker overnight at 37°C. The plates should be stored at −80°C.

3.6. Sequencing

1. The gland-cell LD-PCR cDNA library is subtracted with cDNA probes from the LD-PCR cDNA library made from the intestinal cells to remove common expressed housekeeping and structural genes. Macroarray 6,144 white colonies of the gland-cell LD-PCR cDNA library onto sterile Hybond-XL nylon membranes (70 × 105 mm) on LB plates with ampicillin using a macroarray machine (Fig. 1) (see Note 15).
2. Incubate macroarrayed membranes on the plates at 37°C for 8–10 h or until colonies are visible on the membranes. Treat the membranes as follows: 5 min on 10% SDS, 5 min on denaturing solution, 5 min on neutralizing solution, and 5 min on 2× SSC. Air dry and bake the membranes at 80°C in the oven for 2 h (see Note 16).
3. Hybridize the membrane with 20 ng of ^{32}P-labeled intestinal cDNA probes using standard colony-hybridization/Southern blot procedures to identify common expressed housekeeping and structural genes. An example is shown in Fig. 1 (see Note 17).
4. Transfer nonhybridizing colonies to 96-well MICROTEST III Tissue Culture plates containing 150 μL of 10% glycerol LB medium with ampicillin and incubate the plates by shaking at 250 rpm on an orbital shaker overnight at 37°C to generate a new gland-specific cDNA library. The plates should be stored at −80°C.
5. Grow up 1 mL overnight liquid cultures in a 96-well culture block as follows: Use a flame-sterilized 96 pin replicator to transfer frozen gland-cell cDNA library colonies from 96-well stock plates to 96-well deep culture blocks (Millipore's Montage Plasmid $Miniprep_{96}$ kit) with each well containing 1 mL LB plus ampicillin.
6. Perform plasmid miniprep on the cultures using Millipore's Montage Plasmid $Miniprep_{96}$ kit (see Note 18), according to the manufacturer's instructions.
7. Sequence the gland-cell cDNA clones using an Applied Biosystems' BigDye Terminator v3.1 cycle sequencing kit with the sequencing primer (5′-GGTATCAACGCAGA

GTACGCG-3′) for 5′-end single pass cDNA sequencing and collect sequences on an ABI 3700 autosequencer. Next, the bioinformatics program SignalP 3.0 (http://www.cbs.dtu.dk/services/SignalP/index.php) is used to predict N-terminal hydrophobic signal-peptide sequences for secretion (see Note 19). Deduced proteins with transmembrane helices predicted by TMHMM (http://www.cbs.dtu.dk/services/TMHMM/) are dropped from further consideration. All cDNA clones encoding signal peptides without trasmembrane helices are then screened by using high-throughput in situ hybridizations to confirm their expression in the esophageal gland cells.

3.7. In Situ Hybridization

1. Wash pre-parasitic second-stage juveniles three times with 1 mL of 1× PBS buffer and fix the nematodes in 1 mL of 10% formalin buffered in phosphate for 2 days at room temperature. Mixed parasitic stages previously stored in 10% formalin buffered in phosphate are also processed for in situ hybridization.
2. Randomly cut the fixed-nematodes on glass slides with a razor blade until approximately 90% of the nematodes are cut (see Note 20). Transfer nematodes to an Eppendorf tube.
3. Wash the cut nematodes three times with 1 mL of 1× PBS buffer and then incubate the nematode sections in 1× PBS buffer containing 0.5–1 mg/mL proteinase K with gentle shaking on a rotator at 37°C for 1 h (see Note 21).
4. Wash nematode sections once with 1 mL of PBS buffer and remove the supernatant.
5. Freeze pellet of nematode sections at –80°C for 15 min.
6. Add 1 mL of cold (–80°C) methanol, let the tube stay at room temperature for 1–3 min, and then vortex it to resuspend the nematode pellet. Remove the methanol by centrifugation at 8,000 × *g* for 30 s followed by aspiration.
7. Add 1 mL of cold acetone (–80°C) and let the tube set at –80°C for 10 min (can be set up to 1 h), and then remove the acetone by centrifugation/aspiration.
8. Wash nematode sections once with 1 mL of DIG-hybridization buffer to remove the residual acetone.
9. Pre-hybridize nematode sections in 1 mL of fresh DIG-hybridization buffer for 30 min at 50°C with gentle shaking on a rotator in the oven (see Note 22).
10. Aliquot the pre-hybridized nematode sections (300–400 per well) to wells of 96-well MultiScreen plate and then add 2 μL of denatured PCR DIG-labeled DNA probes (see Note 23) to each well. Solutions are changed by aspiration using a Millipore Multiscreen vacuum manifold in the following steps.
11. Incubate the plate at 37–42°C (see Note 24) for 16 h with gentle agitation.

12. Wash twice 10 min with 100 μL of 2× SSC/0.1% SDS at room temperature. Again wash twice for 5 min with 100 μL of 0.1× SSC/0.1% SDS at 50°C.
13. Wash the nematodes briefly with 150 μL of 1× maleic acid buffer.
14. Incubate the nematodes in 50 μL of 1× blocking solution at 37°C for 30 min with gentle agitation.
15. Incubate the nematodes for 2 h at 37°C in 100 μL of antibody solution with gentle agitation.
16. Wash three times for 15 min with 100 μL of 1× maleic acid buffer at 37°C, and then wash briefly with 100 μL of 1× detection buffer.
17. Add 100 μL of staining solution to each well. Let the plate set overnight in the dark at room temperature without agitation.
18. Stop the staining reaction by washing the nematodes twice in 0.01% Tween-20 in sterile H_2O.
19. Observe nematode specimens with a compound light microscope. An example of gene expression in the esophageal gland cells is shown in Fig. 1.

4. Notes

1. Fish sperm DNA (e.g., sonicated salmon sperm DNA) is denatured at 100°C for 10 min and cooled on ice water. The components of the DIG-hybridization buffer are dissolved in d.s. H_2O, filtered to sterilize and remove particulate matter, divided into aliquots, and then stored at −20°C. The buffer remains stable at this temperature for up to a few months.
2. The desired size of DIG-labeled cDNA probes for this protocol is 200–300 bp in length. We use a computer program to design appropriate gene-specific forward and reverse primers for each gene. The PCR products synthesized in the first PCR are electrophoresed on a 1.2% agarose/EtBr gel in 1× TAE buffer to test if they are predicted ones in size.
3. Approximate number of days allowed for nematode growth depends on the season of the year and how well the greenhouse holds day or nighttime temperature. During summer months, approximately 9 days should be expected and in the winter months, 14 days might be the average.
4. There is a brief window before the agarose will solidify immobilizing the nematodes. Nematodes need to be aligned quickly, and depending on which side of the microscope stage the microinjector is attached, the nematode will be rolled so that the esophageal gland cells are positioned on the side so

that the microinjection needle can enter the glands directly (Fig. 1). Nematodes are aligned with a fine needle made by gluing a #000 insect pin (38 mm × 0.25 mm) into the end of a wood applicator stick (Fisher Scientific #01-340).

5. The needle is attached by tubing to an aquarium air pump to prevent the tip of the needle from becoming plugged during the beveling process. The needle must have a sharp 45° angled tip, which should be confirmed under a microscope.
6. Maintaining a slight positive pressure on the micropipette to keep agarose from plugging the tip as the needle moves toward the nematode will result in most of the buffer being injected from the micropipette. Approximately 1–2 μL of buffer should still be in the micropipette when the cytoplasm is aspirated from the esophageal gland cells.
7. The potential for mRNA degradation is minimized by immediate mixing of the gland cell contents with the mRNA extraction buffer present in the needle during aspiration and immediately freezing of gland cell contents after aspiration.
8. To ensure this protocol is successful, all of the solutions, tips, tubes, and pipettes should be RNase-free. Dynabeads Oligo$(dT)_{25}$ (5 μg/μL) should be resuspended thoroughly before use. The beads per microgram can bind up to 2 ng of mRNA, depending on the tissue or cell type. We use 10 μL of the beads (50 μg in total) for isolation of mRNA from ten contents of the gland cells.
9. This washing step is to completely remove the LiDS from the Dynabeads/mRNA complex, preventing LiDS from inhibiting downstream enzymatic reactions.
10. The amount of mRNA in the esophageal gland cells per nematode is extremely low (possibly only picogram in weight). Therefore, we do not estimate the quality and quantity of the purified mRNA by gel analysis or UV spectrophotometry at this step.
11. Compared to the traditional SMART protocol we originally used (5–8), Clontech's Super SMART cDNA synthesis technology (see manufacturer's user manual) enables one to synthesize more high-quality cDNA from nanogram quantities of gland-cell mRNA by increasing the reaction volumes and performing an additional column purification step. For new users, we recommend that both a water control and a human placenta RNA control be used for first-strand cDNA synthesis in the parallel experiments.
12. Optimal parameters for LD-PCR may vary with different templates and thermal cyclers. To determine the optimal number of cycles for your sample and conditions, we strongly recommend that you test a range of cycles (15–35 cycles).

13. The unincorporated nucleotides (primers) and small (<100 bp) PCR products decrease the efficiency of ligation reactions, influencing the quality of gland-cell cDNA libraries. They must be completely removed.
14. To test the quality of the gland-cell LD-PCR products for constructing gland-cell cDNA library, we use 10× dilutions of the purified LD-PCR products as template to apply several previously identified gland-expressed genes (if available), which represent genes with features including single or multiple copies, short or long, known function or pioneer, and expression in SvG or DG at the different parasitic stages.
15. As an alternative to the intestinal cDNA subtraction, it is possible to directly sequence 10,000 clones of a gland-cell LD-PCR cDNA library after plating. In addition, recent advances in sequencing using the 454 sequencing platform allows high-quality transcriptome analysis of gland cells by direct sequencing 7.5–10 μg of gland-cell LD-PCR products using sequencing primer (5′-GGTATCAACGCAGAGTA CGCG-3′). However, the potential challenge with the latter approach is a time-consuming step to recover gland-specific cDNA clones of interest from nematode/gland cDNA pools after sequencing.
16. The DNA cross-linked membranes remain stable for up to a few months when stored under dry condition at room temperature. For the long-term storage they can be stored in a sealed plastic bag at −20°C.
17. While we use a radioactive-labeling method in this experiment, nonradioactive labeling methods using haptens, such as digoxigenin (DIG), can also be applied to label intestinal cDNA probes for colony-hybridization to identify common expressed genes. Briefly, intestinal cDNA probes will be labeled from 20 ng of intestinal first-strand cDNA or purified LD-PCR products by PCR using 5′ PCR primer II A with a PCR-DIG probe synthesis system (Roche Applied Science). About 15 ng of DIG-labeled probe per milliliter will be used for hybridization. Hybridization will be performed at 40°C in DIG Easy Hyb solution (Roche Applied Science) overnight and washed twice with 0.5× SSC/0.1% SDS solution at 68°C. Probes that hybridize to gland cDNA clones on Southern blots will be detected by alkaline phosphatase-conjugated anti-DIG antibody and disodium 3-(4 methoxyspiro{1,2-dioxetane-3,2′-(5′-chloro)tricyclo[3.3.1.1^{3,7}]decan}-4-yl)pheryl phosphate (CSPD) (Roche Applied Science) chemiluminescent substrate reaction. The membrane will be exposed to high performance chemiluminescence film (Amersham Pharmacia Biotech) at room temperature for 1.5 h.

18. The Montage Plasmid Miniprep$_{96}$ kit employs standard alkaline lysis methods, but is adapted for 96 well plates, utilizing 96-well filtration plates to capture and wash the plasmid DNA. Upon completion, the typical yield is 2–10 μg of plasmid DNA in 35 μL of deionized water per well.
19. The sequencing primer is designed based on the sequence of SMART II A Oligo. SMART cDNA synthesis technology enriches the production of full-length cDNA sequences, allowing us to identify the majority of N-terminal hydrophobic signal-peptide sequences of the gland-expressed parasitism genes encoding secretory proteins through 5′-end single pass cDNA sequencing.
20. The nematodes are cut on the slide with a razor blade. The cut nematodes are checked under a microscope to monitor cutting effectiveness and prevent excessive cutting or having too many uncut nematodes.
21. We use the procedures to treat 50–100 μL of root-knot nematodes at the pre-parasitic and parasitic stages. The parameters for proteinase treatment may require optimization, depending on the number of nematodes-specific developmental stages, and species of plant-parasitic nematode.
22. This step is not really necessary; use it if you have problems with nonspecific binding of the probe.
23. As a negative control for each gene, hybridization with the sense DIG-DNA probe should produce no staining of the treated nematodes. We often use the antisense DIG-DNA probes to do high-throughput in situ hybridizations.
24. Optimal hybridization temperature will vary, depending on probe size and specificity. The best hybridization temperature for hybridization with our DIG-hybridization buffer containing 50% formamide is 20–25°C below the T_m value of an DNA:RNA hybrid, which is calculated according to the GC content and the length of hybrid in base pairs (L) according to the following equation: $T_m = 49.82 + 0.41\ (\%\ G + C) - (600/L)$.

References

1. Hussey, R. S. and Grundler, F. M. (1998) Nematode parasitism of plants, in *Physiology and biochemistry of free-living and plant-parasitic nematodes*, (Perry, R. N. and Wright J., eds.), CAB International, Wallingford, UK, pp. 213–243.
2. Mitchum, M. G., Hussey, R. S., Davis, E. L., and Baum, T.J. (2007) Application of biotechnology to understand pathogenesis of nematode plant pathogens, in *Biotechnology & Plant Disease Management* (Punja, Z. K., DeBoer, S., and Sanfacon, H. eds.) CAB International, Wallingford, UK, pp. 58–86.
3. Davis, E. L., Hussey, R. S., Mitchum, M. G., and Baum, T. J. (2008) Parasitism proteins in nematode-plant interactions. *Curr. Opin. Plant Biol.* **11**, 360–366.
4. Wang, X. H., Allen, R., Ding, X. F., Goellner, M., Maier, T., de Boer, J. M., Baum, T. J., Hussey, R., and Davis, E. L. (2001) Signal peptide-selection

of cDNA cloned directly from the esophageal gland cells of the soybean cyst nematode Heterodera glycines. *Mol. Plant Microbe Interact.* **14**, 536–544.

5. Gao, B. L., Allen, R., Maier, T., Davis, E. L., Baum, T. J., and Hussey, R. S. (2001) Identification of putative parasitism genes expressed in the esophageal gland cells of the soybean cyst nematode *Heterodera glycines*. *Mol. Plant Microbe Interact.* **14**, 1247–1254.
6. Gao, B. L., Allen, R., Maier, T., Davis, E. L., Baum, T. J., and Hussey, R. S. (2003) The parasitome of the phytonematode *Heterodera glycines*. *Mol. Plant Microbe Interact.* **16**, 720–726.
7. Huang, G. Z., Gao, B. L., Maier, T., Allen, R., Davis, E. L., Baum, T. J., and Hussey, R. S. (2003) A profile of putative parasitism genes expressed in the esophageal gland cells of the root-knot nematode *Meloidogyne incognita*. *Mol. Plant Microbe Interact.* **16**, 376–381.
8. Huang, G., Dong, R., Maier, T., Allen, R., Davis, E. L., Baum, T. J., and Hussey, R. S. (2004) Use of solid-phase subtractive hybridization for the identification of parasitism gene candidates from the root-knot nematode *Meloidogyne incognita*. *Mol. Plant Pathol.* **5**, 217–222.
9. Hussey, R. S. and Barker, K. R. (1973) A comparison of methods of collecting inocula of *Meloidogyne* spp., including a new technique. *Plant Dis. Rep.* **57**, 1025–1028.
10. De Boer, J. M., Yan, Y., Smant, G., Davis, E. L. and Baum, T. J. (1998) In-situ hybridization to messenger RNA in *Heterodera glycines*. *J. Nematol.* **30**, 309–312.

Chapter 10

Construction of *Pseudomonas syringae* pv. tomato DC3000 Mutant and Polymutant Strains

Brian H. Kvitko and Alan Collmer

Abstract

Redundancy between *Pseudomonas syringae* pv. tomato DC3000 virulence factors has made their characterization difficult. One method to circumvent redundancy for phenotypic characterization is to simultaneously delete all redundant factors through the generation of polymutant strains. Described here are methods by which single and polymutant strains of DC3000 can be generated through the use of the small mobilizable sucrose counter-selection vector pK18*mobsacB*, FRT-flanked antibiotic marker cassettes, and Flp recombination.

Key words: *Pseudomonas syringae*, Gene deletion, Polymutant, Recycling antibiotic resistance, Sucrose counter-selection, Flp recombinase, Conjugation

1. Introduction

The Gram-negative plant pathogen *Pseudomonas syringae* pv. tomato DC3000 (DC3000) uses a large and well-defined repertoire of type III effector proteins to cause disease in tomato and *Arabidopsis thaliana* (1). These effectors are collectively essential for pathogenicity, but are individually dispensable for virulence. Characterizing the functions and virulence contributions of individual type III effectors has been made difficult by the apparent redundancy of function between unrelated type III effectors.

The theme of redundancy in DC3000 virulence factors extends beyond type III effector repertories. It has also been demonstrated for type III translocators and specialized lytic transglycosylases, and genomic analysis has suggested potential interplay between several bacterial toxins in plant disease (2–4).

John M. McDowell (ed.), *Plant Immunity: Methods and Protocols*, Methods in Molecular Biology, vol. 712,
DOI 10.1007/978-1-61737-998-7_10,

One strategy that has been successfully employed to dissect functional redundancy is to create polymutant strains that combine deletions of all members in a functionally redundant group. When a redundant group is deleted, a synergistic combinatorial phenotype can be revealed that greatly simplifies analyses of individual group members. Using traditional methods of marked mutation, the number of mutations that can be introduced into a single strain is limited by the number of available antibiotic markers. To move past the marker barrier in DC3000, we have adopted and adapted several genetic tools for recycling antibiotic markers and generating unmarked mutants (3, 5).

Although several tools have been used by our group for the generation of DC3000 mutants, we have found that the most effective tool for DC3000 deletion and polydeletion strain construction is the small, mobilizable, sucrose counter-selection vector pK18*mobsacB* (6). To that end, we have sequenced pK18*mobsacB* and deposited the full sequence in Genbank (accession FJ437239). The procedure for making a deletion of your favorite gene (*yfgX*) has two major phases: First, making the pK18*mobsacB*::*ΔyfgX* deletion construct, and second, recombining that deletion into DC3000 using a selection/counter-selection strategy. There is also the optional step of marking the deletion construct with an FRT-flanked antibiotic marker cassette and removing that FRT-flanked marker after recombination into DC3000. These optional steps add robustness to the procedure by allowing selection at both the single and double crossover steps, while still facilitating recycling of the antibiotic markers. The major steps of the protocol are diagramed in Fig. 1. We have used these basic procedures to delete genomic regions ranging in size from 500-bp to 22-kb, and to combine over seven separate deletions into a single DC3000 strain.

2. Materials

2.1. Primer Design

1. DC3000 genome sequence, (Genbank accessions; chromosome AE016853, pDC3000A AE016855, pDC3000B AE016854).
2. pK18*mobsacB* sequence (accession FJ437239).

2.2. Building the pK18mobsacB ::ΔyfgX Deletion Construct

1. Takara Ex Taq Polymerase Premix (Takara Mirus Bio, Otsu, Shiga, Japan).
2. DC3000 genomic DNA (~1,000 ng/μl) prepared using the Wizard Genomic DNA Purification Kit (Promega, Madison, WI).

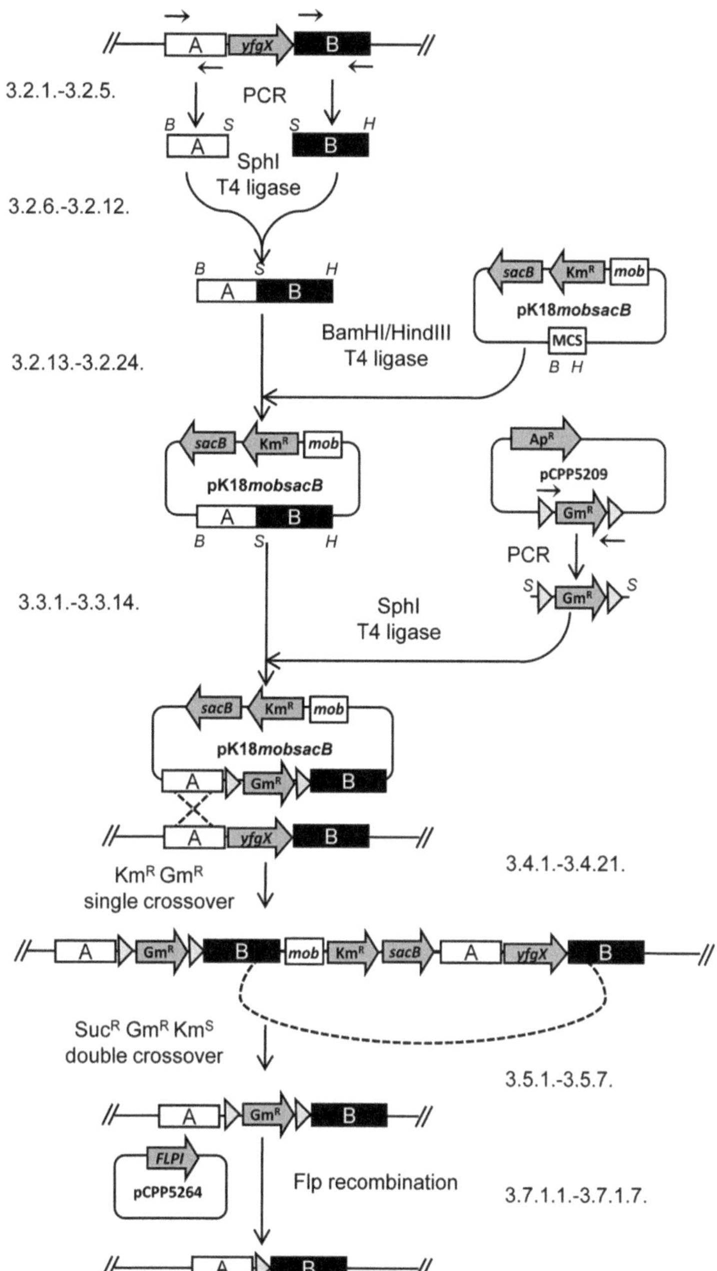

Fig. 1. Flowchart of major steps in deletion of *yfgX*, including optional steps of cloning an FRT-flanked cassette and Flp recombination. The steps of the protocol diagramed in the flowchart are noted. Primers are indicated with small arrows. Directed FRT sites are indicated with triangles. B = *Bam*HI, *S* = *Sph*I, *H* = *Hin*dIII. *Bam*HI, *Sph*I, *Hin*dIII, and the FRTGm^R cassette are presented in the flowchart only for the sake of clarity. Other restriction enzymes or the FRTSp^R cassette can be used in their place. Note that either flanks A or B have the potential to be the sites of crossovers.

3. Oligonucleotides (10 pmol/μl).
   Flank A forward, flank A reverse.
   Flank B forward, flank B reverse.
   A seq forward, B seq reverse.
   M13F TGTAAAACGACGGCCAGT.
   PK18M13R AACAGCTATGACATGA (see Note 1).
4. Sterile dH_2O.
5. Sterile 0.65, 1.0, and 2.0 ml RNase/DNase-free microcentrifuge tubes.
6. Agarose 1™ (Amresco, Solon, OH).
7. Ethidium bromide (1%).
8. 1× TBE buffer (Tris-Base 10.8 g, boric acid 5.5 g, 0.5 M EDTA 4 ml, fill to 1 L with dH_2O).
9. 10× DNA sample loading buffer (0.25% bromophenol blue, 0.25% xylene cyanol, 0.1% SDS, and 30% glycerol).
10. O'GeneRuler 1-kb DNA ladder (Fermentas, Glen Burnie, MD).
11. Gel DNA Recovery Kit and Clean-up and Concentrator Kit (Zymo Research, Orange, CA).
12. Restriction enzymes, 10× restriction buffers and 10× BSA (NEB, Ipswitch, MA).
13. T4 ligase and 5× T4 ligase buffer (NEB, Ipswich, MA).
14. Alkaline phosphatase (CIP) (NEB, Ipswitch, MA).
15. pK18*mobsacB* plasmid DNA (~200 ng/μl).
16. LB liquid and agar solidified media (Tryptone 10.0 g, yeast extract 5.0 g NaCl 10.0 g, 1 M NaOH 1.0 ml, fill to 1 L with dH_2O add 18 g agar for solidified media, autoclave).
17. Sterile 14-ml plastic snap-cap culture tubes.
18. 1,000× kanamycin (Km) 50 mg/ml, filter-sterilized in dH_2O.
19. 1,000× 5-bromo-4-chloro-3-indolyl-beta-D-galactopyranoside (X-gal) 20 mg/ml in DMF (see Note 2).
20. Competent *Escherichia coli* DH5α or any standard blue/white screening plasmid cloning strain.
21. Qiaprep® Spin Miniprep Kit (Qiagen, Valencia, CA).

2.3. (Optional) Introduce an FRT-Flanked Antibiotic Marker Cassette Between ΔyfgX Flanks A and B

1. Oligonucleotides (10 pmol/μl).
   FRT cassette forward.
   FRT cassette reverse.
2. Plasmid DNA (~200 ng/μl) from either pCPP5209 (Accession EU024549) or pCPP5242 (Accession EU024551) (3).
3. pK18*mobsacB*::*ΔyfgX* plasmid DNA (~200 ng/μl).

4. 1,000× Gentamicin (Gm) 5 mg/ml filter-sterilized in dH_2O.
5. 1,000× Spectinomycin (Sp) 50 mg/ml filter-sterilized in dH_2O.

2.4. Conjugate the pK18mobsacB ::ΔyfgX Deletion Construct into DC3000 to Get Single Crossover Merodiploid Transconjugants

1. Competent *E. coli* S17-1.
2. pK18*mobsacB*::Δ*yfgX* plasmid DNA (~200 ng/μl) marked or unmarked variants.
3. DC3000 strain to mutate.
4. LM liquid and solid media (Tryptone 10.0 g, yeast extract 6.0 g, NaCl 0.6 g, $MgSO_4 \times 7H_2O$ 0.4 g, K_2HPO_4 1.5 g (see Note 3) fill to 1 L with dH_2O add 18 g agar for solidified media, autoclave).
5. KB liquid and solid media (Bacto™ Proteose peptone No. 3 20.0 g, $MgSO_4 \times 7H_2O$ 0.4 g, glycerol, K_2HPO_4 1.5 g (see Note 3) fill to 1 L with dH_2O add 18 g agar for solidified media, autoclave).
6. 1,000× Rifampin (Rf) 50 mg/ml in DMSO.
7. 1,000× Ampicillin (Ap) 100 mg/ml filter-sterilized in dH_2O.
8. Immobilon-NY+filter cut into postage stamp size squares with the blue backing paper still attached and sterilized (Millipore, Billerica, MA).

2.5. Counter-Select the Integration with Sucrose to Recover Double-Crossover Mutants

1. 1 M filter-sterilized sucrose.
2. Sterile toothpicks.

2.6. Screen for the yfgX Deletion Mutant by PCR

1. Wizard Genomic DNA Purification Kit (Promega, Madison, WI).
2. Oligonucleotides (10 pmol/μl).
   Out A forward.
   Out B reverse.

2.7. Removal of the FRT-Flanked Marker from a Marked Deletion Strain

2.7.1. Removing FRT-Flanked Cassette Using Flp Recombination

1. Marked deletion strain mutant of DC3000.
2. *Escherichia coli* DH5α pCPP5264 (3) (see Note 4).
3. *Escherichia coli* HB101 pRK2013 (see Note 5).
4. 1,000× tetracycline (Tc) 20 mg/ml in 50% ethanol (see Note 2).

2.7.2. Removing an FRT-Flanked Marker by Secondary Mutagenesis

1. Marked deletion strain mutant of DC3000.
2. *Escherichia coli* S17-1 transformed with unmarked pK18*mobsacB*::Δ*yfgX*.

3. Methods

The following sections describe the protocols to make deletions in DC3000 using the small, mobilizable, sucrose counter-selection vector pK18*mobsacB*. The final *yfgX* deletion in the DC3000 genome looks exactly like the center join of the pK18*mobsacB* deletion construct. Slightly different paths should be taken with this protocol depending on the number of deletions desired in the final strain and the proximity of those genes to one another. If only a single mutation is desired, a deletion construct marked with an FRT-flanked antibiotic resistance cassette can be used and the FRT-flanked marker can be left in place. If two to five deletions in one strain are desired and the genes/regions to be deleted are not located too close to one another, the FRT-flanked cassettes can be removed by Flp-mediated recombination leaving behind FRT site scars. What defines "too close" is unfortunately hard to know universally. Any two genes without some essential gene between them may be too close together. If six or more deletions are to be combined in one strain, or if some of the genes to be deleted are located near one another, the genes can be deleted using unmarked deletion constructs or the FRT-flanked cassettes can be removed by conducting a second mutagenesis with unmarked deletion constructs. The rationales for these recommended paths are explained in Note 29.

3.1. Primer Design

1. Eight primers are designed for each deletion construct. Forward and reverse primers to amplify flank A upstream to *yfgX*, forward and reverse primers to amplify flank B downstream to *yfgX*. Two sequencing (seq) primers within flanks A and B, which prime internally for sequencing the deletion construct. Lastly, two outside (out) primers which are located just external to the flank primers for deletion strain confirmation. The Tm of all eight primers should ideally be within 3°C of one another.
2. Flanks A and B are designed to have a 100–500-bp difference in size. Generally, the flanks are 1-kb and 1.5-kb. Either flank A or B can be designed as the larger of the two flanks (see Notes 6 and 7).

   The flank amplifying primers should be designed according to the following templates:

   Flank A forward: 3–4 nt toe (external restriction site 1) 18–25 nt identity

   Flank A reverse: 3–4 nt toe (internal restriction site) 18–25 nt identity

   Flank B forward: 3–4 nt toe (internal restriction site) 18–25 nt identity

   Flank B reverse: 3–4 nt toe (external restriction site 2) 18–25 nt identity (see Note 8)

[PK18M13R] *Eco*RI *Sma*I *Bam*HI *Xba*I *Sal*I *Pst*I *Sph*I *Hin*dIII [M13F] *Pvu*I *Nhe*I

Fig. 2. pK18*mobsacB* multiple cloning site. Standard primer binding sites are bracketed. Restriction sites are shaded according to standard NEB buffers reported to give 100% activity. Gray = NEB2, Black = NEB3, White = NEB4. Underlined restriction sites are present in FRTSp[R] and/or FRTGm[R] cassettes.

3. Choose your restriction enzyme sites so that they are in the same sequential order as in the multiple cloning site of pK18*mobsacB*, and use sites for which the enzymes have compatible reaction buffers (Fig. 2).

   Examples:

   (external restriction site 1, *Eco*RI), (internal restriction site, *Sma*I), (external restriction site 2, *Sph*I)

   (external restriction site 1, *Bam*HI), (internal restriction site, *Sph*I), (external restriction site 2, *Hin*dIII)

   The chosen restriction sites should not be present within the flanks. Though, a conveniently located natural restriction site, for example, a natural *Hin*dIII site 1.4-kb downstream of the *yfgX* stop codon, could be used in place of a primer-introduced restriction site. Underlined restriction sites in the multiple cloning site are present in FRTSp[R] and/or FRTGm[R] cassettes and should be avoided for use as the internal restriction site if you intend to conduct the optional steps of introducing a cassette into the deletion construct (see Notes 9–11).
4. Additional care should be taken in the design of flank A reverse and flank B forward if *yfgX* is in an operon. To manage the risk of introducing polar effects, the flanks should be designed so that flank A includes the first two or three codons of *yfgX* followed by an in-frame and translatable internal restriction site (for example, *Sma*I, CCCGGG = Pro,Gly) and flank B carries the same restriction site followed in-frame by the last two or three codons, including the stop codon of *yfgX*. This way the deletion strain expresses a small peptide in place of YfgX.
5. Seq A forward and seq B reverse primers should be designed to bind near the middles of flanks A and B and should prime toward *yfgX*. Aim for 18–25 nt each.
6. Out A forward and out B reverse primers should be located about 50-bp upstream from flank A forward and downstream of flank B reverse. Aim for 18–25 nt each.

3.2. Building the pK18mobsacB ::ΔyfgX Deletion Construct (See Note 12)

1. Gather the Ex Taq premix, forward and reverse flank primers and genomic DNA template for the PCR reaction.
2. Set up a separate PCR reaction for each flank in 0.65-ml microcentrifuge tubes according to the following recipe.

   50 μl Ex Taq premix

   2 μl forward PCR primer

   2 μl reverse PCR primer

2 µl genomic DNA

44 µl sterile dH_2O

3. Divide the PCR reaction into two 50 µl aliquots to run in the thermocycler. The PCR works better with the smaller sample volume. Recombine the 50 µl aliquots after the cycling is complete. The following thermocycler program is recommended for PCR amplification.

   5 m at 95°C

   Repeat 30 cycles

   1 m at 94°C

   1 m at (lowest primer Tm –3°C)

   1 m/kb at 72°C

   10 m at 72°C

   Hold at 15°C

4. Use 5 µl of the PCR reaction to confirm product amplification by agarose gel electrophoresis. The band should be strong and run at the correct size. There should be no obvious secondary bands in the reaction (see Note 13).

5. Use the Zymoclean Clean and Concentrate kit to clean the remaining 95 µl volume. Elute with a 6 µl volume.

6. Restriction digest the 5 µl eluted volume of each flank product separately in a 50 µl reaction volume with the internal restriction site enzyme according to the following recipe. Incubate the reaction for at least 1 h at the appropriate temperature for the restriction enzyme (most likely 37°C).

   5 µl purified flank DNA

   5 µl 10× NEB buffer

   5 µl 10× BSA

   1 µl NEB internal restriction enzyme

   34 µl dH_2O

7. Purify the DNA with the Zymoclean Clean and Concentrate kit. Elute with a 6 µl volume.

8. Mix the two purified flanks and ligate them with T4 ligase according to the following recipe. Incubate the reaction overnight at 14°C.

   5 µl flank A DNA

   5 µl flank B DNA

   6 µl NEB 5× T4 ligase buffer

   1 µl NEB T4 ligase

   13 µl dH_2O

9. Using fresh agarose, cast a 1% agarose gel using a comb size which produces wells sufficient to fit the 30 μl ligation reaction volume.
10. Gel-purify the desired product from the ligation mixture by agarose gel electrophoresis of the entire ligation volume. Use fresh TBE running buffer. You should see five bands. Three ligation products, A+A, A+B, B+B, and unligated flank A and flank B.
11. Cut the A+B band out of the gel with a clean razor blade. Most likely, this is the second largest band.
12. Purify the DNA from the agarose block using the Zymoclean Gel Recovery kit. Elute with a 6 μl volume.
13. Restriction digest the ligated A+B flanks and the pK18*mobsacB* vector in a total reaction volume of 50 μl with the external restriction site enzymes according to the following recipes. Incubate the reaction for at least 1 h at the appropriate temperature for the restriction enzyme (most likely 37°C).

    Insert restriction digest

    5 μl purified flank A+B DNA

    5 μl 10× NEB buffer compatible with both enzymes

    5 μl 10× BSA

    1 μl NEB external restriction enzyme 1

    1 μl NEB external restriction enzyme 2

    33 μl dH_2O

    Vector restriction digest

    1 μl pK18*mobsacB*

    5 μl 10× NEB buffer compatible with both enzymes

    5 μl 10× BSA

    1 μl NEB external restriction enzyme 1

    1 μl NEB external restriction enzyme 2

    1 μl NEB Calf Intestinal Phosphatase

    36 μl dH_2O
14. Purify the DNA with the Zymoclean Clean and Concentrate kit. Elute with a 10 μl volume.
15. Set up 20 μl ligation reactions and controls. Incubate the reaction overnight at 14°C.

    Restriction control

    3 μl purified vector DNA

    4 μl NEB 5× T4 ligase buffer

    13 μl dH_2O

Vector control

3 µl purified vector DNA

4 µl NEB 5× T4 ligase buffer

1 µl NEB T4 ligase

12 µl dH_2O

Ligation

3 µl purified vector DNA

9 µl purified insert DNA

4 µl NEB 5× T4 ligase buffer

1 µl NEB T4 ligase

3 µl dH_2O

16. Purify the DNA from the ligations with the Zymoclean Clean and Concentrate kit. Elute with a 6 µl volume.
17. Prepare six LB Km X-gal plates.
18. Transform the purified controls and ligation into *E. coli* compatible with blue/white screening (DH5α, TOP10, etc) using electroporation or chemically competent heat-shock transformation by standard protocols (7).
19. Plate 1/10 of each transformation reaction and a concentrated remainder onto labeled LB Km X-gal plates. Incubate the plates overnight at 37°C. You should have significantly more colonies on the ligation plate than on the control plates and the majority of those colonies should be white or pale blue (see Note 14).
20. Choose six white colonies and start 2-ml overnight LB Km cultures at 37°C.
21. Miniprep the plasmid DNA from the cultures using the Qiaprep Spin Miniprep kit from Qiagen.
22. Conduct a restriction digest screen on 2 µl of plasmid DNA with both external restriction site enzymes in a 20 µl volume according to the following recipe. Incubate the reaction for at least 1 h at the appropriate temperature for the restriction enzyme (most likely 37°C).

    2 µl plasmid DNA

    2 µl 10× NEB buffer compatible with both enzymes

    2 µl 10× BSA

    1 µl NEB external restriction enzyme 1

    1 µl NEB external restriction enzyme 2

    12 µl dH_2O
23. Check by agarose gel electrophoresis that the plasmid digest releases a fragment the same size as the A+B fragment. Discard any plasmids which do not.

24. Send candidate clones for sequencing with M13F, PK18M13R and the two sequencing primers designed in step 5 in Subheading 3.1 to check for any point mutations in the flanks and confirm that the pK18*mobsacB*::Δ*yfgX* deletion construct is correct.

3.3. (Optional) Introduce an FRT-Flanked Antibiotic Marker Cassette Between Δ*yfgX* Flanks A and B (See Note 15)

1. Design FRT cassette primers to PCR amplify either the FRTSp^R or the FRTGm^R cassettes from pCPP5242 or pCPP5209, respectively, with primers designed according to this template (see Note 16):

   FRT cassette forward primer

   3–4 nt toe (internal restriction site) GTGTAGGCTGGAGC TGCTTC

   FRT cassette reverse primer

   3–4 nt toe (internal restriction site) CATATGAATATCC TCCTTA
2. Conduct a PCR reaction as described in steps 2–4 in Subheading 3.2, using 1 μl of pCPP5242 or pCPP5209 plasmid DNA as template.
3. Purify the PCR product with the Zymoclean Clean and Concentrate kit. Elute with a 6 μl volume.
4. Restriction digest the purified FRT cassette and the pK18*mobsacB*::Δ*yfgX* deletion construct in a total reaction volume of 50 μl with the internal restriction site enzyme. Incubate the reaction for at least 1 h at the appropriate temperature for the restriction enzyme (most likely 37°C).

   Insert reaction

   5 μl purified FRT cassette DNA

   5 μl 10× NEB buffer

   5 μl 10× BSA

   1 μl NEB internal restriction enzyme

   34 μl dH_2O

   Vector reaction

   1 μl pK18*mobsacB*::Δ*yfgX*

   5 μl 10× NEB buffer

   5 μl 10× BSA

   1 μl NEB internal restriction enzyme

   1 μl NEB Calf Intestinal Phosphatase

   37 μl dH_2O
5. Purify the DNA with the Zymoclean Clean and Concentrate kit. Elute with a 10 μl volume.

6. Set up 20 μl ligation reactions and controls as in step 15 in Subheading 3.2.

   Incubate the reaction overnight at 14°C.
7. Purify the DNA from the ligations with the Zymoclean Clean and Concentrate kit. Elute with a 6 μl volume.
8. Prepare 6 LB Km plates additionally augmented with either Sp or Gm.
9. Transform the purified controls and ligation into competent DH5α or any standard *E. coli* cloning strain using electroporation or chemically competent heat-shock transformation by standard protocols (7).
10. Plate 1/10 of each transformation reaction and a concentrated remainder onto labeled LB Km plates augmented with either Sp or Gm. Incubate the plates overnight at 37°C. Colonies should only grow on the ligation plates, indicating that the deletion construct now carries the FRT cassette between Flank A and B.
11. Choose three Km^R and marker antibiotic resistant colonies and start 2-ml overnight LB Km cultures at 37°C.
12. Miniprep the plasmid DNA from the cultures using the Qiaprep Spin Miniprep kit from Qiagen.
13. Conduct a restriction digest screen on 2 μl of plasmid DNA with both external restriction site enzymes in a 20 μl volume as in step 22 in Subheading 3.2.
14. Check by agarose gel electrophoresis that the digest releases a fragment the same size as the A + FRT cassette + B fragment. Discard any plasmids which do not (see Note 17).

3.4. Conjugate the pK18mobsacB::ΔyfgX Deletion Construct into DC3000 to Get Single Crossover Merodiploid Transconjugants

1. Prepare an LB Km plate.
2. Transform 1 μl of a 50× dilution (~ 5 ng) of miniprep pK18*mobsacB*::*ΔyfgX* deletion construct DNA into *E. coli* S17-1 by electroporation or chemically competent heat-shock transformation using standard protocols (7). The marked or unmarked pK18*mobsacB*::*ΔyfgX* variants can be used.
3. Plate 1/10 of the transformation reaction onto a labeled LB Km plate. Incubate overnight at 37°C (see Notes 18–20).
4. From isolated colonies, start a 5-ml LM Rf Ap overnight culture with shaking at 30°C of the DC3000 recipient strain and a 5-ml LM Km overnight culture with shaking at 37°C of the *E. coli* S17-1 pK18*mobsacB*::*ΔyfgX* donor strain. Either the marked or unmarked variants of pK18*mobsacB*::*ΔyfgX* can be used depending on your particular strategy (see Note 21).
5. Harvest the cells from the overnight cultures by centrifugation. We use a floor model centrifuge, 10 min at 2,000 × *g* for this step. Discard the supernatant.

6. Resuspend the cells in 1.9 ml of fresh LM by vortexing. Transfer the cultures to labeled 2-ml microcentrifuge tubes.
7. Harvest the cells in a microcentrifuge 1 min at 16,000 × *g*. Discard the supernatant.
8. Resuspend the cells in a final volume of 0.2 ml LM.
9. In a third microcentrifuge tube, mix 50 μl each of the DC3000 and *E. coli* suspensions.
10. Using ethanol-sterilized forceps, position three sterile filter squares on a LM plate, with one square each for the controls and one for the conjugation. Use the forceps to push the filter square onto the surface of the plate. Label the plate near the filter squares with the respective strains.
11. Using a micropipetor, apply 50 μl each of the DC3000 and *E. coli* suspensions as drops to their respective filter squares. Add all 100 μl of the conjugation mixture to its filter square (see Note 22).
12. Allow the drops of suspension to dry.
13. Invert the plates and incubate for 2 days at 30°C.
14. Prepare and label three KB Rf Ap Km plates, one for each filter square (see Note 23).
15. Using ethanol-sterilized forceps, remove the filter squares from the LM plate and insert them into sterile 14-ml plastic snap-cap tubes, ethanol sterilizing between samples (see Note 24).
16. Add 2 ml of KB to each tube and secure the cap. The filter square can be knocked down into the media by tapping the tube repeatedly on the bench. Resuspend the bacteria from the filter square by vortexing. Try to get all the bacteria off of the filter square.
17. Transfer the resuspended cells to labeled 2-ml microcentrifuge tubes.
18. Harvest the cells in a microcentrifuge 1 min at 16,000 × *g*. Discard the supernatant.
19. Resuspend the cells in 0.2 ml of KB.
20. Plate the entire volume of resuspended cells onto the KB Rf Ap Km plates with an ethanol-sterilized spreader.
21. Incubate at room temperature or 30°C for three to 6 days. Rf^R Ap^R Km^R colonies should only grow on the plate spread with the conjugation mixture.

3.5. Counter-Select the Integration with Sucrose to Recover Double-Crossover Mutants

1. Prepare control and counter-selection plates. The control plate is KB Rf. To make the counter-selection plate add three parts prewarmed 1 M sucrose to seven parts melted KB to get approximately 10% final concentration of sucrose, add Rf as normal. If you used the marked variant of pK18*mobsacB*::*ΔyfgX*

in your conjugation, include the marker antibiotic in both the control and sucrose counter-selection plates.

2. Pick two Rf^R Ap^R Km^R colonies from the conjugation plate with a sterile toothpick and suspend the cells in 1.9 ml KB in a 2-ml microcentrifuge tube.
3. Use an ethanol-sterilized spreader to plate 50 µl of the suspension onto the KB Rf control plate and the KB Rf 10% sucrose plate.
4. Incubate at room temperature or 30°C for 2–4 days. The KB Rf control plate should produce a lawn of bacteria, but the KB Rf Suc plates should produce somewhere between 50 and 500 colonies (see Note 25). In the case of the unmarked variant of pK18*mobsacB*::*ΔyfgX*, these colonies represent a mixture of deletion mutants and wild-type revertants. In the case of the marked variant of pK18*mobsacB*::*ΔyfgX*, the included marker antibiotic selects for the deletion strain mutants.
5. Prepare a KB Rf and a KB Rf Km plate.
6. Using sterile toothpicks patch the Suc^R colonies onto the KB Rf and KB Rf Km plates. Patch at least 12 colonies.
7. Incubate at room temperature or 30°C for 2–4 days. The Km^S colonies have evicted the pK18*mobsacB* backbone. Any colony that is Km^R is most likely a spontaneous Suc^R strain rather than a double recombinant and should not be used in any further work.

3.6. Screen for the yfgX Deletion Mutant by PCR (See Notes 26–28)

1. Start 2-ml KB Rf overnight cultures from at least 6 Suc^R Km^S patches.
2. Purify genomic DNA using the Promega wizard genomic DNA kit for Gram-negative bacteria.
3. Gather the Ex Taq premix, forward and reverse flank primers and genomic DNA template for the PCR reaction.
4. In 0.65 microcentrifuge tubes, set up a separate PCR reaction for each clone to be screened as well as a wild-type control reaction using DC3000 genomic DNA according to the following recipe.

   10 µl Ex Taq premix

   2 µl A out forward

   2 µl B out reverse

   1 µl 1/10× diluted genomic DNA

   5 µl sterile dH_2O
5. Run the PCR reaction in a thermocycler according to the following program.

5 m at 95°C

Repeat 30 cycles

1 m at 94°C

1 m at (lowest primer Tm –3°C)

1 m/kb at 72°C

(Base the extension time on the size of the largest product)
10 m at 72°C

Hold at 15°C

6. Check the total volume of the PCR reaction by agarose gel electrophoresis for bands of the appropriate sizes for the wild-type and the deletion strain mutant.

3.7. (Optional) Removal of the FRT-Flanked Marker from a Marked Deletion Strain

You may choose to remove the antibiotic marker from your marked deletion strain either to make that antibiotic available for other purposes or to make additional mutations in the same strain. The removal of the FRT-flanked marker can be done in two ways: Flp-mediated recombination of the FRT-flanked cassette, or a second mutagenesis with the unmarked variant of pK18*mobsacB*::*ΔyfgX*. After FRT flank removal by either method, the strain should be reconfirmed by PCR using methods similar to those described in steps 1–6 in Subheading 3.6 (see Note 29).

3.7.1. Removing FRT-Flanked Cassette Using Flp Recombination

1. Similar to step 4 in Subheading 3.4, from isolated colonies, start a 5-ml LM Rf Ap overnight culture with shaking at 30°C of the DC3000 marked deletion recipient strain, a 5-ml LM Km overnight culture with shaking at 37°C of the *E. coli* HB101 pRK2013 helper strain, and a 5-ml LM Tc overnight culture with shaking at 37°C of the *E. coli* DH5α pCPP5264 donor strain.
2. Process the overnight cultures and set up the LM plate as in steps 5–21 in Subheading 3.4 with these modifications. There is an extra control filter square for the helper strain. Mix 20 μl of each of the three strains for the conjugation and spot 20 μl of each control strain and all 60 μl of the conjugation. Plate the controls and conjugation on KB Rf Ap Tc.
3. Patch the Tc^R pCPP5264 transformed clones on KB Rf and KB Rf with marker antibiotic looking for sensitivity to the marker antibiotic. Sensitivity indicates loss of the FRT-flanked marker and its recombination into an FRT site scar (8) (see Note 30).
4. Start 5-ml KB Rf overnight culture with shaking at 30°C of two or three marker antibiotic sensitive colonies. Without Tc selection pCPP5264 is rapidly lost.
5. Apply 20 μl of overnight cultures to KB Rf plates near the plate's border and spread the bacteria by three-phase streak to get isolated colonies.

6. Incubate 2–3 days at room temperature or 30°C.
7. Patch plate at least 12 isolated colonies for Tc sensitivity to confirm loss of pCPP5264.

3.7.2. Removing an FRT-Flanked Marker by Secondary Mutagenesis

1. Repeat steps 4–7 in Subheading 3.4 using your marked deletion strain as the recipient strain and *E. coli* S17-1 carrying unmarked pK18*mobsacB*::*ΔyfgX* as the donor strain. When you get to step 6 in Subheading 3.5, patch plates for Km^S and also patch for sensitivity to the marker antibiotic. Colonies that are sensitive to both Km and the marker antibiotic are likely to be unmarked deletion mutants.

4. Notes

1. pK18*mobsacB* has a 1-bp deletion in the M13R binding site so it requires a unique primer.
2. X-gal and Tc are light sensitive. Store either in amber microcentrifuge tubes or in a light-free environment.
3. For LM and KB liquid media do not add K_2HPO_4 prior to autoclaving or it will precipitate. Mix up a 100× phosphate stock, K_2HPO_4 75 g in 500 ml dH_2O and filter sterilize. Add to the media after it cools.
4. pCPP5264 is an unstable Flp expression plasmid.
5. pRK2013 is a *tra* helper plasmid.
6. The flanks are designed with a difference in size because they are joined by restriction digest followed by T4 ligation of the linear products. If both flanks were the same size, you could not distinguish undesired flank A + A or flank B + B products from the desired A + B product.
7. Alternative joining methods such as SOEing (*s*plicing by *o*verlap *e*xtension) PCR could be used instead, in which case the flanks could be designed to be identical in size (9).
8. The purpose of the 3–4 nt toe is to provide a place for the restriction enzyme to sit. Many enzymes do not cut efficiently near the end of a DNA molecule. Any base combination can be used here and it can be altered to increase or decrease the Tm of the primer, but try to avoid palindromic sequences which can lead to primer-dimers.
9. Using *Pvu*I or *Nhe*I for cloning removes the M13F priming site, so a different sequencing primer needs to be used in step 24 in Subheading 3.2.
10. SmaI digests at 30°C, which can complicate double digests. XmaI is a non-blunt isoschizomer of SmaI that digests at 37°C.
11. *Xba*I and *Nhe*I produce compatible cohesive ends.

12. If ligating the two fragments together prior to cloning into pK18*mobsacB* does not work, or if a different cloning strategy is preferred, the primer design described here can also be used to clone each fragment sequentially into pK18*mobsacB* using similar restriction digest and cloning strategies to those described.
13. If the bands are not strong or if there are secondary bands, try to optimize your PCR or possibly redesign the flank amplification primers. If secondary bands are a problem, you could alternatively gel purify the correct band. To gel purify, cast a 1% agarose gel using a comb which produces large wells to fit the 95 μl reaction volume. Use fresh agarose and running buffer. Cut out the band of the correct size with a clean razor blade. Purify the DNA from the agarose block using the Zymoclean gel recovery kit. Elute with a 6 μl volume.
14. Incubation of the plates at 4°C strengthens the blue color.
15. Introducing FRT-flanked cassettes into the deletion mutant strain is likely to create polar effects on downstream genes. FRT cassettes can be cloned in such a way that the FRT site scar can be read through in-frame, but to do so, the FRT cassette forward primer has to be redesigned as 3–4 nt toe (internal restriction site) CGCTGGAGCTGCTTCGAA, and the FRT cassette has to be cloned in the appropriate orientation.
16. These primers are based on the common primers 1 and 2 from Datsenko et al. (10).
17. The FRT cassette is cloned in a nondirectional manner. You may want to determine the orientation of the FRT cassette by restriction digest or PCR screen.
18. *Escherichia coli* S17-1 is a mating strain with an integrated RP4 plasmid, so it can mobilize plasmids such as pK18*mobsacB* without the aid of a *tra* helper plasmid. The reason *E. coli* S17-1 was not transformed directly is that it is not blue white compatible and is Sp^R, which would interfere with the cloning of the $FRTSp^R$ cassette.
19. *Escherichia coli* transformed with pK18*mobsacB* form slightly transparent colonies which are also sensitive to sucrose. These strains often grow at a slower pace.
20. *Escherichia coli* S17-1 and pK18*mobsacB* transformed strains do not transition well to 4°C. It is better to keep these strains at room temperature for up to 2 weeks and restreak them fresh from glycerol when needed.
21. When starting overnight, cultures of DC3000 be generous with the amount of cells added to the broth.
22. To prevent cross-contamination, when applying the bacterial suspensions be very cautious not to drip onto the plate, and

try to approach each filter square from its own angle of attack, not crossing over other filter squares. Hold the micropipetor tip close to, but not touching, the filter square, and depress the plunger slowly. Try to keep the entire volume of the bacterial suspension on the filter square.

23. Ampicillin is included to provide additional selection against the *E. coli* donor.
24. The best way to do this is to grab the middle of an edge of a filter square, being careful not to touch the forceps to the bacteria. Insert a corner of the filter square into the mouth of the snap cap tube and twist the filter square into the tube. Ideally, you want the square flat against the wall of the tube with the bacteria facing inward.
25. DC3000 colonies grown on KB 10% sucrose plates are shiny, mucoid, semitransparent, more strongly domed, and develop an orange color as they age.
26. If the intended deletion has a readily testable phenotype, you may want to conduct a phenotypic screen prior to the PCR screen to get some indication as to which strains are likely to be mutant or wild-type. For example, a *fliC* deletion mutant does not produce a flagellum, resulting in the loss of motility on a 0.2% agar swim plate.
27. The most straightforward PCR confirmation screen is to use out A forward and out B reverse. These primers are located outside the flank amplification primers, so they can screen against deletion construct integration into the incorrect genomic location. Ideally, this PCR reaction produces distinct band sizes for both wild-type and deletion mutant strains and confirms the status of both borders simultaneously. However, there are two potential problems. If you have created a marked deletion strain, depending on the size of the gene that was deleted, the mutant and wild-type may not be easily distinguishable by size. Also, if the deleted gene or genomic region was 5 kb or longer, the wild-type band may be difficult or impossible to amplify. In all the alternative PCR confirmation strategies, two PCR reactions are required to confirm the status of each border.

    A out forward/B seq reverse and A seq forward/B out reverse.

    A out forward/flank A reverse and flank B forward/B out reverse

    For marked deletion strains or deletion strains with the FRT site scar:

    A out forward/FRT check 1 and FRT check 2/B out reverse

    FRT check 1 GAAGCAGCTCCAGCCTACAC

    FRT check 2 CTTCGGAATAGGAACTAAGGAGGATATTC ATATG

28. If a particular mutant has proven difficult to recover using the unmarked variant of pK18*mobsacB*::*ΔyfgX* due to high recovery of wild-type revertants, it can often be forced forward by using the marked pK18*mobsacB*::*ΔyfgX* variant. When the marker antibiotic is included at the sucrose counter-selection step, only the marked mutant population and spontaneous Suc^R colonies grow. If no colonies are produced using the two colonies in 2 ml dilution strategy with the marked variant pK18*mobsacB*::*ΔyfgX*, more concentrated suspensions can be spread on the sucrose counter-selective marker antibiotic selective plates. Dense suspensions of approximately 10^8 CFU/ml have been used in this manner to isolate particularly difficult mutants.
29. The use of Flp to remove FRT-flanked cassettes is quick and straightforward, but in strains with multiple deletions there are some potential concerns. After Flp recombination of an FRT-flanked cassette, an FRT site scar is left behind in the genome. After several cycles of mutagenesis, these FRT site scars start to build up in the genome. The FRT site scars are still recombinationally active and can facilitate large-scale inversions between indirect FRT sites or deletions between direct FRT sites mediated by Flp. If you intend to make more than five deletions in sequence in the same strain or if you intend to make additional deletions in genes which are located near an existing FRT site scar, you may want to consider using a second mutagenesis with unmarked pK18*mobsacB*::*ΔyfgX* to remove the FRT-flanked marker rather than the use of Flp recombination. If you do make multiple deletions and use Flp to remove the FRT-flanked markers, then the status of each FRT site scar in the genome should be checked by PCR after Flp recombination to ensure that they are intact.
30. The sequence of the FRT site scar has the following structure. GTGTAGGCTGGAGCTGCTTC**GAAGTTCCTATA**CTT*TCTAGA*G**AATAGGAACTTC**GGAATAGGAACTAAGGAGGATATTCATATG. Bold sequences are Flp recognition sites. The sequence in italics is an *Xba*I site. Underlined sequences are primer binding sites (10).

Acknowledgments

The authors would like to thank Dr. Chia-Fong Wei and Dr. Joanne E. Morello for significant contributions to the refinement of this protocol. This work was supported by NSF Plant Genome Research Program grant DBI-0605059, by NSF grant MCB-0544066 and by NSC grant NSC94-2752-B-005-003-APE.

References

1. Lindeberg, M., Cartinhour, S., Myers, C. R., Schechter, L. M., Schneider, D. J., Collmer, A. (2006) Closing the circle on the discovery of genes encoding Hrp regulon members and type III secretion system effectors in the genomes of three model *Pseudomonas syringae* strains. *Mol Plant Microbe Interact* **19**, 1151–8.
2. Oh, H. S., Kvitko, B. H., Morello, J. E., Collmer, A. (2007) *Pseudomonas syringae* lytic transglycosylases co-regulated with the type III secretion system contribute to the translocation of effectors into plant cells. *J Bacteriol* **189**, 8277–89.
3. Kvitko, B. H., Ramos, A. R., Morello, J. E., Oh, H. S., Collmer, A. (2007) Identification of harpins in *Pseudomonas syringae* pv. *tomato* DC3000, which are functionally similar to HrpK1 in promoting translocation of type III secretion system effectors. *J Bacteriol* **189**, 8059–72.
4. Lindeberg, M., Myers, C. R., Collmer, A., Schneider, D. J. (2008) Roadmap to new virulence determinants in *Pseudomonas syringae*: insights from comparative genomics and genome organization. *Mol Plant Microbe Interact* **21**, 685–700.
5. Wei, C. F., Kvitko, B. H., Shimizu, R., et al. (2007) A *Pseudomonas syringae* pv. tomato DC3000 mutant lacking the type III effector HopQ1-1 is able to cause disease in the model plant *Nicotiana benthamiana*. *Plant J* **51**, 32–46.
6. Schafer, A., Tauch, A., Jager, W., Kalinowski, J., Thierbach, G., Puhler, A. (1994) Small mobilizable multipurpose cloning vectors derived from the *Escherichia coli* plasmids pK18 and pK19: selection of defined deletions in the chromosome of *Corynebacterium glutamicum*. *Gene* **145**, 69–73.
7. Sambrook, J., Russell, D. W. (1989) Molecular Cloning: A Laboratory Manual. Cold Spring Harbor, NY: Cold Spring Harbor Laboratory Press.
8. Kolisnychenko, V., Plunkett, G., 3rd, Herring, C. D., et al. (2002) Engineering a reduced *Escherichia coli* genome. *Genome Res* **12**, 640–7.
9. Horton, R. M. (1995) PCR-mediated recombination and mutagenesis. SOEing together tailor-made genes. *Mol Biotechnol* **3**, 93–9.
10. Datsenko, K. A., Wanner, B. L. (2000) One-step inactivation of chromosomal genes in *Escherichia coli* K-12 using PCR products. *Proc Natl Acad Sci USA* **97**, 6640–5.

Chapter 11

A Straightforward Protocol for Electro-transformation of *Phytophthora capsici* Zoospores

Edgar Huitema, Matthew Smoker, and Sophien Kamoun

Abstract

Genome sequencing combined with high-throughput functional analyses has proved vital in our quest to understand oomycete–plant interactions. With the identification of effector molecules from *Phytophthora* spp. we can now embark on dissecting the mechanisms by which effectors modulate host processes and thus ensure parasite fitness. One of the key limitations, however, is to genetically modify *Phytophthora* and assess gene function during parasitism. Here, we describe a straightforward protocol that allows rapid transformation of *Phytophthora capsici*, an emerging model in oomycete biology. *P. capsici* is a broad host range pathogen that can infect a wide variety of plants under lab conditions making it a suitable model for detailed studies on oomycete–host interactions. This protocol relies on electroporation-assisted uptake of DNA in to motile zoospores and allows the rapid identification and characterization of genetically stable transformants.

Key words: Oomycetes, *Phytophthora capsici*, Genetic transformation, Zoospore, Electroporation

1. Introduction

With the advent of genome sequencing and re-sequencing technologies as well as the availability of tools suited for high-throughput functional analyses, much progress has been made in understanding oomycete–plant interactions (1–3). It is now clear that all *Phytophthora* spp. examined to date, harbour vast arsenals of secreted proteins (effectors) that modulate host processes (4). One of the major objectives, therefore, is to determine the roles of these proteins and understand their function during infection. Strategies towards reaching these goals include (1) identification of host proteins that could be targeted and modified by each effector, (2) silencing or over-expression of effector genes to determine the impact on virulence, and (3) localization

John M. McDowell (ed.), *Plant Immunity: Methods and Protocols*, Methods in Molecular Biology, vol. 712, DOI 10.1007/978-1-61737-998-7_11, © Springer Science+Business Media, LLC 2011

of *Phytophthora* (effector) proteins during infection. One key and rate-limiting step in performing these analyses is to genetically modify *Phytophthora*. Various protocols for DNA transformation of *Phytophthora* are available, but they are typically tedious, time consuming, and with variable reproducibility.

Genetic transformation of *Phytophthora* requires uptake of DNA and subsequent integration into the genome. DNA uptake takes place only in cells that lack intact cell walls and requires the formation of pores in the cell plasma membrane. Transformation of *Phytophthora* protoplasts using polyethylene glycol and $CaCl_2$ has historically been the method of choice (5). Here, we describe a rapid and efficient method towards the transformation of *Phytophthora capsici*. This method takes advantage of the fact that as part of its life cycle, *P. capsici*, similar to most other *Phytophthora* species, produces motile zoospores that lack cell walls and which are amenable to electroporation-assisted transformation.

1.1. Electroporation of P. capsici Zoospores

P. capsici transformation relies on the generation of large quantities of viable zoospores that lack cell walls. Before transformation, *P. capsici* is grown on solid media plates under conditions that allow the formation of sporangia, structures that can be induced to release zoospores in aqueous suspensions. Sporangia are harvested by flooding plates with a cold wash solution, carefully dislodging sporangia and a gentle centrifugation and re-suspension step that concentrate sporangia. Sporangial solutions are then incubated at room temperature while exposed to an external light source, conditions that induce differentiation of sporangia and subsequent release of zoospores. Upon emergence of zoospores, suspensions are mixed with DNA and used directly for electroporation. Finally, electroporated suspensions are mixed with regeneration media and incubated overnight before being subjected to selection on plates.

Electroporation of zoospores using this protocol is feasible in other *Phytophthora* species in principle but is dependent upon the production of large number of zoospores, which does not occur with all strains. Optimal conditions for growth, sporulation, electroporation and subsequent selection should be determined first before embarking on transforming other *Phytophthora* species. The protocol described here can produce up to 40 *P. capsici* transformants per electroporation.

2. Materials

1. Unclarified 10% V8 agar: 100 ml V8 juice, 900 ml H_2O, 1 g of $CaCO_3$, 0.05 g β-sitosterol, 15 g agar/litre (see Note 1).
2. *P. capsici* strain LT1534 (see Note 2).

3. 1× modified Petri's solution (6): 0.25 mM $CaCl_2$, 1 mM $MgSO_4$, 1 mM KH_2PO_4, 0.8 mM KCl/litre (see Note 3).
4. 15–20 μg of linearized plasmid DNA (see Note 4).
5. Gene pulser: BioRad Xcell & compatible Electroporation cuvettes, with gap width 4 mm (see Note 5).
6. Light microscope.
7. Haemocytometer slide.
8. Regeneration medium: 60 g organic rye grains, 20 g/l sucrose, 100 mM mannitol, 1 mM KCl, and 2.5 mM $CaCl_2$ (see Note 6).
9. Selection medium: Unclarified 10% V8 agar with 50 μg/ml Carbenicillin, 25 μg/ml Vancomycin and 50 μg/ml G418 (geneticin) (see Note 7).

3. Methods

3.1. Culturing *P. capsici*

1. Grow *P. capsici* on large (15 cm) Petri dishes by placing 2–4 plugs of previously grown *Phytophthora* cultures onto the surface of non-selective unclarified V8 agar and seal with parafilm (see Note 8).
2. Incubate the plates in continuous light for 4 days at 25°C (see Note 9).
3. Remove the parafilm and incubate the cultures in the same conditions for at least 3 more days (see Note 9).

3.2. Preparation of Regeneration Media

1. Incubate 60 g of organically grown rye grains in 3% sodium hypochlorite for 10 min with light agitation (see Note 10).
2. Decant and rinse the grains in running tap water for 10–15 min until the smell of bleach is completely absent (see Note 10).
3. Cover the washed grains with distilled water (800 ml) and incubate at room temperature for 24 h.
4. Transfer the suspension to a food blender and grind the grains for 1 min.
5. Incubate the suspension for 2 h at 68°C.
6. Centrifuge the suspension for 10 min at 3,300×*g* at room temperature, and collect supernatant.
7. Filter supernatant through four layers of muslin and make volume up to 1 l (see Note 11).
8. Add 20 g sucrose and autoclave.

9. Upon use, take an aliquot of the required volume. (10 ml/ electroporation) and add: 100 mM mannitol, 1 mM KCl, and 2.5 mM $CaCl_2$ (see Note 6).

3.3. Electroporation of *P. capsici* Zoospores

1. Flood a plate with 14 ml of ice-cold 1× modified Petri's solution and gently rub the mycelial mat with a sterile glass spreader to dislodge the sporangia.
2. Tip the re-suspended sporangia solution onto the next plate and repeat step 1 until 3–4 plates have been harvested.
3. Wash all the plates with an extra small volume of cold Petri's (5 ml) to harvest as many sporangia as possible. Between 5×10^5 and 1×10^6/ml sporangia should be obtained. Collect all suspensions into a sterile 15 ml tube on ice.
4. Divide the harvested sporangia solution into ice-cold 2-ml microcentrifuge tubes (one tube per electroporation) and centrifuge at 3,300 × *g* for 5 min at room temperature.
5. Remove excess supernatant to leave approximately 750 μl and gently re-suspend the sporangial pellet.
6. Place microcentrifuge tubes containing re-suspended sporangia onto a light box and incubate. Mix suspensions gently from time to time in order to keep sporangia in suspension.
7. Evaluate the release and number of zoospores carefully every 5 min by placing a 10 μl droplet of suspension onto a haemocytometer and counting the number of swimming spores under a light microscope.
8. Add DNA when zoospore numbers start to increase dramatically.
9. As soon as there are 1×10^6/ml zoospores, pipette the suspensions very carefully (see Notes 12 and 13) into pre-chilled 4 mm gap electroporation cuvettes.
10. Electroporate the suspensions with the following settings; Voltage, 550 V, capacitance, 50 μF, resistance, 200 Ω. This should give a time constant between 4.8 and 6.0 ms.
11. Add 800 μl of ice cold, well-aerated regeneration medium, taking care to pipette slowly and carefully (see Note 14).
12. Transfer zoospore mix into 9 ml of ice-cold regeneration medium in a 15 ml centrifuge tube with lid. Lay the tube on its side and place onto a rocking platform at 18°C in the light and leave for 1 h to recover.
13. Add antibiotics: 25 μg/ml vancomycin and 50 μg/ml carbenicillin.
14. Place tubes back onto the rocking platform and incubate overnight.

15. Centrifuge 15 ml tubes for 5 min, 403 × *g* at room temperature and remove the supernatant to leave between 3 and 4 ml per tube.
16. Disrupt the hyphal mats by pipetting vigorously. Start with a 1 ml pipette tip that has had its tip (approximately 3 mm from its end) cut off (using a hot sterile scalpel) and then with an intact 1 ml tip.
17. Plate the hyphal suspension onto selective V8 15 cm plates (at least 2 plates per electroporation) by dispensing the liquid across the plate. Limit mechanical manipulations on the plate surface as much as possible and allow excess liquid to air dry or soak into the plates (see Note 15).
18. Seal the plates with parafilm and incubate the selective plates (inverted) at 25°C. Geneticin resistant colonies appear after 3–5 days and should be subcultured on fresh selective plates when possible.

4. Notes

1. Both calcium carbonate ($CaCO_3$) and β-sitosterol do not completely dissolve in the media. Upon addition of all ingredients, as well as after sterilization, stir the media for 5–10 min. β-sitosterol is required for the production of viable zoospores. Omitting this component leads to the release of zoospore clumps that fail to separate.
2. *P. capsici* strain LT1534 is our preferred strain for transformation since this isolate grows vigorously on V8 and sporulates profusely using standard conditions. In addition, LT1534 was used in the *P. capsici* genome sequencing project.
3. It is recommended to make a sterile 50× concentrated solution of Petri's (12.5 mM $CaCl_2$, 50 mM $MgSO_4$, 50 mM KH_2PO_4, 40 mM KCl/litre) and dilute in sterile dH_2O prior to transformation. It is critical to use ice-cold Petri's solution for both harvesting and re-suspending sporangial suspensions.
4. We routinely use the pTOR vector for *P. capsici* transformation (7). pTOR constructs are maintained and isolated from *Escherichia coli* cells grown in Luria–Bertani (LB) media supplemented with Carbenicillin (100 μg/ml). The pTOR cassette contains the NPTII gene for the selection of *Phytophthora* transformants on G418 (geneticin), a constitutive oomycete promoter (*ham34*) followed by a multiple cloning site to insert your gene(s) of interest and the *ham34* terminator. Although in other *Phytophthora* spp. linearized DNA is used

for transformation, we found that *P. capsici* transformation can be carried out successfully with undigested plasmid preparations.

5. We have successfully used the Genepulser Xcell system (Biorad) and compatible cuvettes for transformation experiments. It is recommended, however, to test other systems as we have successfully used the Micropulser for a limited number of experiments, and we have never performed side by side comparative experiments to find optimal transformation conditions.
6. For detailed instructions of regeneration media preparation, please refer to Subheading 3.1. It is imperative that mannitol, KCl, and $CaCl_2$ are added on the day of transformation.
7. For strain LT1534, we supplement V8 agar with 50 μg/ml of G418 to select transformants. Before experiments are initiated with other *P. capsici* strains, we recommend establishing optimal G418 concentrations empirically. Carbenicillin and Vancomycin are added to limit the occurrence of bacterial contaminations during the zoospore preparation and electroporation process.
8. We use parafilm to prevent culture plates from drying.
9. We grow *P. capsici* at 25°C in LMS incubators fitted with interior illumination. Interior light intensity roughly corresponds to natural light conditions. Upon removal of parafilm and exposure to light, sporulation (formation of sporangiophores bearing sporangia) is initiated. Sporulation is evident on plates when "powdery rings" are formed around the *Phytophthora* mycelial mat.
10. We strongly recommend using organically grown rye for the preparation of regeneration media. The use of pesticides on non-organically grown rye could impact the growth of *Phytophthora* and introduce variation between rye grain batches. In addition, after sterilization of grains, it is paramount that all traces of bleach are removed to obtain optimal regeneration and growth.
11. We commonly use 3–4 layers of cheesecloth to filter media supernatants. After centrifugation and filtration, media may still contain low levels of particulates that precipitate over time. This does not hamper regeneration.
12. Zoospores are very sensitive to physical stimuli. For example, excessive bouncing of zoospores onto solid surfaces results in immediate encystment. Therefore, care should always be taken when handling (e.g. mixing) zoospore suspensions.
13. Zoospore health is key for successful transformation. Actively swimming zoospores indicate healthy cultures, whereas rapid

(spontaneous) encystment is often observed under suboptimal conditions.

14. Aerate regeneration media before use to ensure optimal conditions for recovery. Once combined with electroporated zoospore suspensions, mix carefully to limit physical disruption of regenerating cysts.
15. We found that physically spreading young and regenerated mycelia onto solid surfaces (agar plates) resulted in reduced transformation efficiency. We therefore dry selection plates considerably before plating regenerated samples. We apply excess liquid onto the plates, spread the suspension by tilting the plates and allow the plates to soak up moisture.

Acknowledgments

We thank Dr. Kurt Lamour for *P. capsici* strain LT1534 that was used for the development of the transformation protocol and Dr. Brett Tyler for sharing an earlier version of a *Phytophthora* zoospore electroporation protocol.

References

1. Kamoun, S. (2007) Groovy times: filamentous pathogen effectors revealed. *Current Opinion in Plant Biology*, **10**:358–365.
2. Lamour, K., Win, J., and Kamoun, S. (2007) Oomycete genomics: new insights and future directions. *FEMS Microbiology Letters*, **274**:1–8.
3. Kamoun, S. (2009) The secretome of plant-associated fungi and oomycetes. *In* H. Deising (Ed.), *Plant Relationships, 2nd Edition, The Mycota V.* Springer-Verlag, Berlin, Heidelberg.
4. Morgan, W., and Kamoun, S. (2007) RXLR effectors of plant pathogenic oomycetes. *Current Opinion in Microbiology*, **10**:332–338.
5. Judelson, H. S., Tyler, B. M., and Michelmore, R. W. (1991) Transformation of the oomycete pathogen, *Phytophthora infestans. Molecular Plant–Microbe Interactions*, **4**:602–607.
6. Erwin, D. C., and Ribeiro, O. K. (1996) *Phytophthora* Diseases Worldwide. American Phytopathological Society, St. Paul, MN.
7. Torche, S. (2004) Isolation et etudes d'expression de genes potentiellement lis a la pathogenicite chez *Phytophthora porri*. Ph.D. Thesis. University of Friborg, Switzerland.

Chapter 12

Propagation, Storage, and Assays with *Hyaloperonospora arabidopsidis*: A Model Oomycete Pathogen of *Arabidopsis*

John M. McDowell, Troy Hoff, Ryan G. Anderson, and Daniel Deegan

Abstract

The oomycete pathogen *Hyaloperonospora arabidopsidis* is a natural pathogen of *Arabidopsis thaliana* and a laboratory model for (1) understanding how *Arabidopsis* responds to pathogen attack; (2) comparative and functional genomics of oomycetes; and (3) the molecular basis and evolution of obligate biotrophy. Here, we describe procedures for propagation and long-term storage of *H. arabidopsidis*, which address complications arising from its biotrophic lifestyle that precludes growth on synthetic media. We also describe four assays that provide information on different facets of the *H. arabidopsidis*–*Arabidopsis* interaction.

Key words: Disease, Resistance, *R* gene, Cytology, Oomycete, Haustoria, Trypan blue, Sporangiophore, Diaminobenzidine, Aniline blue

1. Introduction

Hyaloperonospora arabidopsidis (formerly *Peronospora parasitica* and *Hyaloperonospora parasitica*) is a naturally occurring pathogen of *Arabidopsis thaliana* (1, 2). *Hyaloperonospora arabidopsidis* causes downy mildew disease specifically on *Arabidopsis*, and is congeneric with species that cause economically significant diseases on cultivated Brassicas. Over 800 species causing downy mildew disease have been identified. As a group, the downy mildews parasitize a large number of crops and are estimated to account for ~20% of the $5 billion global fungicide market (3, 4). Two downy mildews of monocots are listed among seven plant pathogens considered to be major US bioterror threats (Agricultural Bioterrorism Protection Act, 2002).

John M. McDowell (ed.), *Plant Immunity: Methods and Protocols*, Methods in Molecular Biology, vol. 712,
DOI 10.1007/978-1-61737-998-7_12,

Downy mildews are related to destructive plant pathogens in the *Phytophthora*, *Pythium*, and *Albugo* genera. Together, these genera are classified as oomycetes. This group shares superficial morphological resemblances with true fungi (e.g., filamentous growth, haustoria, aerial spore-producing structures) and were first aligned with fungi. However, DNA sequencing and other lines of evidence clearly place oomycetes apart from fungi in the kingdom Stramenopila, which includes brown algae and diatoms (5). Thus, oomycetes and fungi have independently evolved to colonize plants, and have convergently evolved very similar infection structures.

Hyaloperonospora arabidopsidis can reproduce sexual and asexual spores. Asexual spores are produced from branched structures (sporangiophores, Fig. 1) that emerge from stomata and produce spores to be dispersed by wind or water. In addition, sexual spores form inside infected organs and are released into the soil upon decomposition of the infected plant. The infection cycle commences when a spore attaches to the surface of a leaf, stem, or root. An infection thread emerges from the spore, penetrates between epidermal cells, and differentiates into a branched web of hyphae that grow through the spaces between cells inside the host organ (Fig. 1). Direct associations form between hyphae

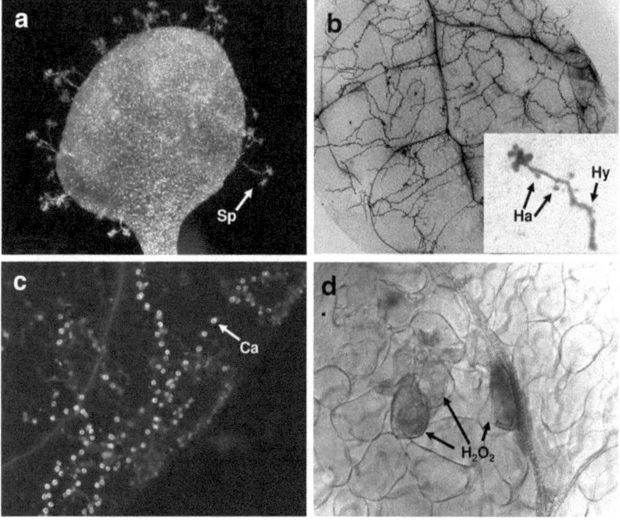

Fig. 1. Examples of assays. (**a**) Susceptible *Arabidopsis* cotyledon with abundant production of sporangiophores (Sp) at 7 days after inoculation. (**b**) Susceptible *Arabidopsis* leaf stained with Trypan Blue at 48 h after inoculation (hai). Hyphae are visible as branched, threadlike structures throughout the leaf. The inset contains an individual interaction site stained at 24 hai, in which a single hypha (Hy) with haustoria (Ha) is visible. (**c**) Susceptible *Arabidopsis* leaf stained with Aniline Blue, highlighting callose (Ca) encasements of haustoria that are typical even in susceptible plants. (**d**) Resistant *Arabidopsis* leaf stained with DAB at 24 hai. DAB precipitate can be visualized as a brown deposit, indicative of the location of H_2O_2 production.

and host cells through structures called haustoria, which project into the plant cell through the host cell wall (Fig. 1). Haustoria play a key role in exploitation of the plant by actively importing carbon, water, and inorganic nutrients (6, 7). In addition, haustoria are sites through which effectors are translocated into host cells (8).

Hyaloperonospora arabidopsidis and other downy mildews are biotrophic, extracting nutrients from living host tissue. Downy mildews are "obligate" pathogens that cannot be cultured on synthetic media and are incapable of surviving apart from their hosts. Little is currently known about the molecular mechanisms that support biotrophic life strategies.

Because *H. arabidopsidis* is a natural pathogen of *Arabidopsis*, it has become one of the two most widely used model pathogens (along with *Pseudomonas syringae*) for studies of *Arabidopsis* defense networks (2). Over 30 naturally occurring "*RPP* genes" (resistance to *P. parasitica*) have been genetically defined (1, 2). Eleven have been positionally cloned and shown to encode intracellular NB-LRR proteins (9). In addition to *RPP*-dependent resistance, *Arabidopsis* deploys basal defenses that inhibit growth of virulent *H. arabidopsidis* strains (e.g., (10)).

Although the *H. arabidopsidis–Arabidopsis* pathosystem has been used primarily to study host defenses, it is now being developed as a model to explore the mechanisms by which biotrophs manipulate their hosts. Experimental tools include a genetic linkage map, large collection of ESTs, and a BAC library that facilitated molecular cloning of two avirulence loci (11–14). Most recently, a high-quality draft genome sequence has been generated (manuscript in preparation), and this has fueled examination of *H. arabidopsidis* RXLR effector genes, as well as other aspects of *H. arabidopsidis* biology. In particular, comparative genomic studies are leveraging the availability of several *Phytophthora* genomes to investigate evolution of pathogenicity in the oomycetes; for example, how did obligate biotrophy in the downy mildews evolve from a *Phytophthora*-like, hemibiotrophic ancestor?

For all of the above reasons, *H. arabidopsidis* is being utilized as by an increasing number of laboratories. However, use of *H. arabidopsidis* in the laboratory is somewhat complicated by its obligate lifestyle; it cannot be grown on synthetic media and must be cultured on living, susceptible *Arabidopsis* plants. As with most other pathogens, careful control of environmental parameters is important for robust propagation and accurate disease/resistance assays. Below, we describe procedures for propagation, long-term storage, and revival of *H. arabidopsidis* cultures. We also describe four assays that report different aspects of interaction between *H. arabidopsidis* and *Arabidopsis.*

2. Materials

2.1. Propagation

1. Dedicated space for experiments with *H. arabidopsidis* (see Note 1).
2. Dedicated plant growth chamber(s) (see Note 2).
3. Sterile hood.
4. Spray bottle with water.
5. Spray bottle with 70% ethanol.
6. Pots (2 and 4 in.), soil (see Note 3), trays, clear plastic domes, 10″ × 20″ flats, magenta boxes, wet paper towels.
7. *Arabidopsis* plants of the appropriate genotype.
8. 50-mL Falcon tubes, forceps, small scissors, magnifying glass (see Note 4).
9. Sterile water.
10. Hemocytometer.

2.2. Storage and Revival

1. Ice bucket with ice, forceps, small scissors, magnifying glass, and 1.5-mL centrifuge tubes.
2. Freezer at –80°C.
3. Sterile water and pipetting device.

2.3. Sporangiophore Assays

1. All materials listed in Subheading 2.1.
2. Computer software (e.g., Microsoft Excel) to calculate averages and standard error and to assess statistical significance using Student's *t*-test.

2.4. Trypan Blue Staining

1. Temperature block or water bath at 90°C.
2. 2-mL Plastic centrifuge tubes or other appropriately sized container (see Note 5).
3. Trypan Blue solution: 50 mL dH_2O, 50 mL phenol, 50 mL lactic acid, 50 mL glycerol, and 100 mg Trypan Blue (see Note 6). Store in a brown bottle at room temperature in an approved chemical fume hood.
4. 95% Ethanol.
5. Chloral hydrate solution (see Notes 7 and 8): Add 1 mL H_2O per 2.5 g chloral hydrate. Allow to dissolve overnight at room temperature in the fume hood (see Note 9). Store at room temperature, following appropriate regulatory guidelines.
6. 50% Glycerol in deionized water, autoclaved, and stored at room temperature.
7. Microscope slides, No. 2 cover slips (24 mm × 50 mm), clear nail polish.

8. Compound microscope with camera, computer, and image processing software (see Note 10).

2.5. Destaining Without Chloral Hydrate

1. Lactophenol solution: 50 mL phenol, 50 mL lactic acid, and 50 mL glycerol (see Note 6).
2. 95% Ethanol, to be mixed in 2:1 ratio with lactophenol solution.

2.6. Diaminobenzidine Staining

1. Diaminobenzidine (DAB) solution: 25 mg DAB, 25 mL diH_2O, 20 μL hydrochloric acid (HCl) in a brown bottle, dissolve in a 42°C water bath for 6–8 h with occasional swirling (see Note 11).
2. Multiwell plates (96-well plates work well for seedlings).
3. Container for storage in the dark; wet paper towels.
4. ELG solution: 3:1:1 mixture of 95% ethanol:lactic acid:glycerol.
5. Chloral hydrate solution as in Subheading 2.4.

2.7. Aniline Blue and Autofluorescence

1. Temperature block or water bath at 90°C.
2. 2-mL Plastic centrifuge tubes or other appropriately sized container.
3. Lactophenol solution as in Subheading 2.5.
4. 95% Ethanol, to be mixed in 2:1 ratio with lactophenol solution.
5. Aniline Blue solution: Add 2.61 g K_2HPO_4 to 80 mL dH_2O. Adjust pH to ~8.4 with HCl. Add 0.01 g Aniline Blue, then add dH_2O to a final volume of 100 mL. Store in brown bottle at room temperature.
6. 50% Glycerol in deionized water, autoclaved, and stored at room temperature.
7. Microscope slides, No. 2 cover slips (24 mm × 50 mm), clear nail polish.
8. Compound microscope with camera, computer, and image processing software (see Note 10).

3. Methods

Although *H. arabidopsidis* cannot be grown on synthetic media, it can reproduce vigorously on susceptible *Arabidopsis* with a life cycle of 5–7 days. Sporangia (asexual spores) can be readily collected in water and sprayed onto fresh plants to produce more inoculum. Below are procedures that take advantage of these attributes for propagation of *H. arabidopsidis* cultures.

3.1. Propagation

1. Follow the following procedures to prevent "escape" of *H. arabidopsidis*:
   (a) Keep infected plants covered unless they are inside the hood.
   (b) Clean the hood and all instruments with 70% ethanol when finished.
   (c) Do not carry infected plants outside the dedicated room.
   (d) For disposal, infected plants should be transferred to an autoclave bag. This bag should be autoclaved, and disposed of in garbage.
   (e) After you have propagated *H. arabidopsidis* do not go into any other plant growth space until you have showered and changed clothes.
2. Use susceptible plants that are 2–4 weeks old for propagation (see Notes 12 and 13). We use the Ws (Wassilewskija) ecotype for propagation of Emco5, Col-0 (Columbia) for production of Noco2, Oy-0 (Oystese) for Emoy2, and Ksk (Keswick) for Hiks1.
3. These plants should not be planted densely (e.g., you should sow 10–20 in a 4-in. pot), to avoid creating an environment favorable for infection by opportunistic soil fungi.
4. A 50-mL Falcon tube or Erlenmeyer flask both work well for collecting inoculum depending on the final volume desired.
5. Fill the tube with sterile distilled H_2O (room temperature) to the desired inoculum volume. As a guideline, 5-mL covers a 4-in. pot very thoroughly, 30-mL covers one 10″×20″ flat. Lower volumes might suffice if sprayed carefully.
6. Remove one infected pot at a time from the growth chamber to collect sporulating leaves. Work quickly to collect the leaves, because spores will begin to spontaneously detach from the sporangiophores soon after the pot is moved from the high humidity environment in the growth chamber to the relatively low ambient humidity in the hood.
7. Cut off several sporulating plants and place them in the water. Avoid plants with extensive chlorotic or necrotic patches – they could be contaminated with other pathogens. Also, avoid transferring soil into the water because this could introduce contaminants (see Note 14).
8. Cover the tube and shake or vortex at medium-high setting for ~3–5 s to release the spores into the water. Remove the plant material from the tube with forceps.
9. Repeat the above steps until you think that the desired amount is collected.
10. Check spore concentration with a hemocytometer (see Note 15). Harvest more leaves or add more water to get the desired

concentration (our standard concentration is 5×10^4 spores/mL for experiments. For propagation, higher concentrations are permissible but not necessary).

11. To avoid clogging the sprayer, remove leaves and large debris by pouring the suspension through a filter of cheesecloth. Transfer inoculum to a Falcon tube with the top cut off (or to a beaker), so that the Preval sprayer tube can reach the bottom of the Falcon tube and thereby suck up all of the suspension (see Note 16).
12. Spray the suspension onto plant leaves with a Preval sprayer. Rotate the pot to spray from all four sides. Stop spraying when runoff is imminent.
13. Cover plants with a plastic dome immediately after spraying. Alternatively, individual pots can be enclosed in magenta boxes or Ziploc bags. The inside of the container should be misted with water, and/or wet paper towels should be placed in the container to maintain high humidity.
14. Grow infected plants under short days (10 h) and cool temperatures (18–20°C). Soil should be damp but not saturated.
15. Uncover plants on the day following inoculation so that water droplets can evaporate from the leaf surface (see Note 17).
16. Leave the cover off until 6 days after inoculation, then replace the cover, taking measures as described above to promote high humidity and thereby induce sporulation (see Note 18). Make sure that the openings of the cover are sealed (e.g., with tape) to maintain high humidity and prevent escape of sporangia.
17. Sporangiophores should be visible the day after the cover is replaced (Fig. 1). Collect inoculum and repeat the cycle. As a general rule, one pot produces enough inoculum to infect at least four pots of equivalent size. Substantially more inoculum can be harvested if care is taken.

3.2. Storage and Revival

Hyaloperonospora arabidopsidis cultures can be stored and revived even after several years using the procedure described below. This capacity for storage and revival allows the experimenter to suspend propagation during times in which inoculum is not needed. Once a culture has been established in your lab, it is highly advisable to make several aliquots for long-term storage, for use if your cultures expire or become contaminated.

1. Cut off heavily sporulating intact seedlings or leaves and put them into an 1.5- or 2.0-mL tube (without water or glycerol).
2. Cool the tube on ice for ~0.5 h, and then place in −80°C freezer.

3. Frozen stocks are revived by thawing the tube on ice for 30 min, collecting spores in ~0.2–0.5 mL H_2O, and inoculating recipient plants either by drop inoculation with a pipettor or spray inoculation with a Preval spray gun.
4. It is possible that only a small number of plants will sporulate, depending on viability of the frozen stock. If this is the case, then collect those few leaves in a small volume of water (i.e., 0.2 mL) adjust concentration to 5×10^4 spores/mL, and drop-inoculate individual leaves with 2–5 drops, depending on the size of the leaf (see Notes 19 and 20).

3.3. Sporangiophore Assays

Hyaloperonospora arabidopsidis asexual reproduction can be quantified by counting sporangiophores (spore-producing structures) on cotyledons or true leaves. This assay is a widely used measure of the degree to which the pathogen has colonized the plant. When performed with care and sufficient replication, this assay can accurately report subtle differences in the timing and magnitude of plant defense responses (e.g., (15)). Our standard assay is performed on cotyledons, but the procedure below can be utilized on true leaves of any size.

1. Think carefully about controls. It is particularly advisable to include at least one susceptible genotype with which you have obtained reliable results in the past, to serve as an indicator of inefficient infection. For example, we always include wild-type Col-0 in experiments involving the Emco5 isolate. This isolate typically produces 12–15 sporangiophores per cotyledon on Col-0.
2. We typically grow plants for experiments under 10-h day length, 23°C, until the time of inoculation.
3. Using a Preval sprayer, inoculate cotyledons of 7-day-old plants with a spore suspension (5×10^4 spores/mL in dH_2O, collected as described in Subheading 3.1) to the point of imminent run off.
4. Place pots with inoculated seedlings in a magenta box or other transparent container to maintain ~100% humidity. Mist the lid of the container with water, or add a paper towel soaked with water. The edges of the container do not need to be sealed.
5. Place the container in a growth chamber at 20/18°C (day/night), 10-h day length.
6. Remove the lid from the container the next day to allow water on the cotyledons to dry. Leave the plants uncovered until ~18–24 h before the plants are to be scored. Then, replace the lid of the container and seal the edges with tape to maintain high humidity and prevent escape of sporangiophores.

7. Assess asexual sporulation at 7 days postinoculation by counting conidiophores on both sides of the cotyledons and scoring the individual cotyledons (see Note 21).
8. Score at least 50 cotyledons per genotype and calculate the averages and standard error. Statistical significance can be assessed with a Student's *t*-test. Results are displayed in a table or bar graph.

3.4. Trypan Blue Staining

Trypan Blue is a cytological stain that adheres preferentially to pathogen structures and to plant cells with damaged plasma membranes. As such, staining with Trypan Blue is excellent for visualizing the timing and extent of plant cell death (e.g., the hypersensitive response) as well as the extent to which vegetative pathogen growth has occurred inside infected plant organs. In addition, pathogen ultrastructure (i.e., hyphae and haustoria, Fig. 1) can be visualized at high magnification, if care is taken for efficient destaining. Finally, the protocol below is applicable to assays with *P. syringae* or any situation in which it is desirable to visualize plant cell death.

1. Inoculate plants with *H. arabidopsidis* as described in Subheading 3.4.
2. Preheat temperature block or water bath to 90°C.
3. Make sufficient alcoholic Trypan Blue solution by mixing one part Trypan Blue solution (see Subheading 2.1) with two parts 95% ethanol. Aliquot this solution into 2-mL centrifuge tubes or other appropriate sized tubes, so that there is enough solution to completely cover the plant tissue.
4. Place whole seedlings (or separated leaves/roots) into the tube. Avoid carrying soil into the tube.
5. Place tubes in 90°C temperature block for 3 min (see Note 22) and then incubate at room temperature for five additional minutes.
6. Remove alcoholic Trypan Blue solution and discard as phenol waste. Immediately add chloral hydrate solution to cover. This solution may require several changes over approximately 48 h – change if the solution becomes dark blue. Placing the tubes in a 55–65°C water bath for a few hours may help with clearing.
7. Discard chloral hydrate solution according to regulatory guidelines at your institution. Add 50% glycerol to cover the sample. Allow at least 0.5 h for equilibration before proceeding to step 8.
8. Mount the samples on slides and seal the cover slips in place with clear nail polish. It is possible to mount 10–20 seedlings on a single slide.

9. Visualize interaction sites with the microscope. Individual interaction sites can be scored quantitatively, as described in ref. 16.

3.5. Destaining Without Chloral Hydrate (Adapted from Ref. 17)

This protocol provides an alternative for tissue destaining that does not require chloral hydrate. However, destaining is significantly less efficient with this procedure. It is possible to distinguish plant cell death lesions, but pathogen structures usually cannot be clearly resolved.

1. Stain plants with alcoholic lactophenol Trypan Blue as described in steps 1–5 of Subheading 3.4.
2. Following the 5-min incubation at room temperature (Subheading 3.4, step 5), replace the Trypan Blue staining solution with a mixture of one part alcoholic lactophenol (without Trypan Blue, Subheading 2.5) and two parts 95% ethanol.
3. Allow plants to destain overnight.
4. If no destaining has occurred after an overnight incubation, then heat the samples to 90°C for 1 min.
5. Replace the destaining solution with 50% glycerol and prepare slides for visualization as described in steps 7–9 of Subheading 3.4.

3.6. Diaminobenzidine Staining (Adapted from Ref. 18)

This protocol enables visualization of H_2O_2 production, as can occur during plant immune responses such as effector-triggered immunity. It is possible to distinguish the cellular location of H_2O_2 production at very high resolution (Fig. 1). In addition, hyphae and haustoria from *H. arabidopsidis* can also be visualized using Nomarski optics.

1. Inoculate plants with *H. arabidopsidis* as described in Subheading 3.4.
2. Add DAB staining solution to wells in a multiwell plate.
3. Collect plants at the appropriate time point by gently extracting each plant from the soil, and immediately removing the root and placing the cut stem in DAB staining solution.
4. Place inside a light-blocking container, with wet paper towels to maintain high humidity.
5. Incubate in the dark for 6 h at room temperature.
6. Immerse samples in ELG solution until all chlorophyll is removed.
7. Place samples in chloral hydrate for 1–1.5 days.
8. Place sample in glycerol and mount on slides as described in steps 7–9 of Subheading 3.4.

3.7. Aniline Blue/ Autofluorescence (Adapted from Ref. 19)

This procedure can be used to visualize two cytological makers of plant defense: Production of autofluorescent metabolites and production of callose, a 1,4-beta-glycan that is an integral component of papillae that are thought to serve as a structural barrier against pathogen invasion.

1. Inoculate plants with *H. arabidopsidis* as described in Subheading 3.4.
2. Preheat temperature block or water bath to 90°C.
3. Make sufficient alcoholic lactophenol solution by mixing one part lactophenol solution with two parts 95% ethanol. Aliquot this solution into 2-mL centrifuge tubes or other appropriate sized tubes, so that there is enough solution to completely cover the plant tissue.
4. Place whole seedlings (or separated leaves/roots) into the tube. Avoid carrying soil into the tube.
5. Place tubes in 90°C temperature block for 3 min (see Note 22) and then incubate at room temperature overnight. Change the lactophenol solution if necessary, until the tissue is clear. Discard as phenol waste.
6. Replace the lactophenol with 50% ethanol, and soak for 15 min.
7. Replace the 50% ethanol with dH_2O and soak for 15 min.
8. To observe autofluorescence, mount the samples in 50% glycerol and proceed to step 9.
9. For callose staining, replace the water with Aniline Blue solution, and soak for 1 h at room temperature.
10. Autofluorescence and callose can both be observed with a DAPI/UV filter on any compound microscope equipped for epifluorescence.

4. Notes

1. It is ideal to have segregated lab space and plant growth chambers that are dedicated to experiments with *H. arabidopsidis*. For example, our lab has an ~10″×20″ room that contains a hood and two refrigerator-sized growth chambers that are used exclusively for pathogen experiments. We believe that escapes are unlikely under normal laboratory growth conditions for *Arabidopsis*, which are typically 3–5° higher and with much lower humidity than is optimal for downy mildew disease. However, physical segregation, at the very least, provides peace of mind for the investigator and for her/his

neighboring colleagues who might otherwise worry about contamination of their experiments. On a related note, we find that powdery mildew outbreaks, caused by true fungi, are common in Virginia and North Carolina. Powdery and downy mildews are easily distinguished based on fruiting body structure. If powdery mildews are endemic in your area, it is worthwhile to learn the morphological distinctions between these two diseases, to guard against contamination and to convince yourself and your colleagues that the "outbreak in the greenhouse" did not originate in your laboratory.

2. We have found Percival Scientific CU36L chambers to be effective, durable, and space efficient. It is not necessary to humidify the chamber, because sufficient humidification is provided by enclosing the plants with plastic domes or other containment.
3. We use Sunshine Mix 1 with good results. We are biased against soil formulations with sand because we suspect that the sand might increase the risk of fungal contamination. However, we have not empirically tested this suspicion.
4. Visors with built-in magnifying glasses (e.g., Optivisor, Donegan Optical Company) provide convenient, hands-free use and sufficient magnification for most applications. Hand-held magnifying glasses and dissecting microscopes can also be used.
5. 2-mL "Bullet tubes" are suitable for 10–20 seedlings. For larger samples, Falcon tubes can be used.
6. This solution contains phenol. Use only in an approved chemical fume hood, with appropriate protective gear and clothing.
7. Choral Hydrate is a narcotic. Use only in an approved chemical fume hood, with appropriate protective gear and clothing.
8. Chloral hydrate is a Class IV controlled substance in the USA. Therefore, it may be necessary to obtain licenses from the Federal Drug Enforcement Agency and the DEA in your state. The most important aspects of licensure are to carefully inventory the stocks and to tightly restrict access to this substance for all unauthorized personnel. After the licenses are granted, they must be renewed yearly for a fee (~$100–120 for each). Annual or biannual inspections may also take place. We find that the inconvenience and expense are well worthwhile, because we have tried several alternatives for destaining/clearing plant tissue, and none work as well as chloral hydrate. However, we describe an alternative approach, with limited applicability, in Subheading 3.5.
9. This can be accelerated by heating (e.g., in a 65°C water bath). However, this should be done inside a fume hood.

10. We use a Zeiss Axioimager M1 and a Zeiss Axiocam MRm.
11. DAB is highly toxic. Use it only in an approved chemical fume hood, with appropriate protective gear and clothing.
12. The best practice is to use naturally susceptible *Arabidopsis* ecotypes for propagation, because any given ecotype is likely resistant to most other pathogens and therefore less likely to enable contamination of the culture by other plant-associated microbes. We occasionally use *Arabidopsis* mutants with enhanced disease susceptibility (e.g., *eds1*) when we need to produce copious amounts of spores for large-scale experiments (e.g., forward genetic screens). Such mutants provide significantly higher spore production compared to wild-type, susceptible ecotypes. However, these mutants can be susceptible to other pathogens (e.g., soil-borne fungi), and therefore could facilitate contamination of the culture.
13. It is possible to simultaneously maintain two or more cultures, but care must be taken to avoid cross-contamination. Ideally, each isolate should be grown in separate chambers, and propagated on different days. Propagating on ecotypes that are susceptible to one isolate but not the other that is being propagated can further reduce chances of cross-contamination. Conversely, use of *eds* mutants (see Note 12) should be avoided while maintaining multiple *H. arabidopsidis* isolates, because these mutants are typically susceptible to most or all isolates and could easily become cross contaminated.
14. Because propagation is not conducted under axenic conditions, the potential always exists for contamination of the culture with fungi or bacteria that are introduced via the soil and then amplified during subsequent cycles of propagation. This risk can be minimized but not eliminated by maintaining healthy plants prior to inoculation, and following steps to avoid contamination described in this chapter. Importantly, downy mildew disease does not produce symptoms other than light to moderate chlorosis and sporangiophore production (which appears "downy"). Aberrant plant symptoms (e.g., rotting or dry, papery lesions) in a significant proportion of infected plants are a warning that the culture could be contaminated. If contamination is detected, we terminate the contaminated culture and start a fresh culture from frozen stock.
15. This step is performed according to the instructions on the hemocytometer. For accurate measurements, it is important to swirl the tube just before you withdraw the aliquot for counting, because spores will quickly settle to the bottom of the tube.
16. Inoculate plants immediately after you have finished collecting the inoculum. Spores will begin to germinate in water within

0.5–1 h after they are collected, and the culture will lose infectivity.

17. It is important to allow the water from the inoculation to evaporate for two reasons: first, the water could promote the growth of contaminating microbes on leaf surface. Second, sporulation is inhibited by water on the cotyledon/leaf surface.
18. High humidity is required for efficient sporulation. It is possible to induce sporulation as early as 5 days after inoculation by covering the plants (and thereby raising the humidity). We prefer to propagate in 7-day cycles.
19. When reviving a culture, it is more effective to inoculate a relatively small number of leaves with a relative high dose of inoculum, even if this means that only a few leaves are inoculated. In our experience, inoculating a relatively large number of leaves with a relatively low (dilute) inoculum often results in zero sporulating leaves.
20. It is not advisable to conduct experiments using frozen material directly, because long-term storage reduces spore viability. Plan on at least two cycles of propagation (Subheading 3.1) following revival, to produce fresh spores in sufficient quantities for accurate experiments.
21. It is difficult to accurately count sporangiophores when 15 or more are present on one side of a cotyledon. If the objective of the experiment is to distinguish subtle alterations in susceptibility, then it would be advisable to consider a procedure by which spores are counted, e.g., as in ref. 10. We are currently developing an quantitative PCR technique to assay pathogen DNA as a proxy for pathogen biomass.
22. This step is potentially dangerous: If overheated, the tube could explode or the solution could boil over when the lid is removed. To minimize this risk, make sure that the temperature block or water bath is no higher than 90°C. In addition, cap the tubes loosely or make several pinholes in the tube cap, using a needle heated with a Bunsen burner. Finally, wear appropriate protective eyewear, gloves, and clothing, and conduct all steps inside a certified chemical safety hood.

Acknowledgments

We thank Devdutta Deb for critical reading and the National Science Foundation for funding.

References

1. Holub, E. B. (2008) Natural history of *Arabidopsis thaliana* and oomycete symbioses, *Eur J Plant Pathol* **122**, 91–109.
2. Slusarenko, A. and Schlaich, N. (2003) Downy mildew of *Arabidopsis thaliana* caused by *Hyaloperonospora parasitica* (formerly *Peronospora parasitica*), *Mol Plant Pathol* **4**, 159–70.
3. Gisi, U. (2002) *in* "Advances in downy mildew research" (Spencer-Phillips, P., Gisi, U., and Lebeda, A., Eds.), pp. 119–60, Kluwer Academic Publishers, Dordrecht.
4. Clark, J. and Spencer-Phillips, P. (2000) *in* "Encyclopedia of microbiology" (Lederberg, J., Ed.), Vol. 2, pp. 117–29, Academic Press, Inc, San Diego.
5. Dick, M. (2001) Straminipilous fungi: Systematics of the peronosporomycetes including accounts of the marine straminipilous protists, the plasmodiophorids and similar organisms, Kluwer Academic Publishers, Dordrecht/Boston/London.
6. Voegele, R. T. and Mendgen, K. (2003) Rust haustoria: Nutrient uptake and beyond, *New Phytol* **159**, 93–100.
7. Staples, R. (2001) Nutrients for a rust fungus: The role of haustoria, *Trends Plant Sci* **6**, 496–98.
8. Whisson, S. C., Boevink, P. C., Moleleki, L., et al. (2007) A translocation signal for delivery of oomycete effector proteins into host plant cells, *Nature* **450**, 115–8.
9. Holub, E. B. (2001) The arms race is ancient history in *Arabidopsis*, the wildflower, *Nat Rev Genet* **2**, 516–27.
10. Feys, B. J., Moisan, L. J., Newman, M. A., and Parker, J. E. (2001) Direct interaction between the *Arabidopsis* disease resistance signaling proteins, *eds1* and *pad4*, *EMBO J* **20**, 5400–11.
11. Rehmany, A. P., Grenville, L. J., Gunn, N. D., et al. (2003) A genetic interval and physical contig spanning the *Peronospora parasitica* avirulence gene locus *Atr1nd*, *Fungal Genet Biol* **38**, 33–42.
12. Bittner-Eddy, P., Allen, R., Rehmany, A., et al. (2003) Use of suppressive subtrative hybridization to identify downy mildew genes expressed during infection of *Arabidopsis thaliana*, *Mol Plant Pathol* **4**, 501–8.
13. Rehmany, A. P., Gordon, A., Rose, L. E., et al. (2005) Differential recognition of highly divergent downy mildew avirulence gene alleles by *RPP1* resistance genes from two *Arabidopsis* lines, *Plant Cell* **17**, 1839–50.
14. Allen, R. L., Bittner-Eddy, P. D., Grenvitte-Briggs, L. J., et al. (2004) Host-parasite coevolutionary conflict between *Arabidopsis* and downy mildew, *Science* **306**, 1957–60.
15. McDowell, J. M., Cuzick, A., Can, C., et al. (2000) Downy mildew (*Peronospora parasitica*) resistance genes in *Arabidopsis* vary in functional requirements for *NDR1*, *EDS1*, *NPR1* and salicylic acid accumulation, *Plant J* **22**, 523–9.
16. Aviv, D. H., Rusterucci, C., Holt, B. F. III, et al. (2002) Runaway cell death, but not basal disease resistance, in *lsd1* is SA- and *nim1/npr1*-dependent, *Plant J* **29**, 381–91.
17. Rate, D. N., Cuenca, J. V., Bowman, G. R., et al. (1999) The gain-of-function *Arabidopsis* acd6 mutant reveals novel regulation and function of the salicylic acid signaling pathway in controlling cell death, defenses, and cell growth, *Plant Cell* **11**, 1695–708.
18. Thordal-Christensen, H., Zhang, Z., Wei, Y., and Collinge, D. (1997) Subcellular localization of H_2O_2 in plants. H_2O_2 accumulation in papillae and hypersensitive response during the barley–powdery mildew interaction, *Plant J* **11**, 1187–94.
19. Dietrich, R. A., Richberg, M. H., Schmidt, R., et al. (1997) A novel zinc-finger protein is encoded by the *Arabidopsis lsd1* gene and functions as a negative regulator of plant cell death, *Cell* **88**, 685–94.

Chapter 13

Assaying Effector Function in Planta Using Double-Barreled Particle Bombardment

Shiv D. Kale and Brett M. Tyler

Abstract

The biolistic transient gene expression assay is a beneficial tool for studying gene function in vivo. However, biolistic transient assay systems have inherent pitfalls that often cause experimental inaccuracies such as poor transformation efficiency, which can be confused with biological phenomena. The double-barreled gene gun device is an inexpensive and highly effective attachment that enables statistically significant data to be obtained with one-tenth the number of experimental replicates compared to conventional biolistic assays. The principle behind the attachment is to perform two simultaneous bombardments with control and test DNA preparations onto the same leaf. The control bombardment measures the efficiency of the transformation while the ratio of the test bombardment to the control bombardment measures the activity of the gene of interest. With care, the ratio between the pair of bombardments can be highly reproducible from bombardment to bombardment. The double-barreled attachment has been used to study plant resistance *(R)* gene-mediated responses to effectors, induction and suppression of cell death by a wide variety of pathogen and host molecules, and the role of oömycete effector RXLR motifs in cell reentry.

Key words: Transient expression, Biolistic, Double-barreled gene gun, Effectors, *R* gene, Avirulence, Soybean, Tobacco, BAX, PAMP-triggered immunity, Effector-triggered immunity, Suppression of PCD

1. Introduction

Biolistic transformation has been used extensively to create stable and transiently genetically modified plant, animal, fungal, and bacterial cells (1). This process has been used in a variety of applications, such as the creation of genetically modified crops (2), the delivery of DNA vaccines into animals (3), and the transformation of mitochondria and chloroplasts (4, 5). Biolistic transformation has also been used to study the transient expression of individual gene products in vivo (6, 7). Biolistic transient gene

John M. McDowell (ed.), *Plant Immunity: Methods and Protocols*, Methods in Molecular Biology, vol. 712,
DOI 10.1007/978-1-61737-998-7_13, © Springer Science+Business Media, LLC 2011

expression has been particularly useful for the characterization of plant pathogen effector proteins (6–8). This application uses cells that are biolistically transformed with a reporter gene and the gene(s) of interest using a particle delivery system such as a gene gun. The most commonly used reporter gene encodes beta-glucuronidase (GUS); when expressed in a living cell, GUS produces a blue precipitate (indigo blue) in the region of the transformed cells in the presence of X-Gluc and an oxidizing agent, such as potassium ferri(III) cyanide that accelerates the reaction (9). In the presence of genes that encode inducers of cell death, fewer viable cells expressing GUS are produced and so fewer blue tissue patches (spots) are observed in comparison to a control (a process called ablation). Taking this approach one step further, when a GUS reporter together with an inducer of cell death are used as the control, the ability to suppress the cell death of a protein encoded by a third added gene can be measured. In summary, biolistic delivery can be used for transient assays of effector avirulence activity (induction of *R* gene-dependent plant cell death) and virulence activity (suppression of *R* gene-dependent plant cell death). Additionally, biolistic delivery has also been used to study the structure and function of host-targeting sequences in oömycete effector proteins (7, 8).

When using conventional bombardment assays, large sets of replicates are required to produce precise and accurate results due to the high variability between individual assays, especially in three-gene assays. This variability is attributable to fluctuations in transformation efficiency caused by inconsistent particle acceleration, as well as the natural variations in tissue physiology among samples (e.g., individual leaves). In the most commonly used form of the gene gun, a large helium burst produced by a rupture disk that breaks at an approximate pressure is used to induce particle acceleration (10). The strength of this burst appears to be quite variable between bombardments. The double-barreled attachment was designed to circumvent the variability associated with the gene gun.

The concept behind the double-barreled attachment (Fig. 1c) is to allow for the simultaneous bombardment of two different preparations of particle-bound DNA, thereby controlling for variations in transformation efficiency from bombardment to bombardment (Fig. 1b). The first preparation is most commonly a control consisting of a reporter gene (GUS) that measures the number of transformed cells. The second preparation consists of the reporter gene and a gene of interest that is expected to alter the number of viable transformed cells, for example, an avirulence gene that encodes an inducer of cell death. The simultaneous bombardment of the two preparations avoids the variability associated with particle acceleration, since both particle preparations are bombarded at the same time. The ratio between the variable

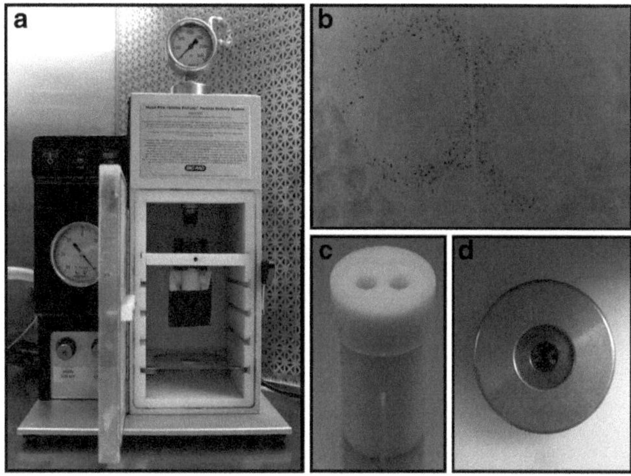

Fig. 1. Double-barreled gene gun attachment. (**a**) Double-barreled gene gun attachment setup inside of PDS-1000/He. (**b**) Leaf after co-bombardment with two control preparations, incubation, staining, and destaining. (**c**) Double-barreled gene gun attachment (not shown is the dividing screen to separate overlap between bombardments). (**d**) Alignment of macrocarrier with preparations in the PDS-1000/He with double-barreled gene gun attachment.

bombardment and the control is highly reproducible from assay to assay. Therefore, a treatment bombardment that produces more or fewer blue spots than a control bombardment is more likely to result from a biological phenomenon than from variation in transformation efficiency. The greatly increased consistency allows statistically significant results to be obtained in less than one-tenth the number of assays performed using conventional biolistic transformation. Therefore, it becomes feasible to measure much smaller variations in cell killing, such as variations resulting from mutations in a death-inducing or death-suppressing gene.

The double-barreled gene gun attachment we have developed fits into the existing Bio-Rad PDS-1000/He (Fig. 1a) (see Note 1). The attachment replaces the internal spacer rings nested inside of the support manifold. The single opening that particles travel through in the PDS-1000/He is replaced by a double opening (i.e., two barrels) that channel the control and variable preparations of particles into separate regions of the sample (usually a leaf) (see Note 2). To date, the double-barreled gene gun attachment has been used for a variety of published studies on plant *R* gene-mediated responses to effectors, induction and suppression of cell death by a variety of pathogen and host molecules, and functional studies of the RXLR-dEER motif from oömycete effectors (7, 8). However, this attachment can in principle be used to minimize variability in any gene function assay that utilizes biolistic delivery.

2. Materials

2.1. Hardware and Disposables

1. PDS-1000/He (Bio-Rad Inc., Hercules, CA).
2. Double-barreled attachment (Fig. 1c).
3. Dissecting microscope with minimum 10× magnification.
4. Hand held counter.
5. Soil (Miracle Grow Potting Mix).
6. Trays (Griffin Greenhouse).
7. Pots (Griffen Greenhouse).
8. Soybean seeds.
9. Qiagen maxi prep kit (Qiagen Inc., Valencia, CA).
10. M-10 tungsten microcarriers (Bio-Rad Inc., Hercules, CA).
11. Macrocarrier (Inbio Gold Inc., Victoria, Australia).
12. Macrocarrier holder (Bio-Rad Inc., Hercules, CA).
13. Rupture disks 650 PSI (Soybean and Tobacco) (Inbio Gold Inc., Victoria, Australia).
14. Stopping screens (Inbio Gold Inc., Victoria, Australia).
15. Petri dishes.
16. Whatman filter paper 70 mm (Fisher Scientific, Pittsburgh, PA).
17. Nescofilm (parafilm).
18. 100% Methanol.

2.2. Solutions

1. Spermidine: 25 μL aliquots (Sigma-Aldrich Inc., St. Louis, MO) in 0.6 mL eppendorf tubes. Make 30 aliquots and store at 4°C in a container with drierite (W.A. Hammond Drierite Co. LTD, Xenia, OH).
2. 2.5 M sodium chloride (NaCl)*: Dissolve 7.305 g of NaCl in 30 mL of sterile deionized water (dH_2O) and increase the final volume to 50 mL.
3. LB Plates: 5 g tryptone, 5 g sodium chloride, 2.5 g yeast extract, 7.5 g agar, raise the volume to 500 mL with water. Autoclave the media for 40 min on a liquid cycle. Cool the media until it can be held then pour approximately 25 mL into sterile Petri dishes in a sterile hood. Allow plates to solidify and store at 4°C until use.
4. 0.2 M potassium phosphate monobasic* (KH_2PO_4): Dissolve 12 g anhydrous KH_2PO_4 (Sigma-Aldrich, St. Louis, MO) in 400 mL of sterile dH_2O. Raise the volume to 500 mL. Filter Sterilize*.

5. 0.2 M potassium phosphate dibasic (K_2HPO_4)*: Dissolve 14.2 g anhydrous K_2HPO_4 (Sigma-Aldrich, St. Louis, MO) in 400 mL of sterile dH_2O. Raise the volume to 500 mL.
6. 0.1 M potassium ferricyanide(III)** ($K_3[Fe(CN)_6]$): Dissolve 1.65 g $K_3[Fe(CN)_6]$ (Sigma-Aldrich, St. Louis, MO) in 50 mL sterile dH_2O. Highly light sensitive.
7. 0.1 M potassium hexacyanoferrate(II) trihydrate** ($K_4[Fe(CN)_6]3H_2O$): Dissolve 2.11 g of $K_4[Fe(CN)_6]3H_2O$ (Sigma-Aldrich, St. Louis, MO) in 50 mL sterile dH_2O. Highly light sensitive.
8. 0.5 M sodium ethylenediaminetetraacetic acid (Na_2EDTA)*: Dissolve 18.612 g of Na_2EDTA (Fischer Scientific, Pittsburgh, PA) in 50 mL of sterile dH_2O. Raise the volume to 100 mL.
9. Staining Solution (500 mL): Prepare 200 mL of 0.2 M potassium phosphate buffer pH 7.0* in a sterile bottle with 124 mL 0.2 M K_2HPO_4* and 76 mL 0.2 M KH_2PO_4*. Add 177 mL sterile dH_2O to the 200 mL stain solution. Add 2 mL of 0.1 M $K_3[Fe(CN)_6]$** and 2 mL of 0.1 M $K_4[Fe(CN)_6]3H_2O$** to the 377 mL stain solution. Wrap the stain solution bottle in aluminum foil as it is light sensitive. Add 8 mL 0.5 M Na_2EDTA to the 381 mL stain solution. Mix the solution thoroughly. Dissolve 400 mg of X-Gluc (5-bromo-4-chloro-3-indolyl-beta-D-glucuronic acid; Gold BioTechnology, St. Louis, MO) in 8 mL of DMSO, mix, and immediately add to the 389 mL stain solution. Mix the stain solution well. X-Gluc is highly light sensitive. Add 100 mL of 100% Methanol to the 397 mL stain solution and mix. Add 3 mL of 10% Triton X-100 (Sigma-Aldrich, St. Louis, MO) to the 497 mL stain solution and mix thoroughly. Store the solution at 4°C. It is stable for up to 1 year when stored in the dark.

*Filter sterilized (0.22 µm).

**Contact with acids liberates very toxic cyanide gas. Read Material Safety Data Sheet (MSDS) before handling chemical.

3. Methods

3.1. Experimental Design

This section describes assays for induction of cell death and suppression of cell death. The assay for induction of cell death consists of 16 co-bombardments per experimental replicate. A co-bombardment comprises two simultaneous bombardments using the double-barreled gene gun attachment. If the inducer of cell death being used is independent of plant genotype (for example, DNA encoding the mouse BAX that induced cell death equally well on all tested soybean lines), then 16 co-bombardments (16 pairs) are recommended to achieve a high level of statistical reliability (see Note 3). If the leaf is of sufficient size, then two co-bombardments may be fired onto the same leaf, with the preparation containing BAX being fired through one barrel and the control preparation being fired through the other barrel. The leaf is then rotated 180° so that the preparations are bombarded on the alternate side of the leaf for the second bombardment (Fig. 2a). This is done for the first four co-bombardments. For the next four co-bombardments, the BAX preparation and control preparation should be fired through the other barrel (Fig. 2a). The preparations are reversed so that not all control preparations are fired on only one side of the leaf throughout the experiments. By switching the preparations the control and variable bombardments are delivered to all sides of the target area an even number of times throughout the experiment. These eight co-bombardments should then be repeated to give a total of 16 co-bombardments. These 16 co-bombardments are equivalent to one experimental replicate. Ablation due to the plant cell death response can then be calculated as 1 – (treatment spot number)/(control spot number).

If the inducer of cell death being used is dependent on plant genotype, for example, the *Phytophthora sojae Avr1b* protein that is dependent of the presence of the soybean *R* gene *Rps1b* to trigger cell death, then leaves from plants lacking the *Rps1b* gene may be used as an additional control. Following the *Rps1b* example, eight co-bombardments would be performed on the leaves expressing *Rps1b*, and eight co-bombardments would be performed on leaves lacking *Rps1b*, arranged as follows: four co-bombardments on *Rps1b* leaves followed by four co-bombardments on *Rps* leaves with treatment (*Avr1b*) and control fired through the two barrels. The treatment and control barrels are then switched and the eight co-bombardments are repeated (Fig. 2b). Ablation due to the plant cell death response can then be calculated as 1 – [(treatment spot number in presence of gene)/(control spot number in presence of gene)]/[(treatment spot number in absence of gene)/(control spot number in absence of gene)].

For assays to test whether an effector can suppress cell death (e.g., triggered by BAX), we recommend carrying out both a direct and an indirect assay (see Note 3). In the direct assay, one barrel contains a suppressor preparation (GUS + cell death trigger + suppressor) and the other barrel contains the cell death preparation (GUS + cell death trigger + empty vector control). In the indirect assay, two assay results are compared to determine the relative ablation: a suppressor preparation (GUS + cell death trigger + suppressor) versus a control preparation (GUS + control) are compared to the cell death preparation (GUS + cell death trigger + control) versus the control preparation (GUS + control).

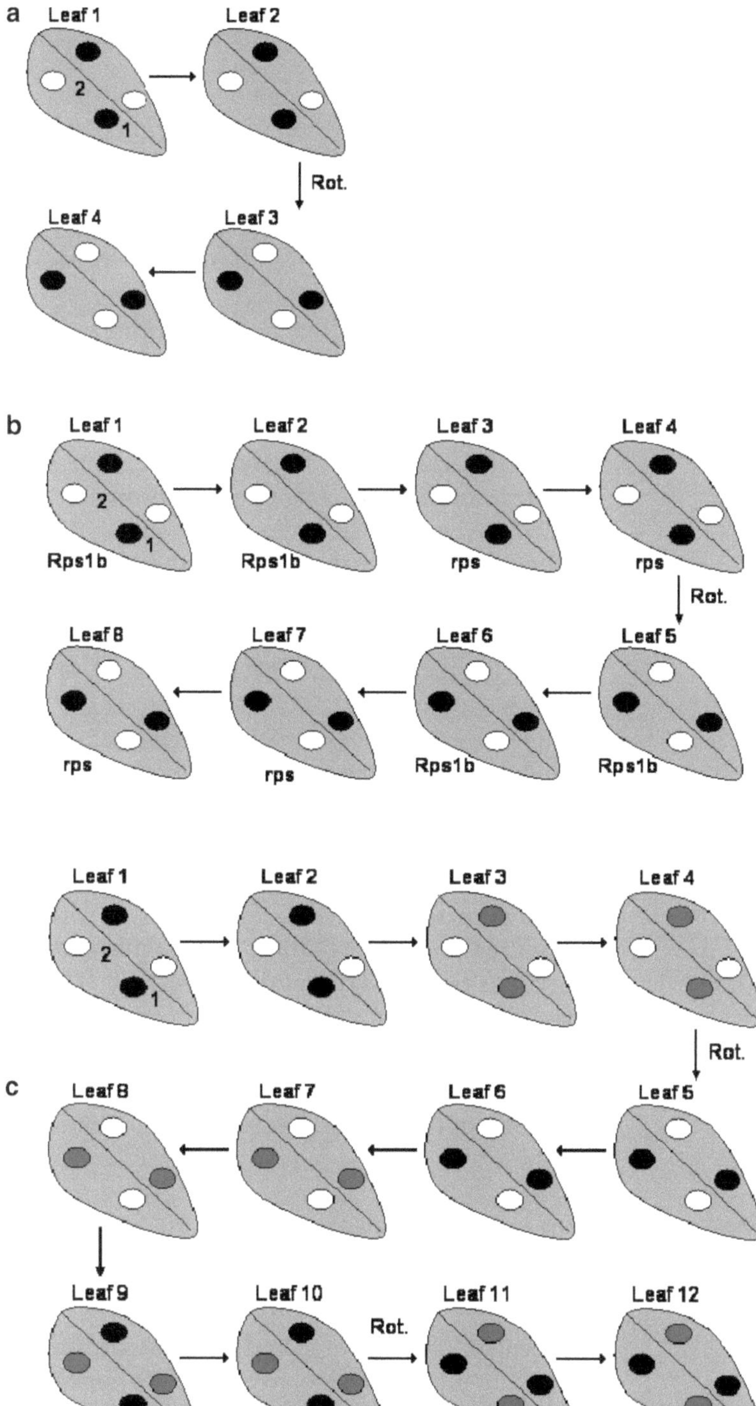

Fig. 2. Order of co-bombardments for various types of experiments. (**a**) Inducer of cell death independent of plant genotype. (**b**) Induction of cell death based on plant genotype, using the *Avr1b-Rps1b* interaction as an example. (**c**) Suppression of cell death. White circles are bombardments of the control preparation; black circles are bombardment of preparations containing an inducer of cell death. Gray circles are bombardments of preparations containing the inducer of cell death and the suppressor of cell death. 1 signifies that first co-bombardment. 2 signifies the second co-bombardment. The two co-bombardments on a leaf are switched by rotating the leaf. Rot. signifies rotating/switching the preparations so that they are fired through the other barrel.

16 co-bombardments are required for the two parts of the indirect assay, while 8 co-bombardments are required for the direct measurement (Fig. 2c). Preparations are fired through alternate barrels every eight co-bombardments in the indirect assay and every four bombardments in the direct assay. Leaves are rotated as described in the inducer of cell death assay between the two co-bombardments for each leaf.

The procedure described below is optimized for assays to test the function of effectors from *P. sojae* on leaves of soybean (*Glycine max*).

3.2. Plasmid Design

1. Plasmids are designed for transient expression under the control of the cauliflower mosaic virus 35s promoter and terminator. pUC19 serves as the vector backbone so that the bombardment plasmid is small in size and therefore easier to prepare.
2. Genes can be cloned into the bombardment vector using these template primer designs. Forward primer 5′AGAT <u>CCCGGG</u>GGAGCAATGAGAT**ATG**.... 3′. The reverse primer consists of 5′....**TGA**<u>GGTACC</u>catgc 3′ (7). Underlined sequences represent restriction sites used for cloning. Bold sequences represent actual gene sequence.
3. The PCR product and bombardment vector should be digested with XmaI and KpnI and then ligated (7).

3.3. Soybean Growth

1. Approximately 2 weeks before the assay is to be performed, plant ten pots of soybeans in the greenhouse, with five plants per pot. Set the day length for 14 h at 28°C and the night length for 10 h at 25°C. Water as required.
2. Select the first (monofoliate) true leaves for bombardment 14–18 days after planting. If the monofoliate leaves are too old or too young, the transformation efficiency will drop drastically. Leaves must be in very good health for bombardment.

3.4. DNA Purification

1. Prepare two 4-L flasks with 500 mL of LB and autoclave for 60 min.
2. Add sterile ampicillin to reach a final concentration of 100 μg/mL.
3. Inoculate flasks from a freshly streaked plate of *Escherichia. coli* containing the plasmid and gene of interest.
4. Grow culture in a shaker at 240 rpm at 37°C for approximately 14 h.
5. Harvest cells at 6,000 × *g* for 15 min.
6. Perform purification using Qiagen Maxi Prep kits. 500 mL of culture pellet should receive 10 mL of P1, 10 mL of P2, 10 mL P3, and should follow the maxi prep protocol. Columns can be used twice. Columns must be washed twice

with QBT between each purification if they are being reused (see Note 4).

7. Dried DNA should be resuspended in water and set to a final concentration between 3 and 5 μg/μL. Concentration can be determined using a spectrophotometer; we use a nanodrop-1000. Measurement should be taken of a 1–20 dilution since high concentrations fall outside the linear conversion range.
8. DNA should be portioned into aliquots of roughly 25 μL and stored at −20°C.

3.5. Setup and Calibration of PDS-1000/He

1. Replace spacer rings and stop-screen support inside of the fixed nest with the double-barreled attachment.
2. Attach the rupture disk retaining cap. Attach the macrocarrier holder and the macrocarrier cover lid to the macrocarrier launch assembly. Set the spacing between the rupture retaining cap and the macrocarrier launch assembly using the 0.25 in. gap adjustment tool so that the adjustment tool just fits between the two.
3. Place the target shelf on the lowest setting in the PDS-1000/He.
4. The chamber vacuum must be able to reach 23 PSI.
5. Set the pressure on the He-tank gage to exceed no more than 200 PSI of the rupture disk pressure. Rupture disk = 650 PSI, tank pressure = 825–850 PSI.
6. Adjust the helium metering valve so that upon firing the rupture disk takes 15 s to fire.

3.6. Double-Barreled Attachment

1. Remove internal spacer rings and stop-screen support ring nested inside the fixed nest with retaining spring (see Note 5).
2. Place the double-barreled attachment into the fixed nest with retaining spring.
3. Lock the double-barreled attachment and nest with retaining spring into the adjustable brass nest. Place the adjustable brass shelf on the top shelf inside the PDS-1000/He.

3.7. Tungsten

1. Add 90 mg of tungsten to 1 mL 95% EtOH in a 1.5 mL eppendorf tube and vortex for 1 min, followed by centrifugation at 13,000 × *g* for 10 min. Remove all of the ethanol by gently pipetting. Repeat once (see Note 6).
2. Add 1 mL of sterile dH_2O to the dried tungsten and vortex for 1 min followed by centrifugation at 13,000 × *g* for 15 min. Remove all of the dH_2O by gently pipetting. Repeat once (see Note 7).

3. Add 1 mL of sterile 50% glycerol to the dried tungsten and store at –20°C. Can be stored indefinitely. Making 20–50 tubes of tungsten is recommended.

3.8. Bombardment Mixture

1. Thoroughly vortex stored tungsten (2 min) and transfer 100 μL into two 0.6 mL eppendorf tubes (Control and Test). Minimizing surface area in smaller tubes helps reduce degradation and keeps mixture homogenous (see Note 8).
2. Centrifuge tubes for 1 min at 13,000 × g and remove 50 μL of supernatant.
3. 25 μL of volume is reserved for DNA. This step is discussed in depth below in Subheadings 3.9 and 3.10. Once DNA is added, vortex thoroughly for 1 min and keep tubes on ice from here until the end of the actual bombardments (see Note 9).
4. Add 65 μL of 2.5 M NaCl, vortex thoroughly for 1 min, and place back on ice.
5. Add 4 μL of aliquot spermidine to 246 μL sterile dH_2O to make 0.1 M spermidine and vortex thoroughly.
6. Add 25 μL 0.1 M spermidine to the bombardment mixture. Vortex thoroughly for 2 min and put on ice for a minimum of 30 min. Discard 0.1 M spermidine after 1 day if not used again.
7. Particle mixture can be stored at –20°C for up to 8 h.
8. Centrifuge mixture for 1 min at 13,000 × *g* and remove 135 μL of the supernatant. Be very careful not to disturb the tungsten-DNA pellet.
9. Thoroughly vortex mixture for 2–3 min and store on ice. The mixture begins to clump after 3 h. Bombardments must be carried out immediately and usually take 2.5 h for a novice user and 1.5 h for an experienced user.

3.9. DNA Mixture for Induction of Cell Death

Two preparations are created, each with a final volume of 25 μL. Each preparation provides 16 bombardments. The first is a control preparation consisting of 50 μg of DNA encoding GUS and 50 μg of DNA encoding containing an empty vector. The second is an induction of cell death preparation consisting of 50 μg of DNA encoding GUS and 50 μg of DNA encoding an inducer of cell death. If the inducer of cell death is very strong, the amount of inducer of cell death and control pUC19 should be lowered so that an equal amount and ratio of plasmids are present in each preparation.

Example

Control preparation: 8 μL GUS (50 μg) + 10 μL pUC19 (50 μg) + 7 μL water = 25 μL total.

Test preparation 8 μL GUS (50 μg) + 9 μL pUCBomb-*Avr1b* (50 μg) + 8 μL water = 25 μL total.

3.10. DNA Mixture for Suppression of Cell Death

Three preparations are created, each with a total volume of 25 μL. Each preparation provides 16 bombardments. The first is an induction of cell death preparation consisting of 50 μg of DNA encoding GUS, 25 μg of DNA encoding an inducer of cell death (BAX), and 50 μg of control DNA (pUC19). The second is a cell death suppressor preparation consisting of 50 μg DNA encoding GUS, 25 μg DNA encoding an inducer of cell death (BAX), and 50 μg of DNA encoding a suppressor of cell death (*Avr1b*). The control preparation consists of 50 μg of DNA encoding GUS, and 75 μg of control DNA (pUC19).

Example

Control preparation: 8 μL GUS (50 μg) + 15 μL pUC19 (75 μg) + 2 μL water = 25 μL total.

Cell death preparation: 8 μL GUS (50 μg) + 7 μL BAX (25 μg) + 10 μL pUC19 (50 μg) = 25 μL total.

Suppression preparation: 8 μL GUS (50 μg) + 7 μL BAX (25 μg) + 7 μL pUC*Avr1b* (50 μg) + 3 μL water = 25 μL total.

3.11. Bombardment

1. Place 650 PSI rupture disk inside the rupture disk retaining cap and tighten fully.
2. Load 1 μL of each preparation (control and gene of interest) onto a macrocarrier placed inside a macrocarrier holder. Vortex preparations thoroughly for (15–30 s) and pipette 1 μL onto the macrocarrier. The preparations are put immediately back on ice.
3. The macrocarrier holder is placed on top of the fixed nest with the retaining spring and the 1 μL preparations are aligned with each barrel so that they are directly in the center of each barrel opening.
4. Add the macrocarrier lid cover and tighten completely. Visually make sure that the 1 μL preparations are still aligned completely. If a shift has occurred, realign the preparations (Fig. 1d).
5. Place the microcarrier launch assembly into the PDS-1000/He.
6. Remove a leaf from the plant and place it flat on a media Petri dish. Cover the petiole portion of the leaf. The distal portion of the leaf should be shot first then the petiole portion immediately after.
7. Place the Petri dish onto the target shelf and align the sample with the barrels. The major vein of the leaf should be directly under the spacer of the double-barreled attachment and there should be adequate exposed tissue (see Note 10).

8. Close the door to the vacuum chamber and begin vacuum. Wait until vacuum reaches 23 PSI and then set the Vac/Vent/Hold switch to Hold.
9. Hold the fire button until rupture disk ruptures and then quickly release.
10. Set the Vac/Vent/Hold switch to Vent and wait until chamber depressurizes.
11. Change the rupture disk and dispose of the macrocarrier and stopping screen. Clean the barrels with a paper towel soaked in 70% ethanol. Dry double-barreled attachment using compressed air.
12. Repeat above bombardment protocol on the petiole portion of the leaf, with each of the two preparations bombarding the other side of the leaf. Do this by rotating the leaf to produce the opposite alignment of samples.
13. Once the leaf has been co-bombarded twice, place the leaf in a Petri dish containing Whatman filter paper soaked in water so that the stem is in contact with the filter paper. There should be 0.5–1 mL of free water in the Petri dish. Wrap in parafilm.
14. A bombardment experiment usually entails 16 or 24 bombardments (as described in experimental design). All bombardments for an experiment must be carried out successively as soon as possible (see Note 11).

3.12. Incubation, Staining, Destaining, and Counting

1. Store Petri dishes at 25°C (see Note 12) for approximately 72 h.
2. Completely drain water from each Petri dish.
3. Add 1 mL of stain solution to each leaf and blot using a kimwipe until the leaf is completely covered in stain solution. Add more stain if required to cover the entire leaf.
4. Rewrap Petri dishes in nescofilm and incubate at 25°C for 24 h.
5. Soak leaves in methanol and shake at 85 rpm on an orbital shaker until they fully destain. Methanol should be replaced if leaves are not fully destained. Petri dishes must be wrapped in parafilm during the destaining.
6. Perform counting using a dissecting scope and a hand held counter. If the co-bombardment fails due to misalignment, then the bombardment is not counted. An example would be having a bombardment from one barrel miss the leaf or a portion of the leaf. If the target region is not completely occupied by leaf tissue, then the co-bombardment is considered a failure (Table 1).

Table 1
Data analysis table for suppression of BAX by *Avh331*

A	B	C	D	E	F	G
Leaf	Plasmid for petiole area	GUS + petiole area	R for petiole	Plasmid for distal area	GUS + distal area	R for distal
1	BAX (25 μg) + Gus Control + Gus	25 184	–0.852	Control + Gus BAX (25 μg) + Gus	182 3	–1.660
2	BAX (25 μg) + Gus Control + Gus	18 241	–1.105	Control + Gus BAX (25 μg) + Gus	399 194	–0.312
3	Control + Gus BAX (25 μg) + Gus	262 7	–1.517	BAX (25 μg) + Gus Control + Gus	16 170	–1.003
4	Control + Gus BAX (25 μg) + Gus	304 41	–0.861	BAX (25 μg) + Gus Control + Gus		
5	BAX (25 μg) + Avh331 (50 μg) + Gus Control + Gus	156 229	–0.166	Control + Gus BAX (25 μg) + Avh331 (50 μg) + Gus	96 62	–0.187
6	BAX (25 μg) + Avh331 (50 μg) + Gus Control + Gus	138 208	–0.177	Control + Gus BAX (25 μg) + Avh331 (50 μg) + Gus	194 142	–0.135
7	Control + Gus BAX (25 μg) + Avh331 (50 μg) + Gus	194 124	–0.193	BAX (25 μg) + Avh331 (50 μg) + Gus Control + Gus	111 162	–0.163
8	Control + Gus BAX (25 μg) + Avh331 (50 μg) + Gus	272 155	–0.243	BAX (25 μg) + Avh331 (50 μg) + Gus Control + Gus	37 133	–0.547

(continued)

Table 1 (continued)

A	B	C	D	E	F	G
Leaf	**Plasmid for petiole area**	**GUS + petiole area**	**R for petiole**	**Plasmid for distal area**	**GUS + distal area**	**R for distal**
9	BAX (25 μg) + Avh331 (50 μg) + Gus	12	0.637	BAX (25 μg) + Gus	3	0.916
	BAX (25 μg) + Gus	2		BAX (25 μg) + Avh331 (50 μg) + Gus	32	
10	BAX (25 μg) + Avh331 (50 μg) + Gus	146	0.576	BAX (25 μg) + Gus	14	0.815
	BAX (25 μg) + Gus	38		BAX (25 μg) + Avh331 (50 μg) + Gus	97	
11	BAX (25 μg) + Gus	21	0.788	BAX (25 μg) + Avh331 (50 μg) + Gus	72	1.085
	BAX (25 μg) + Avh331 (50 μg) + Gus	134		BAX (25 μg) + Gus	5	
12	BAX (25 μg) + Gus	69	−0.067	BAX (25 μg) + Avh331 (50 μg) + Gus		
	BAX (25 μg) + Avh331 (50 μg) + Gus	59		BAX (25 μg) + Gus		

(A) Table is sorted by order of co-bombarded leaf. (B) DNA for co-bombardment shot in the petiole portion of the leaf. (C) Number of GUS + spots for each co-bombardment in the petiole portion. (D) Calculated *R* value for co-bombardments shot on the petiole portion of the leaf. (E) DNA for co-bombardment shot in the distal portion of the leaf. (F) Number of GUS + spots for each co-bombardment in the distal portion. (G) Calculated *R* value for co-bombardments shot on the distal portion of the leaf. Blacked out cells indicate failed co-bombardments

7. Input results of the counting into a spreadsheet (Table 1). Every two rows in the spreadsheet correspond to a leaf. The area of the co-bombardment falls under either distal or petiole. The spreadsheet rows are ordered in the same order the leaves were bombarded.

3.13. Statistical Analysis

Statistical analysis is performed by using two different statistical tests designed for nonparametric data sets (11). The Wilcoxon signed ranks test is employed for measuring the statistical significance for a direct assay of an inducer or suppressor of cell death. The direct assay directly compares cell killing by two preparations bombarded side by side in the gene gun. Indirect measurements of suppression of cell death or *Avr-R* gene interaction can be assessed for significance using the Wilcoxon rank sum test. The indirect assay directly compares cell killing by two preparations bombarded in two separate experiments that include a common control preparation as reference (Table 1).

In both cases, the log ratio of the number of blue cells is calculated for each pair of bombardments $\log[(1+b_1)/(1+b_2)]$, where b_1 and b_2 are the number of blue cells produced by each member of the paired sample. Each blue cell number is incremented by 1 to avoid zeros and to slightly decrease the influence of very small denominators; this adjustment makes the statistical tests slightly more conservative.

The Wilcoxon rank sum test is a nonparametric test used to determine the statistical difference between two small samples that need not have a normal distribution (11). Figure 3a illustrates the procedure of the test using Excel. The two sample groups BAX (column 4a-1) and *Avh331*+BAX (4a-2) are considered together (4a-3) and ranked from 1 (smallest value) to 16 (largest value) (4a-4). Ties are resolved by averaging the ranks of the tied samples. The sum of the ranks for the two sample groups (4a-5, 4a-6, 4a-7, 4a-8) is calculated separately. In this example, the sum for BAX is 27 and for BAX+*Avh331* is 91. These values are used in the following equations to calculate W_a and W_b: $W_a = (n_1 \times n_2) + ((n_2 \times (n_2 + 1))/2) - R_1$ and $W_b = n_1 \times n_2 - W_a$, where n_1 is equal to the number of values in the larger sample, n_2 is equal to the number of values in the smaller sample, and R_1 is equal to the sum of the ranks from the smaller sample (e.g., $W_a = (8 \times 7) + [7 \times (7+1)]/2 - 27 = 55$. $W_b = (7 \times 8) - 55 = 1$). The larger W value becomes the Wilcoxon's two-sample statistic to determine the statistical significance for sample size n_1 and n_2 using a statistical table (12, Table CC]. In this case, the p value corresponds to <0.001 and the null hypothesis that there is no difference between the two sample groups is rejected.

The Wilcoxon signed ranks test is designed to test for significance in small nonparametric samples (11). It is usually used to compare the differences between paired samples and determines

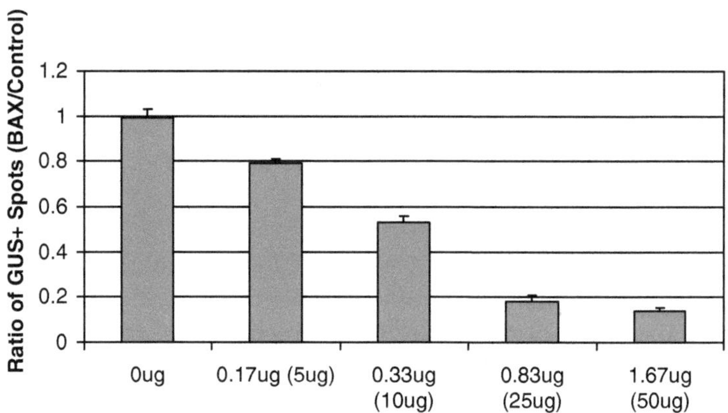

Fig. 3. Ablation of GUS activity due to cell death caused by varying concentrations of BAX. Soybean leaves were bombarded using a double-barreled gene gun attachment that delivered varying concentrations of mouse BAX DNA and GUS DNA preparation to one side of the leaf and a control preparation of empty vector and GUS DNA to the other side. The ratio of blue spots in comparison to the BAX preparation and the control preparation was calculated for varying concentrations of mouse BAX. # μg signifies the amount of BAX DNA bombarded from a barrel during a co-bombardment. (# μg) indicate overall amount of BAX DNA used to make the preparation.

if the magnitudes of the differences greater than zero are significantly different than the magnitudes of the differences less than zero. In this case, the log ratio is considered to be the relevant difference [i.e., $\log(1+b_1)-\log(1+b_2)$]. Figure 3 shows the procedure of the test using Excel. All of the ratios from the eight co-bombardments (eight pairs) are listed in a column. These ratios are the log[(inducer of cell death + 1)/(control + 1)] from each co-bombardment. These ratios are then ordered from smallest to largest regardless of their sign (positive or negative) and then ranked from one (smallest) to eight (largest, for example, if you have 16 co-bombardments, the rank for the largest number is now 16). Ties are resolved by averaging the ranks of the tied samples. If a ratio is negative, the rank gets assigned to a negative value. Figure 3b shows that there are no negative log ratios, and therefore the sum of the negative ranks is zero while the sum of the positive ratio ranks is 28. The sums of the positive and negative ranks are added such that the contribution of the negative ranks decreases the sum. In this case, the negative sum was 0 and the positive value was 28. The smaller sum is assigned to *T*. The number of samples is assigned to *n*. In this example, we had eight values, so $n=8$. The values are then looked up in a Wilcoxon signed ranks table (12, Table DD]. At $n=8$ with a $T=0$, the alpha value is 0.0156. In this example, we can thus reject the null hypothesis that there is no difference between the bombardments from either barrel with a significance value (*p*-value) of 0.0156.

The cut off for significance is usually considered to be a *p*-value of less than 0.05. Note that a *p*-value of 0.05 means there is a 5% chance that a result is called significant when there really is no difference between the samples. Thus, if 20 experiments are carried out with identical DNA preparations, one experiment may be expected to give a p value less than 0.05, just by chance. Therefore, a lower p value (e.g., 0.01) is recommended when large numbers of experiments are carried out. For a rigorous treatment of this multiple testing issue and procedures for accounting for it, see (13, 14). In our hands, the recommended numbers of replicates produce p values less than 0.001, thus multiple testing is rarely an issue.

4. Notes

1. The first embodiment of the double-barreled gene gun attachment consisted of two metal pipes that were fused to a custom made metal ring that fit inside the PDS-1000/He. The current double-barreled gene gun attachment is produced using a mold tool (Fig. 1c). Plans for large-scale commercial manufacturing are in progress. Please contact Dr. Brett Tyler (bmtyler@vt.edu) about acquiring the technology.
2. The leaf width perpendicular to the mid vein at the center of the leaf should measure a minimum 4.5 cm and should range between 4.5 and 6 cm. The length of the leaf should be measured using the mid vein. The length should be a minimum of 6.5 cm and should range between 6.5 and 8.5 cm.
3. The amount of DNA encoding an inducer of cell death should be varied to determine the ideal amount that produces an adequate level of cell death (50–90% reduction in blue spots; an example is shown in Fig. 4). Likewise, the amount of suppressor of cell death should also be determined. We find using double the amount of suppressor works well for the suppression of mouse BAX by Avr1b and Avh331.
4. The terms used to describe the various components of the PDS-1000/He are the exact names used in the user's manual. Pictures of components are included in the manual to avoid confusion. This manual is available online through Bio-rad.
5. Occasionally, a column clogs due to high viscosity of the solution after centrifugation. We do not recommend running these samples through the column since the purification time increases significantly and the quality of the purification diminishes greatly. If the pellet achieved from harvesting is larger than normal, adjust the amount of P1, P2, and P3. Do not decrease the amount of the buffer solutions, only increase them.

a Wilcoxon Rank Sum Test

5a–1 BAX	5a–2 Avh331+BAX	5a–3 Sort	5a–4 Rank	5a–5 BAX Sort	5a–6 Rank	5a–7 Avh331+BAX Sort	5a–8 Rank
−0.852	−0.166	−1.66	1	−1.66	1	−0.547	7
−1.105	−0.177	−1.517	2	−1.517	2	−0.243	9
−1.517	−0.193	−1.105	3	−1.105	3	−0.193	10
−0.861	−0.243	−1.003	4	−1.003	4	−0.187	11
−1.66	−0.187	−0.861	5	−0.861	5	−0.177	12
−0.312	−0.135	−0.852	6	−0.852	6	−0.166	13
−1.003	−0.163	−0.547	7	−0.312	8	−0.163	14
−1.04429	−0.547	−0.312	8	**N=7**	**29**	−0.135	15
9.030552	**−0.22638**	−0.243	9			**N=8**	**91**
	59.37792	−0.193	10				
		−0.187	11				
		−0.177	12	**Wilcoxon**	**55**	**Wilcoxon(2)**	**0**
		−0.166	13				
		−0.163	14	**Alpha 0.001**	**54**		
		−0.135	15				

b Wilcoxon signed ranks test

5b–1 Ratio	5b–2 Rank	5b–3	
−0.067	−1	**N**	**7**
0.637	+2	**Sum of Negative**	**1**
0.737	+3	**Sum of Positive**	**27**
0.788	+4	**T(s)**	**1**
0.815	+5	**T(s) = 1**	**Alpha 0.0156**
0.916	+6	**Log Ratio Avr1k+BAX/BAX**	**0.794**
1.085	+7	**Ratio Avr1k+BAX/BAX**	**6.216**

Fig. 4. Statistical analysis of *Avh331* suppression of BAX triggered cell death. (**a**) Wilcoxon Rank Sum Test. Nonparametric test used to measure difference between two small samples. (5a-1) Co-bombardment *R* values for induction of cell death by BAX in comparison to a control. (5a-2) Co-bombardments *R* values for suppression of BAX triggered cell death by *Avh331* in comparison to a control. (5a-3) *R* values for 5a-1 and 4a-2 combined and sorted smallest to largest. (5a-4) Rank assigned to sorted values. (5a-5 and 5a-6) Sorted *R* values for BAX bombardment and respective rank. (5a-7 and 5a-8) Sorted *R* values for BAX + *Avh331* bombardment and respective rank. (**b**) Wilcoxon Signed Rank Test. Statistical test for significance in small nonparametric samples. (5b-1) Sorted co-bombardment *R* values of suppression of BAX triggered cell death by *Avh331* in comparison to BAX triggered cell death. (5b-2) Signed ranks for co-bombardments. (5b-3) Calculations for statistical analysis.

6. The tungsten pellet requires extra centrifugation to pellet on occasion. Repeat the centrifugation step to further pellet the tungsten. The pellet is not as densely packed as it is in other aqueous solutions. Take care when slowly removing the ethanol from the top of the solution. If a small amount of residual ethanol is present after the second wash, proceed to the water washes as this will help remove any small amounts of ethanol.

7. The tungsten pellet requires extra centrifugation time to pellet on occasion. Repeat the centrifugation step to further pellet the tungsten. The pellet should be dried after water removal. In all cases, a little tungsten is removed during pipetting.

8. The homogeneity of the tungsten solution plays a significant role in the transformation efficiency of the preparations. Less tungsten in one of the two tubes causes the results from the bombardment to be skewed. From a 1 mL preparation of tungsten, we only recommend using the first 800 μL and disregarding the rest. Sometimes, the tungsten pellets at the bottom of the eppendorf tube. This pellet must be fully resuspended before transferring tungsten to the new tube. Flip the tube upside down and flick the tube; if the tungsten comes away from the bottom conical area, the tungsten is appropriately resuspended. Vortex again for 15–20 s and then pipette quickly.
9. Mixing DNA is one of the most important steps involved in this process. DNA at high concentrations is very viscous and error due to pipetting is common. DNA should be pipetted up and down several times slowly. Using DNA at very high concentrations increase variability. The best advice is to maximize volume for the DNA (e.g., 12.5 μL of GUS at 4 μg/μl and 12.5 μL pUC19 at 4 μg/μl is preferable to 5 μL GUS at 10 μg/μl , 5 μL pUC19 at 10 μg/μl , and 15 μL of water).
10. The alignment of the leaf under the bombardment is critical to the success of the assay. Wide leaves that are symmetrical in shape and size across the major vein should be used for bombardment. Bombardments are comparable only across the mid vein, not together on the same side. Transformation efficiency varies between the distal portion of the leaf and the petiole. The minimal media plate used to hold the leaves can dehydrate the bombarded area of the leaf after a short period of time since this is wounded tissue. Minimizing the amount of time, the leaf spends on the minimal media after bombardment is important.
11. The double-barreled gene gun should first be mastered by bombarding a control sample through both barrels. Initially, the user should use one preparation for both barrels. The ratio between the barrels should be approximately 1, but a 10% deviation is considered acceptable. Once successful with this bombardment requirement, the user should prepare two separate control preparations and repeat the assay. This ensures that the user can prepare the solutions for bombardment accurately.
12. Resistance through effector-triggered immunity can be temperature dependent (15). We found that 25°C is an appropriate temperature to trigger cell death through the recognition of Avr1b by Rps1b and for cell death induction by mouse BAX. A temperature that is physiologically relevant for infection should be used for incubation.

References

1. Klein, T. et al. (1992). Transformation of microbes, plants, and animals by particle bombardment. *Nature Biotechnology*, **10**, 286–291.
2. Liang, G.H. and Skinner, D.Z. (2004). Genetically modified crops: their development, uses, and risks. Haworth Press, Binghamton, New York.
3. Barry, M.A. and Johnston, S.A. (1997). Biological features of genetic immunization. *Vaccine*, **15**, 788–791.
4. Daniel, H. et al. (1990). Transient foreign gene expression in chloroplasts of cultured tobacco cells after biolistic delivery of chloroplast vectors. *Proceedings of the National Academy of Sciences of the United States of America*, **87**, 88–92.
5. Remacle, C. et al. (2005). High-efficiency biolistic transformation of Chlamydomonas mitochondria can be used to insert mutations in complex I genes. *Proceedings of the National Academy of Sciences of the United States of America*, **103**, 4771–4776.
6. Qutob, D., Kamoun, S., and Gijzen, M. (2002). Expression of a *Phytophthora sojae* necrosis-inducing protein occurs during transition from biotrophy to necrotrophy. *Plant Journal*, **32**, 361–373.
7. Dou, D., Kale, S.D. et al. (2008). Conserved C-Terminal motifs required for avirulence and suppression of cell death by *Phytophthora sojae* effector Avr1b. *The Plant Cell*, **20**, 1118–1133.
8. Dou, D., Kale, S.D. et al. (2008). RXLR-mediated entry of *Phytophthora sojae* effector Avr1b into soybean cells does not require pathogen-encoded machinery. *The Plant Cell*, **20**, 1930–1947.
9. Jefferson, R.A. (1987) Assaying chimeric genes in plants: the GUS gene fusion system. *Plant Molecular Biology Reporter* **5**, 387–405.
10. Sanford, J.C. et al. (1991). An improved, helium-driven biolistic device. *Technique*, **3**, 3–16.
11. Sokal, R.R. and Rohlf, F.J. (1969) Biometry. W.H. Freeman and Company, San Francisco.
12. Sokal, R.R. and Rohlf, F.J. (1969) Statistical tables. W.H. Freeman and Company, San Francisco.
13. Benjamini, Y. and Hochberg, Y. (1995). Controlling the false discovery rate – a practical and powerful approach to multiple testing. *Journal of the Royal Statistical Society. Series B*, **57**, 289–300.
14. Benjamini, Y. and Yekutieli, D. (2001). The control of the false discovery rate in multiple testing under dependency. *Annals of Statistics*, **29**, 1165–1188.
15. Gijzen, M. et al. (1996) Temperature induced susceptibility to *Phytophthora sojae* in soybean isolines carrying different *Rps* genes. *Physiological and Molecular Plant Pathology*, **48**, 209–215.

Chapter 14

Assays for Effector-Mediated Suppression of Programmed Cell Death in Yeast

Yuanchao Wang and Qian Huang

Abstract

Pathogens such as bacteria, fungi, and oomycetes deliver diverse arrays of virulence or avirulence molecules, defined here as effectors, into the host cells. Effectors enable parasitic colonization by manipulating classes of biochemical, physiological, and morphological processes. An effective strategy to modulate host defense circuitry is to suppress their programmed cell death (PCD) response. Here, we describe a method for analyzing whether effectors function to suppress PCD in yeast. We use Bax and H_2O_2 to induce cell death and mimic some PCD features that naturally appear during the development of multicellular *Saccharomyces cerevisiae* colonies and assay whether plant pathogen effectors can inhibit the process. This technology provides an assay to test whether individual effectors can suppress PCD.

Key words: Yeast, PCD, Effectors, Bax

1. Introduction

Programmed cell death (PCD) is an important mechanism in defense against pathogens in both animals and plants. Bacteria, fungi, and oomycetes deliver diverse arrays of effectors into the host cells to suppress this defense process, and many effectors act as suppressors of PCD (1).

Apoptosis is one form of PCD, some features of which may be evolutionarily conserved in eukaryotic cells. Furthermore, PCD is a tightly regulated, active process, involving production of reactive oxygen species (ROS), increased ion leakage, cytochrome *c* release from mitochondria, and proteolytic activities (i.e., caspases) (2). Inputs from a variety of sources can impact on a core set of enzymes that coordinate destruction of key cellular components necessary for cell survival.

John M. McDowell (ed.), *Plant Immunity: Methods and Protocols*, Methods in Molecular Biology, vol. 712, DOI 10.1007/978-1-61737-998-7_14, © Springer Science+Business Media, LLC 2011

The occurrence of PCD in fungi has received support from numerous studies using the budding yeast *Saccharomyces cerevisiae* (3–5). Several so-called hallmarks of apoptosis can be observed in populations of yeast, plant, and animal cells. For example, oxidative stress from hydrogen peroxide may result through mechanical damage of cells and lead to PCD. Transfer of mammalian Bax into the cytosol of plant or animal cells can induce PCD. Additionally, expression of the mammalian, pro-apoptotic protein Bax in yeast leads to mitochondrial dysfunction, accumulation of ROS, and cell death (5). These results and others suggest that PCD mechanisms may be conserved in yeast, plants, and animals.

Previously, the bacterial (*Pseudomonas syringae*) type III effector AvrPtoB had been reported to suppress cell death in yeast induced by H_2O_2 (6). To address whether effectors from oomycete pathogens target the pathway(s) leading to PCD, we assayed whether the Avr1b-1 and Avh331 proteins from *Phythophthora sojae* can suppress apoptosis in the yeast, *Saccharomyces cerevisiae* (7). Here, we describe our procedure for introducing candidate effector genes under the control of the GAL1 promoter into the yeast strain *W303*, followed by assays for resistance to PCD induced by hydrogen peroxide or overexpression of mammalian *Bax*.

2. Materials

2.1. Yeast Media

1. YPD medium: 20 g Difco peptone, 10 g yeast extract and 20 g agar (for plates only), add H_2O to 950 ml. Adjust the pH to 6.5 if necessary, then autoclave. Allow medium to cool to ~55°C and then add dextrose (glucose) to 2% (100 ml of a sterile 20% stock solution). Adjust the final volume to 1 L.
2. SD medium: 6.7 g yeast nitrogen base without amino acids, 20 g agar, 850 ml H_2O, add 100 ml of the appropriate sterile 100× Dropout (DO) solution. Add glucose, galactose, and/or raffinose as described below, and adjust the final volume to 1 L.
3. SD/Gal, SD/Glu medium: Minimal SD medium with 100 ml of 20% galactose or 20% dextrose stock for final concentration of 2% in 1 L of medium.
4. SD/Gal/Raf medium: Minimal SD medium with 1% (100 ml of 20%) galactose and 1% (100 ml of 20%) raffinose stock for a final concentration of 2% in 1 L.

2.2. Carbon Sources

Filter sterilized or autoclaved:

1. 20% Dextrose (d-glucose). Store at 4°C.
2. 20% Galactose. Store at 4°C (see Note 5).
3. 20% Raffinose. Store at 4°C.

2.3. 100× Dropout Solution

A combination of a minimal SD Base and a DO Supplement will produce a synthetic, defined minimal medium lacking one or more specific nutrients. The specific nutrients omitted depends on the selection medium desired.

Nutrient 100×	Concentration (g/L)
L-Adenine hemisulfate salt	4
L-Histidine HCl monohydrate	2
L-Leucine	6
L-Methionine	2
L-Tryptophan	4
L-Uracil	2

100× dropout supplements (water solution, do not need to adjust the pH) may be autoclaved and stored at 4°C for up to 1 year.

2.4. Transformation of Yeast

1. Herring testes carrier DNA (10 mg/ml): Sonicated herring testes carrier DNA in solution can be purchased separately, or can be prepared using a standard method (8). Just prior to use, denature the carrier DNA by placing it in a boiling water bath for 20 min and immediately cooling it on ice. Use only high-quality carrier DNA; nicked calf thymus DNA is not recommended.
2. PEG/LiAc solution (polyethylene glycol/lithium acetate): The solution should be prepared freshly from stock solution. Ten milliliter of solution is made from 8 ml of 50% PEG4000, 1 ml of 10× TE and 1 ml of 10× LiAc.
3. 50% PEG 3350 (polyethylene glycol, average molecular weight = 3,350; Sigma) prepare with sterile deionized H_2O; if necessary, warm solution to 50°C to help the PEG to dissolve into solution.
4. 100% DMSO (dimethyl sulfoxide).
5. 10× TE buffer: 0.1 M Tris–HCl, 10 mM EDTA, pH 7.5. Autoclave.
6. 10× LiAc: 1 M lithium acetate, adjust to pH 7.5 with dilute acetic acid and autoclave.

2.5. Yeast Strains

The *S. cerevisiae* strain W303 (MATa; ura3-52; trp1delta2; leu2-3, 112; his3-11; ade2-1; can1-100) is used essentially as described by Priault et al. (5).

3. Methods

3.1. Plasmid Construction and Yeast Transformation

1. Expression plasmids containing *P. sojae* effector genes were constructed using pGilda, a shuttle plasmid containing a 0.4 kb GAL1 promoter and histidine selection marker (9).
2. *Eco*RI and *Sal*I restriction sites were respectively added to the 5′ and 3′ ends of *Avr1b-1* by PCR utilizing primers that incorporated these restriction sites. The PCR product was digested and inserted into the multiple cloning site (MCS) of pGilda, yielding pGilda-*Avr1b-1*.
3. pGilda-*Bax* was constructed by inserting mouse BAX (9) into MCS with *Eco*RI at the 5′ end and *Sal*I at the 3′ end, as described above for *Avr1b-1*.
4. Expression vectors for experiments involving co-transformation of yeast with Bax and the effector gene of interest were constructed using pYES2, a shuttle plasmid containing a 0.4 kb GAL1 promoter and uracil selection marker.
5. pYES2-*Bcl2*, pYES2-*Avr1b-1* and pYES2-*Avh331* were made by adding *Eco*RI at the 5′ end and *Sal*I at the 3′ end as described above, and inserting the PCR product into the MCS.
6. Plasmids were transformed into W303 by the lithium acetate method described in Subheading 3.2 (10), generating the following strains: *W303*::pGilda (empty vector); *W303*::*Avr1b-1*; *W303*::*Bax*::PYES2 (empty vector), *W303*::*Bax*::*Avr1b-1*; *W303*::*Bax*::*Avh331*.
7. Surviving colonies were restreaked on SD-his- or SD-his-ura-media at least twice.

3.2. Small-Scale LiAc Yeast Transformation Procedure

1. Inoculate 1 ml of YPD or SD with several colonies, 2–3 mm in diameter (see Note 1).
2. Vortex vigorously for 5 min to disperse any clumps.
3. Transfer this into a flask containing 50 ml of YPD or the appropriate SD medium.
4. Incubate at 30°C for 16–18 h with shaking at 250 rpm to stationary phase ($OD_{600} > 1.5$).
5. Transfer 30 ml of overnight culture to a flask containing 300 ml of YPD. Check the OD_{600} of the diluted culture; if necessary, add more of the overnight culture to bring the OD_{600} up to 0.2–0.3.

6. Incubate at 30°C for 3 h with shaking (230 rpm). At this point, the OD_{600} should be 0.4–0.6 (see Note 2).
7. Place cells in 50-ml tubes and centrifuge at 1,000 × *g* for 5 min at room temperature (20–21°C).
8. Discard the supernatants and thoroughly resuspend the cell pellets in sterile TE or distilled H_2O. Pool the cells into one tube (final volume 25–50 ml).
9. Centrifuge at 1,000 × *g* for 5 min at room temperature.
10. Decant the supernatant.
11. Resuspend the cell pellet in 1.5 ml of freshly prepared, sterile 1× TE/1× LiAc.
12. Add 0.1 μg of plasmid DNA and 0.1 mg of herring testes carrier DNA to a fresh 1.5-ml tube and mix (see Notes 3 and 4).
13. Add 0.1 ml of yeast competent cells to each tube and mix well by vortexing.
14. Add 0.6 ml of sterile PEG/LiAc solution to each tube and vortex at high speed for 10 s to mix.
15. Incubate at 30°C for 30 min with shaking at 200 rpm.
16. Add 70 μl of DMSO. Mix well by gentle inversion. Do not vortex.
17. Heat shock for 15 min in a 42°C water bath.
18. Chill cells on ice for 1–2 min.
19. Centrifuge cells for 5 s at 14,000 rpm (1000 × *g*) at room temperature. Remove the supernatant.
20. Resuspend cells in 0.5 ml of sterile 1× TE buffer.
21. Plate 100 μl of cells on each SD agar plate that will select for the desired transformants.
22. Pick the largest colonies and restreak them on the same selection medium to make master plates. Seal plates with Parafilm and store at 4°C for 3–4 weeks.

3.3. Yeast Viability Assays for H_2O_2 Tolerance

1. Cells of *W303* and the transformants were pelleted, washed, and resuspended in SD medium (0.17% YNB-AA/AS; 0.5% $(NH_4)_2SO_4$) containing 2% galactose and 1% raffinose as carbon sources (SD/gal/raff/±his).
2. After 12 h of induction, cells were diluted to $OD_{600} = 0.05$ (see Note 6).

   H_2O_2 was added to the medium at a final concentration of 3.15 mM and cultures were incubated at 30°C with vigorous shaking for 6 h.

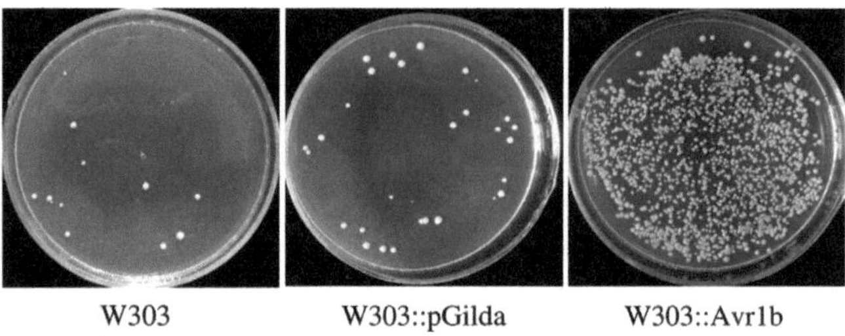

Fig. 1. Avr1b suppresses oxidative stress-induced cell death in yeast.

3. Treated and untreated cells (100 μl) were sampled and spread onto YPD medium with 2% agar, then incubated at 30°C for 48 h.
4. The number of colony forming units (c.f.u.) from treated cells was compared with the c.f.u. from untreated cells and photos were taken (Fig. 1).
5. All the above experiments were repeated in at least triplicate.

3.4. Assays for Suppression of Cell Death Induced by Bax

1. The following strains were cultured in SD medium containing 2% glucose as carbon sources (SD/glu/-his/±ura) for 18 h: W303::Bax::pYes (empty vector); W303::Bax::Bcl2; W303::Bax::Avr1b; W303::Bax::Avh331.
2. Cells were pelleted and washed four to five times with deionized H_2O (see Note 7).
3. Cells were diluted to $OD_{600} = 0.05$.
4. 100 μl of each cell suspension was spread onto plates containing the glucose as the carbon source plate (SD-glu-His-Ura). Expression of Bax will not be induced in these cells.
5. 100 μl of each cell suspension was spread onto plates containing galactose as the carbon source plate (SD-gal-His-Ura) to induce expression of Bax and effector proteins from the GAL1 promoter.
6. Viability was determined by counting colonies from a noninduced sample grown on glucose.
7. Plates were incubated at 30°C for 4 days.
8. The number of colony forming units (c.f.u.) are compared between the different strains. W303::Bax::pYES2 (empty vector) typically produces zero colonies, while W303::Bax::Bcl2 grows well because Bcl2 counters the apoptotic activity of Bax. Figure 2 displays low to moderate growth of W303::Bax::Avr1b-1 and W303::Bax::Avh331, indicating the ability of these effector proteins to suppress cell death induced by *BAX*.
9. All the experiments were repeated in at least triplicate.

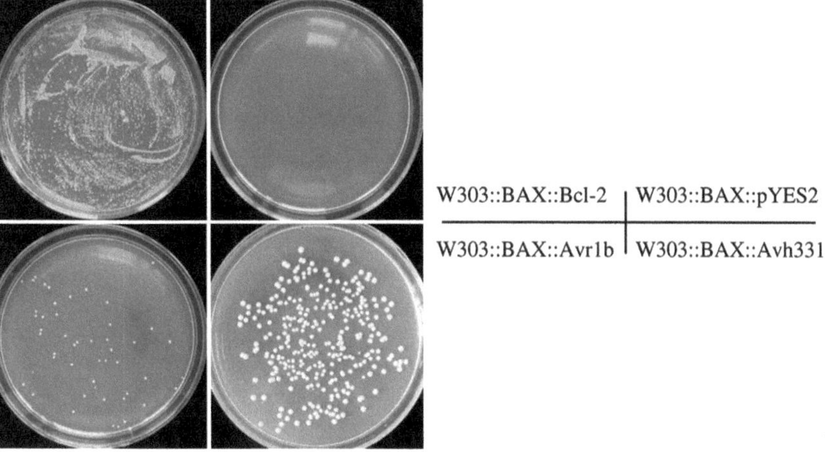

Fig. 2. *PsAvr1b* and *PsAvh331* expression can weakly suppress PCD induced by *BAX*.

4. Notes

1. If you purchase galactose separately, it must be highly pure and contain <0.01% glucose.
2. If you add the sugar solution before autoclaving, autoclave at 121°C for 15 min.
3. Autoclave at 121°C for 15 min; autoclaving at a higher temperature, or for a longer period of time, may cause the sugar solution to darken and will decrease the performance of the medium.
4. For host strains previously transformed with another autonomously replicating plasmid, use the appropriate SD selection medium to maintain the plasmid.
5. For transformations to integrate a reporter vector, use at least 1 μg of linearized plasmid DNA in addition to the carrier DNA.
6. Remember to vortex the tubes repeatedly during the dilution process.
7. After culturing in SD-glu liquid media, wash the yeast cells four to five times with deionized H_2O in order to eliminate the interference of glucose.

Acknowledgments

We would like to thank Brett Tyler and Daolong Dou from Virginia Polytechnic Institute and State University for help and gifts of materials. This work was supported by the National 973 program 2009CB119200.

References

1. Jones, J.D.G., Dangl, J.L. (2006) The plant immune system. *Nature* **444**, 323–329.
2. McKenna, Sharon L., McGowan, Adrian J., Cotter, Thomas G. (1998) Molecular mechanisms of programmed cell death. *Adv. Biochem. Eng. Biotechnol.* **62**, 1–31.
3. Madeo, F., Engelhardt, S., Herker, E., Lehmann, N., Maldener, C., Proksch, A., Wissing, S., Frohlich, K.U. (2002) Apoptosis in yeast: a new model system with applications in cell biology and medicine. *Curr. Genet.* **41**, 208–216.
4. Madeo, F., Herker, E., Wissing, S., Jungwirth, H., Eisenberg, T., Frohlich, K.U. (2004) Apoptosis in yeast. *Curr. Opin. Microbiol.* **7**, 655–660.
5. Priault, M., Camougrand, N., Kinnally, K.W., Vallette, F.M., Manon, S. (2003) Yeast as a tool to study Bax/mitochondrial interactions in cell death. *FEMS Yeast Res.* **4**, 15–27.
6. Abramovitch, R.B., Kim, Y.J., Chen. S., Dickman, M.B., Martin, G.B. (2003) Pseudomonas type III effector AvrPtoB induces plant disease susceptibility by inhibition of host programmed cell death. *EMBO J.* **22**, 60-69.
7. Dou, D., Kale, S.D., Wang, X., Chen, Y., Wang, Q., Wang, X., Jiang, R.H.Y., Arredondo, F.D., Anderson, R.G., Thakur, P.B., McDowell, J.M., Wang, Y., Tyler, B.M. (2008) Conserved C-terminal motifs required for avirulence and suppression of cell death by *Phytophthora sojae* effector Avr1b. *Plant Cell* **20**, 1118–1133.
8. Sambrook, J., Fritsch, E.F., Maniatis, T. (1989) *Molecular Cloning: A Laboratory Manual* (Cold Spring Harbor Laboratory, Cold Spring Harbor, NY).
9. Kampranis, S.C., Damianova, R., Atallah, M., Toby, G., Kondi, G., Tsichlis, P.N., Makris, A.M. (2000) A novel plant glutathione S-transferase/peroxidase suppresses Bax lethality in yeast. *J. Biol. Chem.* **275**, 29207–29216.
10. Ito, H., Fukuda, Y., Murata, K., et al. (1983) Transformations of intact yeast cells treated with alkali cations. *J. Bacteriol.* **153**, 163–168.

Chapter 15

Purification of Effector–Target Protein Complexes via Transient Expression in *Nicotiana benthamiana*

Joe Win, Sophien Kamoun, and Alexandra M.E. Jones

Abstract

Effectors of plant pathogens play important roles in not only pathogenesis but also plant immunity. Plant pathogens use these effectors to manipulate host cells for colonization, and their activities likely influence the evolution of plant immune responses. Analyses of genome sequences revealed that oomycete pathogens, such as *Phytophthora* spp., possess hundreds of RXLR effectors that are thought to be delivered into the host cells and hence function inside the cells by interacting with the host protein complexes. This article describes a co-immunoprecipitation protocol aimed at identifying putative target complexes of the effectors by transiently overexpressing the tagged effectors *in planta*. The identification of the eluted protein complexes was achieved by LC-MS/MS mass spectrometry and peptide spectrum matching.

Key words: Effectors, Co-immunoprecipitation, Oomycetes, *Phytophthora*, Mass spectrometry, Target complexes, Plant–microbe interaction

1. Introduction

The surveillance system of plant immunity relies on detection of molecules generated by an invading pathogen to mount defence responses. Perturbations caused by these pathogen molecules are perceived by resistance gene products in plants and defence procedures against the pathogen are initiated through downstream signalling cascades (1, 2). Although much knowledge has been accumulating in molecular events of recognition and signalling, little is known about pathogen molecules and their cognate functions except for a few cases in prokaryotes. Molecules, usually proteinaceous, secreted by pathogens to manipulate their hosts are known as "effectors", and their activities likely influence the evolution of plant immune responses. It is therefore pertinent to

John M. McDowell (ed.), *Plant Immunity: Methods and Protocols*, Methods in Molecular Biology, vol. 712,
DOI 10.1007/978-1-61737-998-7_15, © Springer Science+Business Media, LLC 2011

study the functions of these effectors *in planta* to gain deeper understanding of plant immunity.

In addition to validated effectors (3, 4), repertoires of effectors can be predicted for eukaryotic pathogens and studied in high-throughput manner from the available genome sequences. For example, by performing motif searches, complete sets of RXLR effectors are identified from the genome sequences of *Phytophthora sojae* and *Phytophthora ramorum*, oomycetes that cause devastating diseases on soybean and oak trees, respectively (5). A set of RXLR effectors have also been identified bioinformatically in potato late blight pathogen *Phytophthora infestans* and other oomycetes (6). RXLR effectors are pathogen-secreted proteins that are thought to translocate and function inside plant cells. It is most likely that effectors interact with plant proteins to condition the host plant for colonization.

Significant insights into how an effector carries out its function can be achieved by examining its interacting partners, including other proteins, inside the host cell (7, 8). An effector of interest can be tagged, expressed *in planta*, and purified in its native form by immunoprecipitation. In doing so, associated proteins are co-immunoprecipitated and these can be identified by mass spectrometry. Here, we describe a medium-throughput co-immunoprecipitation procedure adapted from ref. 7 to identify plant protein complexes interacting with pathogen effectors.

2. Materials

2.1. Preparation of Electro-Competent Agrobacterium Cells

1. *Agrobacterium tumefaciens* strain GV3101::pMP90::pSoup (see Note 1).
2. Luria–Bertani (LB) medium: Dissolve 10 g Bacto Tryptone (Difco, Voigt Global Distribution Inc., Lawrence, KS), 5 g yeast extract (Difco), 10 g NaCl in 1 L water, and adjust pH to 7.00 with NaOH. Autoclave at 121°C and 103.42 kPa for 15 min.
3. 10% (v/v) Glycerol.

2.2. Expression Vector and Transformation of Agrobacterium by Electroporation

1. TRBO high efficiency over expression vector (9).
2. Antibiotic stocks: Tetracycline (5 mg/mL in ethanol (EtOH)), rifampicin (10 mg/mL in methanol), kanamycin (10 mg/mL).
3. Gene Pulser Xcell Electroporation System (Bio-Rad) or similar electroporator.
4. 60% (v/v) Glycerol.

2.3. Agro-Infiltration

1. Plants: *Nicotiana benthamiana*, 4- to 6-week-old.
2. Greenhouse: Average temperature of 20°C and 16/8 h (light/dark) cycle supplemented with sodium vapour lamps.
3. Acetosyringone: Dissolve powder in DMSO to make a stock solution of 500 mM. Store at –20°C.
4. Agrobacterium Infiltration medium: 10 mM $MgCl_2$, 10 mM MES, pH 5.6, 150 mM acetosyringone (Sigma-Aldrich, St. Louis, MO). This should be made fresh from stock solutions.

2.4. Protein Extraction and Co-immuno-precipitation

1. GTEN: 10% glycerol, 25 mM Tris pH 7.5, 1 mM EDTA, 150 mM NaCl.
2. Extraction buffer: GTEN, 2% w/v PVPP, 10 mM DTT, 1× protease inhibitor cocktail (Sigma), 0.1% Tween 20 (Sigma).
3. Anti-FLAG M2 affinity Gel (Sigma).
4. 3× FLAG peptide (Sigma) stock solution at 5 mg/mL.
5. Immunoprecipitation (IP) buffer: GTEN, 0.1% Tween 20 (see Note 2).
6. Elution buffer: IP buffer containing 150 ng/μL 3× FLAG peptides.

2.5. SDS–Polyacrylamide Gel Electrophoresis

1. Separating buffer (4×): 1.5 M Tris–HCl, pH 8.8.
2. Stacking buffer (4×): 0.5 M Tris–HCl, pH 6.8.
3. 10% (w/v) SDS (see Note 3).
4. Forty percent acrylamide/bis solution (37.5:1) (Bio-Rad, Hercules, CA) (see Note 4).
5. *N,N,N,N′*-Tetramethyl-ethylenediamine (TEMED, Bio-Rad).
6. Ammonium persulfate (APS, Bio-Rad): prepare 10% solution in water just before use. The solution can be stored at 4°C for up to 5 days.
7. Mini-PROTEAN Tetra Electrophoresis System (Bio-Rad).
8. Running buffer (10×): Dissolve 30.3 g Tris base, 144 g glycine and 10 g SDS in 700 mL water and make to 1 L with water after everything is dissolved.
9. Sample loading dye (4×): 200 mM Tris–HCl (pH 6.8), 8% (w/v) SDS, 40% (v/v) glycerol, 50 mM EDTA, 0.08% bromophenol blue.
10. 1 M Dithiothreitol (DTT). Divide into 1 mL aliquots and store at –20°C.
11. Prestained Protein Marker, Broad Range (New England BioLabs, Ipswich, MA).
12. Colloidal Coomassie Blue solution, e.g. SimplyBlue™ SafeStain (Invitrogen, Carlsbad, CA).

2.6. Western Blotting

1. Transfer buffer: 25 mM Tris base, 190 mM glycine, 20% (v/v) methanol.
2. Polyvinylidine fluoride membrane (PVDF, Bio-Rad) and 3MM chromatography paper (Whatman, Maidstone, UK).
3. Mini Trans-Blot Cell (Bio-Rad).
4. Tris-buffered saline with Tween (TBS-T) (10×): 1.37 M NaCl, 27 mM KCl, 250 mM Tris–HCl, pH 7.4, 1% (v/v) Tween 20.
5. Blocking solution: 5% (w/v) non-fat milk powder in 1× TBS-T.
6. Primary antibody: anti-FLAG monoclonal antibody (Sigma).
7. Secondary antibody: anti-mouse IgG conjugated to alkaline phosphatase (AP) (Sigma).
8. Colour development reagent: AP Conjugate Substrate Kit (Bio-Rad) (see Note 5).

2.7. Gel Excision and Tryptic Digestion

1. Clean razor blades and spatulas to cut and handle gel slices.
2. 20 mM Ammonium bicarbonate (ABC; NH_4HCO_3, Sigma, Cat. No. A6141).
3. 100% EtOH.
4. 50 mM ABC/100% EtOH (1:1).
5. 10 mM Dithiothreitol (DTT, Sigma, Cat. No. D0632) in 50 mM ABC.
6. 55 mM Iodoacetamide (IAA, Sigma, Cat. No. I6125) in 20 mM ABC.
7. Trypsin (12.5 μg/mL Promega Gold, Cat. No. V5113).
8. 100% TFA (trifluoroacetic acid).
9. 100% FA, formic acid.
10. 100% ACN, acetonitrile.

2.8. In-Solution Digestion

1. Rapigest SF (Waters Corp., #186001861).
2. Reduction buffer: 1 M dithiothreitol (DTT, Sigma, Cat. No. D0632) in 50 mM ammonium bicarbonate (Sigma, Cat. No. A6141).
3. Alkylation buffer: 55 mM iodoacetamide (Sigma, Cat. No. I6125) in 50 mM ABC.
4. Trypsin, sequencing grade, modified (Promega, Cat. No. V5113).

2.9. Mass Spectrometry

1. Nano-flow liquid chromatography system (nanoAcquity, Waters Corp.).
2. LTQ-Orbitrap XL (Thermo Scientific, Bremen, Germany).
3. Pre-column (Symmetry C18 5 μm beads, 180 μm×20 mm column, Waters Corp., #186003514).

4. Analytical column: (BEH 130 C18 1.7 μm beads, 75 μm × 250 mm column, Waters Corp., #186003545).
5. Nano-spray source (Proxeon).
6. HPLC buffer A: MS-grade water with 0.1% formic acid (Water LiChrosolv, VWR International Ltd # 1 15333 2500).
7. HPLC buffer B: MS-grade acetonitrile with 0.1% formic acid (Acetonitrile UV grade, Fisher Scientific UK Ltd, # A0627B17).

2.10. Data Analysis

1. Bioworks/extract.msn (version 3.3.1, Thermo Scientific, San Jose, USA).
2. merge.pl (Matrix Science, UK).
3. Mascot (v 2.2 Matrix Science, UK).
4. Scaffold (v 2.2.03, Proteome Software).

3. Methods

We designed effector proteins for expression by replacing the N-terminal secretory signal peptide with FLAG epitope (MDYDDDDK) which can be used for affinity purification of the tagged protein with anti-FLAG resin. We made use of a gene synthesis service for our effector constructs: coding sequences of the constructs were codon-optimized for *in planta* expression, synthesized, and cloned by GenScript Corp. (Piscataway, NJ) into TRBO.

To prepare material for co-immunoprecipitation, *A. tumefaciens* cells are transformed with TRBO-effector constructs by electroporation. Agro-infiltration enables over expression of FLAG-tagged effector proteins in *N. benthamiana* leaves after T-DNA transfer by agrobacteria. Effector proteins are allowed to accumulate in the leaves for 2–3 days post-infiltration before harvesting. Total proteins are extracted from the leaves and FLAG-tagged effectors are co-immunoprecipitated with anti-FLAG M2 affinity resins. Bound proteins are specifically eluted with 3× FLAG peptides and separated by SDS–PAGE. Proteins in the gel are stained with colloidal Coomassie blue, excised, digested with trypsin, and identified by mass spectrometry.

Samples from co-IP experiments may be prepared for mass spectrometry in two ways; by excision from SDS–PAGE gel or the eluates can be analysed directly with in-solution digestion. Preparing samples in 1D gels offers the advantages of visualizing the proteins before analysis, allowing an estimate of quantity and complexity. Additionally, 1D gels provide a simple and reliable method to pre-fractionate complex protein mixtures, concentrating proteins according to their size and allowing more abundant proteins to be cut away from less abundant proteins. The disadvantages of gel preparation are that it provides ample opportunity for the introduction of significant amounts of keratin and some smaller proteins (<10 kDa) are typically lost from standard gels. Analysing eluates directly can retain the smaller proteins and reduce contamination from skin proteins, but complex mixtures may require further fractionation (for example, by STAGE tips (10)) before analysis.

3.1. Preparation of Electro-Competent GV3101 Cells and Transformation

1. Plate the original culture of GV3101 cells on LB agar plate supplemented with tetracycline (2.5 μg/mL) and grow 2–3 days at 28°C.
2. Pick one colony and grow in 5 mL LB supplemented with tetracycline (5 μg/mL) medium, overnight at 28°C in a shaking incubator.
3. Take 1 mL start culture and grow in 200 mL LB + tetracycline (5 μg/mL) overnight at 28°C in a shaking incubator.
4. Measure the OD_{600}. When OD_{600} reaches 0.5–0.7, divide the culture into two pre-chilled 250-mL centrifuge bottles and keep on ice at least 30 min.
5. Centrifuge the cultures for 15 min at 3,500 × *g* at 4°C.
6. Pour off the supernatant and resuspend the pellets in 50 mL ice cold 10% glycerol.
7. Centrifuge the suspension for 15 min at 3,500 × *g* at 4°C.
8. Discard the supernatant and wash pellets with 50 mL ice cold 10% glycerol again.
9. Discard the supernatant and resuspend each pellet in 200 μL 10% glycerol.
10. Aliquot 50 μL of the resuspended competent cells into pre-chilled 1.5-mL Expender tubes.
11. Snap-freeze the cells in liquid nitrogen and store at –80°C for up to 1 year.

3.2. Transformation of Agrobacterium by Electroporation

1. Thaw competent cells on ice (50 μL per transformation).
2. Add TRBO plasmid DNA (1–2 μL) to the cells, and mix them together on ice by tapping the side of the tube.
3. Transfer the mixture to a pre-chilled electroporation cuvette with 2 mm gap.
4. Carry out electroporation using a Bio-Rad electroporator set to following conditions: capacitance: 25 μF, voltage: 2.4 kV, resistance: 200 Ω. This should yield a time constant (pulse length) of 5 ms upon electroporation.
5. Immediately after electroporation, add 950 μL LB to the cuvette, and transfer the bacterial suspension to a 15-mL culture tube. Incubate for 4 h at 28°C with gentle agitation.

6. Collect the cells by centrifuging briefly, and spread them on an LB agar plate containing kanamycin (50 μg/mL) and rifampicin (100 μg/mL).
7. Incubate the cells for 3–4 days at 28°C.
8. Screen colonies by polymerase chain reaction (PCR) for the presence of TRBO plasmid constructs and pick 2–3 colonies with inserts.
9. Confirm the integrity of inserts by sequencing the PCR products.
10. Grow validated clones in 5 mL LB medium supplemented with kanamycin (50 μg/mL) and rifampicin (100 μg/mL) overnight in a shaking incubator at 28°C.
11. Add 0.5 mL 60% glycerol to 1 mL overnight cultures in a 2-mL cryotube, mix by vortexing, and store at −80°C until use.

3.3. Agro-Infiltration

1. Streak out glycerol stock of Agrobacterium containing the TRBO construct onto LB plates containing 50 μg/mL kanamycin and 100 μg/mL rifampicin and incubate at 28°C for 24 h.
2. Inoculate 10 mL LB broth containing 50 μg/mL kanamycin and 100 μg/mL rifampicin with a colony from the streak and grow overnight at 28°C in a shaking incubator.
3. Centrifuge the cells at 3,500 × *g* and resuspend in Agrobacterium infiltration medium and incubate at room temperature for at least 2 h (overnight incubation also works).
4. Take an OD_{600} of each Agrobacterium culture to be infiltrated. Dilute the culture with the Agrobacterium infiltration medium to achieve OD_{600} 0.3–0.5.
5. Infiltrate middle leaves of 4- to 6-week-old *N. benthamiana* plants using a 1-mL syringe without the needle making sure the whole leaf area is infiltrated.
6. Harvest the leaves 3–4 days post-infiltration for protein extraction.

3.4. Protein Extraction and Co-immunoprecipitation

1. Freeze 3–4 leaves in liquid nitrogen and grind into powder using a mortar and pestle making sure the samples are constantly frozen with liquid nitrogen during grinding.
2. Weigh out 1 g of leaf powder on a pre-chilled aluminium foil and add the powder to 2.0 mL of ice-cold extraction buffer in a 15-mL centrifuge tube. Vortex to mix so that all the powder comes in contact with the extraction buffer. Keep the tube on ice until the powder is thawed completely in the extraction buffer. Vortex to mix thoroughly for 20 s.
3. Centrifuge at 3,000 × *g* for 10 min at 4°C and transfer the supernatant to a 2-mL microcentrifuge tube.

4. Centrifuge at full speed in a microcentrifuge for 10 min at 4°C. Transfer the supernatant to a new tube.
5. To a new 2-mL microcentrifuge tube, add 250 μL of extract (freeze the left-over samples in liquid nitrogen and store at –80°C until use). Bring up the total volume to 2.0 mL with IP buffer. Keep this solution on ice until use.
6. Resuspend the resin well by tapping the side of the vial several times and pipetting the resin up and down using a 1-mL pipette with a cut tip (so that the opening is wide enough to let the resin move through without too much damage).
7. Pipette enough resin (50 μL per sample, e.g. prepare 200 μL resin for four samples) into a 2.0-mL Eppendorf tube and centrifuge at 800×*g* for 1 min and remove the supernatant using a needle attached to a syringe (take care not to aspirate the resin).
8. Resuspend the resin in 5× volumes of IP buffer.
9. Centrifuge at 800×*g* for 1 min and remove the supernatant as above.
10. Repeat above two steps twice more.
11. Resuspend the resin to the original volume with the IP buffer and add 50 μL of resin to the leaf extract prepared above (step 5).
12. Mix the resin and the extract well by turning end-over-end for at least 1 h at 4°C.
13. Centrifuge at 800×*g* for 30 s. Discard supernatant and add 1 mL of fresh IP buffer. Repeat four more times but always leave about 50 μL at the bottom of the tube to avoid aspirating the beads. After the last wash, centrifuge to spin down any liquid on the sides of the tube and aspirate the remaining liquid with a needle attached to a 1-mL syringe.
14. Elute the bound proteins by adding 100 μL IP buffer containing 150 ng/μL 3× FLAG peptide and incubating with gentle shaking for 30 min at 4°C.
15. Transfer the supernatant containing the eluted proteins to a fresh tube using a syringe and needle (take care not to aspirate any resin) (see Note 6).
16. Load 10–20 μL of the sample onto an SDS–PAGE gel for colloidal Coomassie blue staining followed by protein identification using mass spectrometry and western blotting.

3.5. SDS–Polyacrylamide Gel Electrophoresis

1. Set up the gel cassettes on casting stand of Mini PROTEAN electrophoresis system using the glass plates with 1-mm spacers following the manufacturer's instructions.
2. Prepare a 15% separating gel by mixing 2.5 mL 4× separating buffer, 3.8 mL 40% acrylamide/bis solution, 3.5 mL water,

100 μL 10% SDS, 50 μL 10% APS, and 5 μL TEMED. Immediately pour 3.75 mL gel mixture between the glass plates seated on the gel casting stand and overlay with 1 mL water. There should be enough solution for two gels. The gels should polymerize within 30 min.

3. Once the gels have polymerized, pour off the water and rinse the top of the gel with water.
4. Prepare the stacking gel by mixing 1.25 mL 4× stacking gel buffer, 0.5 mL 40% acrylamide/bis solution, 3.2 mL water, 50 μL 10% SDS, 50 μL 10% APS, and 5 μL TEMED. Immediately pour the stack on top of both gels and insert the 10-well combs. The stacking gel should polymerize within 30 min. Allow at least one additional hour for gels to set completely.
5. Prepare 1 L running buffer by mixing 100 mL 10× running buffer and 900 mL water.
6. Carefully remove the combs from the gel and rinse the wells with running buffer using a Pasteur pipette. Remove the gel plates from the casting frames and place in electrode module and buffer tank.
7. Add the running buffer to upper and lower chambers of the electrophoresis unit to the indicated level and load the wells with up to 35 μL sample containing 4× sample loading dye and 100 mM DTT. Samples should be loaded on one gel for western blotting and duplicate sample on the other gel for Coomassie blue staining and subsequent analysis by mass spectrometry.
8. Place the lid on the buffer tank and assembly making sure to align the colour coded banana plugs and jacks. The lid should be securely and tightly positioned on the tank.
9. Connect the cables to the power supply and run at a constant voltage of 200 V until the dye front reaches to the bottom of the gel.

3.6. Western Blotting

Proteins separated by SDS–PAGE are transferred using Mini Trans-Blot Cell (Bio-Rad) to PVDF membrane for detection by anti-FLAG antibody and anti-mouse secondary antibody conjugated to alkaline phosphatase catalysing a colorimetric reaction. Fill the Bio-Ice cooling unit with water and store it at −20°C until ready to use.

1. Remove the gel from the cassette and rinse with water and place in transfer buffer and equilibrate for 15 min on a shaking platform.
2. Cut the PVDF membrane and the 3MM paper to the dimensions of the gel.

3. Soak the membrane, 3MM paper, and fibre pads in transfer buffer for 10 min.
4. Prepare the gel cassette following the manufacturer's instructions.
5. Place the cassette in module and add the frozen Bio-Ice cooling unit. Place in tank and completely fill the tank with buffer.
6. Add a stir bar to the tank to distribute buffer temperature and ion in the tank. Set the speed as fast as possible.
7. Put on the lid, plug the cables into the power supply, and run the blot at constant current of 250 mA for 1 h.
8. Upon completion of the run, disassemble the blotting sandwich and carefully remove the membrane.
9. Wash the membrane in water for 2–3 min with mild agitation, then with TBS-T.
10. Incubate the membrane with blocking solution for 1 h at room temperature.
11. Rinse the membrane with TBS-T.
12. Dilute anti-flag M2 antibody (usually 1:3,000) in 10 mL blocking solution and add to the membrane. Incubate for 1 h on a rocking platform.
13. Rinse twice with TBS-T and then wash three times with TBS-T, each time for 10 min.
14. Add anti-mouse antibody (diluted 1:40,000 in blocking solution), incubate for 1 h.
15. Rinse twice with TBS-T and then wash three times with TBS-T, each time for 10 min.
16. Prepare AP substrate solution by mixing 400 μL 25× development buffer, 9.4 mL water, 100 μL AP colour reagent A, 100 μL AP colour reagent B immediately before use.
17. Immerse the membrane in the AP substrate solution and incubate at room temperature with gentle agitation until colour development reaches desired intensity.
18. Stop the development by washing the membrane in ddH_2O for 10 min with gentle agitation. Air dry the membrane and digitally record the image of the membrane.

3.7. Gel Excision and Trypsin Digestion

At all stages strive to avoid keratin contaminations of gels; wear gloves at all times and use clean trays that have never been used for processes, such as Western blotting.

1. Cut gel slices into 1-mm cubes so that they site at the base of the tube. Do not mash or grind the gel.
2. Wash the gel pieces in 50% EtOH with shaking at 40°C, ensure that the gel slices are totally covered and change the buffer as required until completely destained.

3. Remove destaining solution and cover gel pieces with 100% EtOH. Gel pieces will dehydrate, shrinking and becoming opaque, remove EtOH.
4. Add 50–150 μL of 10 mM DTT in 20 mM ABC (see Note 7).
5. Incubate for 30 min 57°C with shaking.
6. Cool samples to room temperature before adding 55 mM IAA in 20 mM ABC (same volume as above).
7. Incubate 15 min in the dark at room temperature.
8. 2 × 10 min wash in 20 mM ABC/EtOH 1:1, and 10 min dehydration in 100% EtOH.
9. Add 10–50 μL trypsin in 20 mM ABC to the dehydrated gel pieces.
10. Add 20–100 μL 20 mM ABC to keep gel wet (ensure that large amounts of gel are covered).
11. Incubate 4 h or overnight at 37°C.
12. Transfer supernatant to new tube. Add 20–200 μL 100% ACN to gel pieces, vortex and gently centrifuge. Combine ACN with supernatant and speed-vac to remove ACN and reduce total volume to approximately 10 μL for MS analysis.

3.8. In-Solution Digestion

1. Prepare fresh Rapigest to 0.1% (w/v) in 50 mM ABC.
2. Dissolve the protein pellet (often invisible) in 0.1% Rapigest solution, using a cycle of sonication, vortexing and heating to 95°C (5 min) to fully dissolve the pellet.
3. Add DTT to a final concentration of 5 mM.
4. Incubate at 57°C for 30 min with shaking.
5. Cool to room temperature before adding IAA to a final concentration of 15 mM.
6. Incubate 15 min in the dark at room temperature.
7. Add trypsin (1:100 w/w). Incubate samples at 37°C for 4 h or overnight.
8. Acidify samples (TFA to final concentration of 0.5%) to halt digestion and hydrolyse Rapigest. Incubate samples for 30 min at 37°C.
9. Centrifuge acid-treated samples at maximum speed in a bench top centrifuge for 15 min. A whitish pellet may be visible (see Note 8).
10. Immediately after centrifugation take the supernatant to a clean, labelled tube. Do not disturb the pellet.

3.9. Mass Spectrometry

LC-MS/MS analysis is performed using an LTQ-Orbitrap XL mass spectrometer, a UPLC system adapted for nano-flow and a nano-electrospray source. The MS is operated in positive ion mode with a capillary temperature of 200°C, no sheath gas is employed and the source and focusing voltages are optimized for the transmission of angiotensin (see Note 9).

1. Apply peptides to a pre-column connected to a self-packed C18 8-cm analytical column.
2. Elute peptides using a gradient of 2–40% acetonitrile in 0.1% formic acid over 60 min at 250 nL/min (see Note 10).
3. Acquire full scan MS in profile mode from 300 to 2,000 *m/z* at 60,000 resolution using an internal lock mass. Maximal accumulation time for MS1 is set to 1 s to a target fill of 1×10^6.
4. A minimum signal of 3,000 counts is required to trigger data dependent acquisition of MS2 spectra. MS2 are acquired in a data dependent mode consisting of selection of the six most abundant ions in each cycle. Fragmentation is typically collision induced disassociation (CID) with collision energy at 35%, in MS2 ions are accumulated to 5×10^4 and spectra are centroided at acquisition by the MS.
5. Dynamic exclusion parameters allow one repeat hits before the precursor *m/z* is added to an exclusion list for 300 s.

3.10. Data Analysis

1. Extract MS/MS spectra from the raw file using BioWorks (v 3.3.1). We typically use the following parameters to generate dta files; MW range 300–4,500, a threshold of 100 ion counts, precursor ion tolerance of 5 ppm, one group scan and a minimum ion count of 10. MS level is set to automatic and we do not use any of the filtering options such as ZSA.
2. Use the merge.pl script (Matrix Science) to concatenate the dta files into generic mascot format (an mgf file).
3. Search the appropriate database with Mascot and compile the results in Scaffold to combine data from gel bands and to facilitate comparison between samples.

4. Notes

1. We use *A. tumefaciens* strain GV3101 for its ability to transfer T-DNA efficiently to *N. benthamiana* and mainly for its high transformation efficiency in our lab. For more information on use of *A. tumefaciens* and binary vectors, see ref. 11.
2. We use non-denaturing non-ionic detergent Tween 20 to extract the proteins from the leaves. NP-40 is also commonly used but discontinued by the manufacturer. Denaturing ionic detergents may disrupt the protein complexes.

3. SDS powder should be weighed out in a safety cabinet. SDS causes irritation to eyes, respiratory system, and skin on contact.
4. Always wear protective clothing, such as gloves, lab coat, and safety glasses, when handling Acrylamide. Acrylamide is highly toxic and may cause cancer and heritable genetic damage.
5. An alternative to visualize the western blots is to use secondary antibodies conjugated to horseradish peroxidase and chemiluminescent substrate to expose X-ray films. This procedure may offer better sensitivity if the protein expression is very low.
6. Repeat this step if there is contamination with IgG fragments in the eluted samples. Smallest number of resin beads could contribute to the IgG contamination.
7. Alter the volumes used to ensure that the gel pieces are entirely covered. Adjust subsequent volumes to compensate.
8. We have occasionally observed a white residue that floats on the surface of the preparation rather than forming a pellet, we think this is Rapigest and recommend transferring the supernatant to a clean tube and repeating the centrifugation at 4°C. If the solution is not entirely clear, we filter the peptides though a 10-kDa membrane (e.g. microcon from Millipore) before mass spectrometry.
9. Other types of mass spectrometer are suitable for this analysis. Some steps of the data analysis differ accordingly.
10. The gradient used depends on the complexity of the peptide mixture and the HLPC available. We typically use a 60-min gradient for medium complexity mixtures (<100 proteins, with identification rather than coverage as the goal) and increase up to 4 h gradients for highly complex mixtures.

Acknowledgments

We thank John Lindbo (UC Davis) for providing us with the TRBO vector, and Liliya Serazetdinova for technical assistance. This work was supported by the Gatsby Charitable Foundation.

References

1. Dangl, J. and Jones, J. D. G. (2001) Plant pathogens and integrated defence responses to infection. *Nature* **411**, 826–833.
2. Jones, J. D. G. and Dangl, J. (2006) The plant immune system. *Nature* **444**, 323-329.
3. Kamoun, S. (2006) A catalogue of the effector secretome of plant pathogenic oomycetes. *Annu. Rev. Phytopathol.* **44**, 41–60.
4. Kamoun, S. (2007) Groovy times: filamentous pathogen effectors revealed. *Curr. Opin. Plant Biol.* **10**, 358–365.

5. Tyler, B. M., Tripathy, S., Zhang, X., Dehal, P., Jiang, R. H., et al. (2006) *Phytophthora* genome sequences uncover evolutionary origins and mechanisms of pathogenesis. *Science* **313**, 1261–1266.
6. Win, J., Morgan, W., Bos, J., Krasileva, K. V., Cano, L. M., Chaparro-Garcia, A., Ammar, R., Staskawicz, B. J., and Kamoun, S. (2007) Adaptive evolution has targeted the C-terminal domain of the RXLR effectors of plant pathogenic oomycetes. *Plant Cell* **19**, 2349–2369.
7. Sacco, M. A., Mansoor, S., and Moffett, P. (2007) A RanGAP protein physically interacts with the NB-LRR protein Rx, and is required for *Rx*-mediated viral resistance. *Plant J.* **52**, 82–93.
8. Monti, M., Cozzolino, M., Cozzolino, F., Vitiello, G., Tedesco, R., Flagiello, A., and Pucci, P. (2009) Puzzle of protein complexes in vivo: a present and future challenge for functional proteomics. *Expert Rev. Proteomics* **6**, 159–169.
9. Lindbo, J. (2007) TRBO: a high-efficiency tobacco mosaic virus RNA-based overexpression vector. *Plant Physiol.* **145**, 1232–1240.
10. Rappsilber, J., Mann, M., and Ishihama, Y. (2007) Protocol for micro-purification, enrichment, pre-fractionation and storage of peptides for proteomics using stagetips. *Nat. Protoc.* **2**, 1896–1906.
11. Hellens, R., Mullineaux, P., and Klee, H. (2000) Technical focus: a guide to Agrobacterium binary Ti vectors. *Trends Plant Sci.* **5**, 446–451.

Chapter 16

Imaging Fluorescently Tagged *Phytophthora* Effector Proteins Inside Infected Plant Tissue

Petra C. Boevink, Paul R.J. Birch, and Stephen C. Whisson

Abstract

Assays to determine the role of pathogen effectors within an infected plant cell are yielding valuable information about which host processes are targeted to allow successful pathogen colonization. However, this does not necessarily inform on the cellular location of these interactions, or if these effector–virulence target interactions occur only in the presence of the pathogen. Here, we describe techniques to allow the subcellular localization of pathogen effectors inside infected plant cells or tissues, based largely on infiltration of plant tissue by *Agrobacterium tumefaciens* and its delivery of DNA encoding fluorescent protein-tagged effectors, and subsequent confocal microscopy.

Key words: Effector, Fluorescent protein, Oomycete, Oomycete transformation, Agroinfiltration

1. Introduction

Oomycete plant pathogens, like most other plant pathogens, must suppress or otherwise evade detection by the host plant defense surveillance systems. Plants can recognize surface or secreted molecules common to broad groups of pathogens, so-called pathogen-associated molecular patterns (PAMPs). This recognition activates defense responses from the host plant to control infection, referred to as PAMP or pattern-triggered immunity (PTI) (1). Pathogens can counter PTI by producing additional secreted proteins that can target and manipulate all components of PTI, from recognition to signal transduction, to transcriptional and post-translational regulation of defense mechanisms (e.g., ref. 2). Plants have an additional layer of defense mediated by resistance (R) proteins that guard virulence targets such that they initiate a hypersensitive resistance response (HR) when they perceive interaction of an effector with its virulence target. This is known as

John M. McDowell (ed.), *Plant Immunity: Methods and Protocols*, Methods in Molecular Biology, vol. 712,
DOI 10.1007/978-1-61737-998-7_16,

effector-triggered immunity (ETI), and an effector that triggers it is typically referred to as an avirulence (Avr) protein. Avr proteins from oomycetes have been identified by various cloning strategies in recent years, such as map-based cloning, transcriptional profiling, association genetics, and functional screening. To date, reports describing many cloned Avr genes from oomycetes have been published (3–10). All oomycete Avr proteins identified to date have a predicted signal peptide for secretion followed by a conserved RXLR motif and, frequently, a region enriched for acidic amino acids, and ending in EER. The RXLR-EER motifs have been shown to mediate translocation from the pathogen into infected host plant cells (11, 12). The RXLR-EER motifs are then followed by a C-terminal "effector" domain which is surmised to carry out the virulence function of the effector (13, 14, 15).

Sequencing of oomycete genomes has revealed hundreds of candidate RXLR effectors in each of the sequenced genomes (11, 13, 16, 17). High-throughput expression profiling, or quantitative RT-PCR has revealed that many of the effectors in *Phytophthora infestans* may be expressed (11, 18). In addition, oomycetes secrete numerous effectors that are predicted to localize within the plant apoplast (reviewed in ref. 19), such as protease and glucanase inhibitors (20, 21), cell wall degrading enzymes (22), and other cysteine-rich toxin-like proteins (23). While the discovery of potential effectors is no longer a limiting factor due to technical advances in nucleic acid sequencing, expression profiling, and associated bioinformatics, functional characterization of effectors is more challenging. There are numerous avenues to discover the function of a given effector such as identifying its cognate interacting virulence target, or its effect on defense responses such as suppression of PTI or ETI. An additional aspect for consideration in functional characterization of effectors is the location of the effector (and its targets) during infection by the pathogen.

Tagging of specific proteins with fluorescent proteins originating from marine organisms, such as jellyfish or corals, has revolutionized how proteins and the processes that they participate in are revealed in living cells. In terms of localizing effector proteins from plant pathogens, it is possible to tag effectors and express them inside host plant cells by using the bacterium *Agrobacterium tumefaciens* as a delivery vehicle (24). This process is particularly effective for localizing effectors that are known to be delivered inside host cells. Alternatively, fluorescently tagged effectors can be expressed in transgenic pathogens (11, 25, 26), although many pathogens are recalcitrant to stable transformation, such as obligate biotrophs. This chapter introduces and discusses methods for (1) *A. tumefaciens* delivery of tagged effectors, (2) localization of tagged effectors by confocal microscopy, (3) confirmation of host protein interaction by bimolecular fluorescence complementation (BiFC), and (4) delivery and detection of oomycete effectors from transgenic pathogens. Many of the techniques described in this chapter largely depend on the system under study and the equipment used. Many of the technical aspects such as standard gene cloning techniques are already well described elsewhere. As such, this chapter largely focuses on the considerations to be taken into account when attempting to localize effector proteins in host–pathogen interactions.

1.1. Considerations for Fluorophore Choice

Since the discovery and cloning of the gene encoding green fluorescent protein (GFP) there has been a proliferation of fluorescent variants of GFP and other similar fluorescent proteins derived from marine organisms (for example see refs. 27, 28). Many of the more recent variants are for specialist applications. For basic localization studies, most scientists use the more widely used fluorophores that are bright (have high quantum yields), largely monomeric, stable and well tested, e.g., enhanced cyan, green, and yellow fluorescent proteins (eCFP, eGFP, and eYFP; no longer available from Clontech but widely available in the scientific community), mGFP5 (29) and monomeric red fluorescent protein (mRFP) (30); although mRFP does not exhibit particularly bright fluorescence. The very bright red mRFP variant tdTomato (28) is a very useful tag for whole cells or organisms when expressed as a free protein in the cytoplasm, but is not optimal for protein fusions due to its larger size. It is important to use monomeric fluorescent proteins for tagging effectors, since many fluorescent protein variants form dimers or oligomers (dsRed for example) that may lead to aggregation and mislocalization of tagged proteins.

The orientation of the protein fusion, i.e., to which end of the protein of interest the fluorescent tag is fused can be an important factor in the outcome of the experiment. For some proteins, available information on localization signal sequences, functional domains or essential amino acids may be used to determine the most logical fusion protein to construct. It may be, however, that different constructs are necessary to study properties such as localization and interactions separately; for example, if a targeting signal is present in the N terminus but an interacting domain is C-terminal. More complex constructions in which fluorescent tags are inserted between terminal signal sequences and the remainder of the protein are sometimes required.

The possible location of the protein of interest must be taken into account when selecting fluorophores. For example, the apoplast of the commonly used *Nicotiana* species is hostile to GFP and its variants, since a secreted form of GFP gave no apoplastic fluorescence in these plants (31). A few protein fusions to GFP fluoresce well in the *Nicotiana* cell wall/apoplast but it is likely that the fused protein protects the GFP in some way, but in most cases the GFP does not fluoresce. The *Arabidopsis* apoplast appears to be less hostile to GFP (32). mRFP and its variants are stable in the *Nicotiana* and *Solanum* apoplasts (11, 33). For localization studies, the identity of a cellular compartment labeled by a tagged protein is generally sought. This often involves the use of stains, dyes, and fluorescent markers of known (sub)cellular compartments. The choice of fluorophore for locating the protein of interest affects the options available for subsequent identification of the compartment. If the protein of interest is to be co-expressed with a fluorescent protein marker of a cellular compartment or a putative interactor for an assessment of co-localization, the two fluorescent protein tags used must be spectrally separable.

1.2. Cloning of Effectors into A. tumefaciens Cloning Vectors

Numerous plasmid vectors exist for *A. tumefaciens* gene delivery into plant cells, but commonly used vectors for expression in dicots typically incorporate the cauliflower mosaic virus 35S promoter to direct very high levels of gene expression (see Note 1). It is not the intention of this chapter to describe gene cloning techniques. Restriction endonuclease-based cloning is well described (34) and GATEWAY recombination based cloning is best described in the user's manual from the manufacturer (Invitrogen).

1.3. Expression of Fluorescent Proteins from Oomycetes

An alternative to expressing fluorescent-tagged effector proteins within plant cells is to express and secrete them from stable transformants of the source pathogen (Fig. 1c, d). For oomycetes, several *Phytophthora* species and one *Pythium* species have been stably transformed. Other plant pathogenic oomycetes such as species of *Aphanomyces*, and the obligate biotroph downy mildews and white rusts, have been recalcitrant to stable transformation to date.

2. Materials

1. *A. tumefaciens* cells. Glycerol stocks should be maintained at -80°C.
2. Selective agar plates (e.g., LB or YEB + streptomycin for strain LBA4404; see Note 2).
3. Appropriate antibiotic for overnight culture (e.g., streptomycin for LBA4404).
4. Sterile tubes.
5. LB, YEB, and SOC media (YEB is preferable to LB for avoiding clumping of cells) (see ref. 34 for LB and SOC). LB = 10 g/L tryptone, 5 g/L yeast extract, 10 g/L NaCl, pH 7.0. YEB = 1 g/L yeast extract, 5 g/L beef extract, 5 g/L peptone, 5 g/L sucrose, 0.5 g/L $MgSO_4 \cdot 7H_2O$. SOC = 20 g/L tryptone, 5 g/L yeast extract, 0.5 g/L NaCl, 2.5 mM KCl, 10 mM $MgCl_2$, 20 mM glucose, pH 7.0.

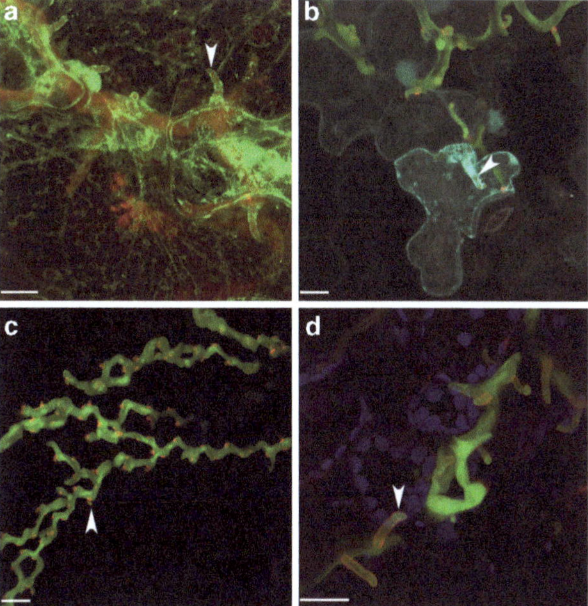

Fig. 1. Confocal imaging of fluorescent protein fusions in *P. infestans*–host plant interactions. (**a**) *P. infestans* expressing cytoplasmic monomeric red fluorescent protein (mRFP) and infecting alternative host, *N. benthamiana*. To label the endoplasmic reticulum (ER) a signal peptide and ER retention signal have been fused to green fluorescent protein (GFP) and this fusion expressed constitutively in transgenic *N. benthamiana*. Note the accumulation of host ER surrounding the invading haustoria (*arrowed*). (**b**) *In planta* expression in *N. benthamiana* of the *P. infestans* translocated RXLR effector Avr3a (minus signal peptide) fused to CFP, following agroinfiltration. Transgenic *P. infestans* expressing cytoplasmic GFP and the Pihmp1 haustorial membrane protein fused to mRFP was inoculated 3 days prior to agroinfiltration to allow establishment of infection. Note the variation in expression levels for the Avr3a::CFP fusion in plant cells, and accumulation of Avr3a::CFP surrounding the site of haustorial penetration (*arrowed*). (**c**) Infection of potato by transgenic *P. infestans* expressing cytoplasmic GFP and an Avr3a::mRFP fusion. Although the Avr3a::mRFP fusion is constitutively overexpressed, it cannot be detected in the much larger host plant cells, presumably due to dilution in the host cytoplasm, but is instead seen to accumulate where haustoria are formed (*arrowed*). (**d**) Higher magnification image of *P. infestans* transformant expressing Avr3a::mRFP. Host chloroplast autofluorescence is false colored *blue*. An example of haustoria are indicated by *arrows*. Scale bars represent 20 μm.

6. Sterile HEPES buffer (10 mM, pH 7.0) or distilled water alone and/or with 10% glycerol, chilled.
7. Liquid nitrogen or dry ice.
8. *Agrobacterium* infiltration buffer (10 mM $MgCl_2$, 10 mM MES pH 5.6, 15 μM acetosyringone).
9. 400–600 mM mannitol or sucrose.
10. Equipment: microcentrifuge (Eppendorf 5415D for example), refrigerated centrifuge (Sorvall RC5C Plus with rotors SLA1500 and SS34 for example, plus associated sterile

centrifuge bottles/tubes), benchtop centrifuge (Eppendorf 5804R for example), electroporator (Bio Rad Micropulser for example), incubator (28°C), incubator (28°C) with shaker, spectrophotometer capable of measuring absorbance at 600 nm (OD_{600}), glasshouse or plant growth cabinet (22°C), confocal microscope (Leica TCS-SP2 AOBS for example).

11. Disposables: sterile 0.5-, 1.5-, and 2.0-mL microcentrifuge tubes, electroporation cuvettes, 1-mL syringes, double sided adhesive tape, microscope slides.

3. Methods

3.1. Preparing Competent A. tumefaciens for Electroporation

This protocol is similar to those described in refs. 35–37.

1. Grow a culture of the *Agrobacterium* strain of choice overnight (5–10 mL YEB plus antibiotic) at 28°C with shaking.
2. Use the overnight culture to inoculate a larger volume of liquid medium (for example, 1-L YEB plus antibiotic).
3. Grow culture in a 28°C incubator with shaking until the cell concentration, measured on a spectrophotometer as absorbance at 600 nm (OD_{600}), reaches between 0.5 and 1 OD_{600} (log-phase culture).
4. Centrifuge cells for 15 min at 5,000 rpm (3,800 × *g*) in Sorvall SLA1500 rotor at 4°C.
5. Resuspend cells in 1 L (i.e., the same volume as the culture) of ice-cold HEPES or water (see Note 3) by gently pipetting up and down with a large-bore pipette (e.g., 25 mL). Keep the centrifuge bottles in ice during the resuspension.
6. Repeat steps 2 and 3; once with water/HEPES and once with 10% glycerol (see Note 3), decreasing the volume to 0.5 L (total of three washes).
7. Finally, resuspend the pellet in approximately 3 mL of 10% glycerol (more may be used if required to ensure that the suspension is liquid enough to pipette).
8. Aliquot bacteria into pre-chilled 0.5-mL microcentrifuge tubes, 40-µL per tube.
9. Snap-freeze in liquid nitrogen or dry ice and store at –80°C.

3.2. Electroporation of A. tumefaciens

1. Thaw an aliquot of electrocompetent *A. tumefaciens* cells on ice.
2. Add plasmid DNA and mix gently, avoiding air bubbles (0.2–0.5 µL of QIAGEN mini-prep plasmid DNA for example; one may first test the efficiency of the cells by electroporating a known quantity of DNA and plating a known proportion of the cells on a plate and counting colonies).

3. Add cells + DNA mixture to an electroporation cuvette that has been pre-chilled on ice.
4. Wipe any moisture from the outside of the cuvette, place cuvette in electroporator, and electroporate using the same settings as for *Escherichia coli* (for example, 1.8 kV for a 0.1-cm gap cuvette).
5. Remove cuvette.
6. Immediately add 1 mL of SOC, LB, or YEB medium and transfer bacteria to a 2-mL microcentrifuge tube.
7. Incubate bacteria at 28°C with shaking for 2–3 h (place tubes horizontally to ensure aeration).
8. Spread aliquots of the bacterial suspension on selective medium (typically YEB plates with kanamycin and rifampicin).

3.3. Transient Expression in Plants

This *Agrobacterium* transient expression ("agroinfiltration") protocol is largely as described in ref. 38. We routinely use *Nicotiana benthamiana* plants for transient expression by agroinfiltration (see Note 2).

1. Grow *N. benthamiana* plants in a greenhouse at 22°C (day temperature) and 18°C (night temperature) with a minimum of 16-h light, plants are best used once they have at least three true leaves (i.e., three fully expanded leaves above the lowest two leaves).
2. Grow overnight *Agrobacterium* cultures of strains containing the binary vector expression constructs in LB or YEB (with appropriate antibiotic) with shaking at 28°C. Some *Agrobacterium* strains may require more than overnight growth.
3. Centrifuge to precipitate bacteria (either 5 min at 12,000 rpm (13,400 × *g*) in a microcentrifuge or, for larger volumes, 15 min at 2,500 rpm (1120 × *g*) in a benchtop centrifuge).
4. Resuspend the pellets in *Agrobacterium* infiltration buffer.
5. Measure the cell density on a spectrophotometer as absorbance at 600 nm (OD_{600}).
6. Dilute the bacterial suspension with the infiltration buffer to adjust the inoculum concentration to the required final OD_{600} (see Note 4). We generally use concentrations of between 0.001 and 0.1 OD_{600}, for imaging experiments, and up to 0.3 for detection in western blot experiments.
7. Lightly touch or scrape leaves with a blade to facilitate infiltration, and infiltrate inoculum into the cut sites gently with a needle-less syringe, using the minimum pressure required, while applying counter-pressure on the opposite side of the leaf with a gloved finger, essentially as described in ref. 39. Where required delimit and label the infiltrated area of the leaf gently with a fine-tip indelible pen.
8. Grow infiltrated plants under normal conditions (as above).

3.4. Confocal Imaging

Imaging of fluorescently tagged effectors and transgenic *Phytophthora* is best performed with a confocal laser scanning microscope, although good low magnification images of the overall growth patterns of a fluorescently tagged *Phytophthora*, for example, may be collected using a standard fluorescence microscope. Precisely how cells expressing fluorescent proteins are imaged by confocal microscopy largely depends on the make and model of the confocal microscope being used (see Note 5). It is recommended that adequate training in operation of the specific microscope is received prior to the commencement of imaging.

For all images shown in Fig. 1, and elsewhere (11, 40), we have used a Leica TCS-SP2 AOBS confocal microscope using HCX APO L 20×/0.5, 40×/0.8, and 63×/0.9 water-dipping lenses. The use of water-dipping lenses allows minimal preparation of leaf material for imaging. Leaves are simply stuck to slides with double-sided adhesive tape. The abaxial side should face upward when agroinfiltrated leaves are being examined as the lower epidermis contains many more expressing cells than the upper epidermis due to the greater number of airspaces in the spongy mesophyll into which the agrobacterial suspension is infiltrated. Cells are imaged through a drop of water placed on the leaf surface into which the lens is dipped. An additional advantage is that the leaf cells remain hydrated by the water droplets.

3.5. Bimolecular Fluorescence Complementation

BiFC or "split YFP" relies on the ability of the two separated halves of the YFP protein to re-associate to form a fully fluorescent protein. This implies that protein–protein interactions can be tested in living cells (15). However, successful application of this strategy does require stringent controls to ensure the validity of results obtained. Controls such as using the two free halves (N-terminal "YN" and C-terminal "YC") of YFP are required to determine the background level of fluorescence from YFP that reforms spontaneously under the expression conditions used (see Note 6). Additional control pairings are shown in Table 1. These control pairings indicate what level of background fluorescence occurs in the expression conditions used and whether there is dimerization of either the effector or the interaction partner that would lead to the reconstitution of YFP fluorescence, and also control for mislocalization due to inappropriate tagging of the proteins that may lead to altered protein stability or interactions.

3.6. Expression of Fluorescent Proteins from Oomycetes

Transformation of *Phytophthora* species is possible by a variety of methods, such as $CaCl_2$/polyethylene glycol (PEG) transformation of protoplasts, electroporation of zoospores or protoplasts, *A. tumefaciens*, and biolistic delivery of DNA. However, most reports where high levels of overexpression have been reported have used $CaCl_2$/PEG transformation of protoplasts

Table 1
Expression constructs for bimolecular fluorescence complementation (BiFC) (split YFP): controls and test fusions

Controls for BiFC	*Purpose*
YFP-effector Effector-YFP	Expression, stability, and localization of effector
YFP-interactor Interactor-YFP	Expression, stability, and localization of interactor
YN-non-interactor 1 + YC-non-interactor 2	Background re-association of YN and YC
YN-effector + YC-effector YN-effector + effector-YC Effector-YN + YC-effector Effector-YN + effector-YC	Self-association of effector
YN-interactor + YC-interactor YN-interactor + interactor-YC Interactor-YN + YC-interactor Interactor-YN + interactor-YC	Self-association of interactor
Test pairings for BiFC	*Specific controls*
YN-effector + YC-interactor YN-effector + interactor-YC Effector-YN + YC-interactor Effector-YN + interactor-YC YN-interactor + YC-effector YN-interactor + effector-YC Interactor-YN + YC-effector Interactor-YN + effector-YC	Combinations as in "test pairings" but using non-interacting forms of either or both proteins (as determined by other techniques or proposed through analysis of sequence motifs)

and plasmids incorporating the *Bremia lactucae Ham34* promoter to direct constitutive overexpression (see Note 7). It is not the intention to describe *Phytophthora* transformation here. Chapter 11 in this volume has described one available method, and other protocols are well documented and can be found at (http://oomyceteworld.net/) and in ref. (41) and citations therein.

Localization of effectors delivered from an oomycete pathogen to the host cytoplasm is yet to be observed using fluorescent protein tagging (Fig. 1c, d). This is presumably due to a dilution of pathogen effectors within the host cell cytoplasm. It is possible to concentrate the host cell cytoplasm and thus identify delivery of effectors. This can be done by plasmolysis using sucrose or mannitol.

3.7. Plasmolysis of Plant Cells

1. Syringe or vacuum-infiltrate the leaf with 400–600 mM mannitol or sucrose (mannitol acts faster but can cause problems with crystal formation).
2. Plasmolysis should occur within 5–30 min (see Note 8). The cytoplasm, bounded by the plasma membrane, moves away from the cell wall. Hechtian strands, where the cytoplasm, ER and cytoskeleton remain connected to the cell wall should be visible.

4. Notes

If constructing an *A. tumefaciens* vector for fluorescent protein tagging of effectors, and depending on the gene cloning system being used, it is useful to incorporate cloning sites at either end of the fluorescent protein gene to allow both N- and C-terminal tagging of effectors. Genes encoding effector proteins typically encode an N-terminal signal peptide for secretion. For localization inside plant cells, the effector gene should be cloned without its signal peptide sequence, such that it is retained in the plant cell upon expression. These gene cloning considerations are most important for RXLR-EER or other effectors directed to the host cell cytoplasm. For other potentially secreted effectors, it may be advantageous to retain the signal peptide for secretion into the apoplast.

The strain LBA4404 has the advantage of being commercially available. It may not, however, be the optimal strain to use. Other strains may be preferable in different plant hosts, different host genotypes, different pathosystems, and different expression requirements. We routinely use *N. benthamiana* for transient expression by agroinfiltration. All strains of agrobacteria commonly used in our laboratory (AGL1, LBA4404, and GV3101) work well in this species.

All of the washes in preparation of electrocompetent *A. tumefaciens* cells may be done with HEPES or water with 10% glycerol.

For experiments requiring co-infection of more than one fluorescent protein fusion construct, bacterial strains containing the constructs are mixed prior to the leaf infiltration, with the concentration of each strain adjusted to the required final OD_{600}.

Leaves may be agroinfiltrated through stomatal openings if they are open (usually in the morning and on overcast days when the plants are sufficiently hydrated). The needle-less syringe with the agroinfiltration mixture is simply pressed as lightly as possible against the leaf with a gloved fingertip on the other side for support.

If the stomata are not open, it is best to minimize damage to the leaf by making a very small incision with a sharp blade tip or needle; just enough to break the cuticle and epidermis (a single touch creating a minute point is sufficient).

It may be expected that effectors may target host proteins that are only present/correctly activated/correctly localized in the presence of pathogen infection. Thus, to most closely simulate the effects of a compatible pathogen infection, it may be advantageous to co-inoculate the *A. tumefaciens* infiltrated site on the plant leaf with either a wild-type or transgenic pathogen cytoplasmically expressing a fluorescent protein with a different emission spectrum (Fig. 1). For example, a GFP-tagged effector expressed in plant cells would be best imaged in conjunction with, for example, an mRFP-expressing pathogen. If this is done, it is important to first test if the chosen pathogen is compatible with *A. tumefaciens* infection, as it may inhibit ingress by some pathogens. Passage of the pathogen through a compatible host, followed by re-isolation, can improve the aggressiveness of many pathogens, especially if these cultures have been held under laboratory conditions for lengthy periods. This strategy can prove useful in improving co-infections with *A. tumefaciens.* We have found that co-infection of *P. infestans* with *A. tumefaciens* can be conducted by inoculating with *P. infestans* immediately after infiltration, and within the infiltrated area. Alternatively, pathogen infection can be initiated some days before infiltration with *A. tumefaciens* at the infection margin such that the *Phytophthora* grows into the agroinfiltrated area (Fig. 1b). The desired outcome of co-infection is to identify those host cells that express the fluorescently tagged effector, while also being infected or contacted by the pathogen. Transient expression after *A. tumefaciens* infiltration generally persists for up to 5 days, and so all imaging of tagged effectors must occur within this time.

It should also be considered that the localization of effectors by *A. tumefaciens* infiltrations involves the presence of a pathogen (*A. tumefaciens*), and that this may influence the outcome of an experiment. Alternatives to *A. tumefaciens* infiltration are transient expression from biolistic delivery of DNA (i.e., microprojectile bombardment) or establishment of stable transgenic plants expressing the tagged effector. Biolistic delivery results in a very scattered distribution of effector-expressing cells and often significant cell damage, while the production of transgenic plants can be very labor-intensive and time-consuming.

The excitation and emission spectrum for a fluorophore expressed in a leaf is not identical to that obtained from a purified preparation of that protein. Ideally, lambda scans should be performed on each of the fluorophores being used in the experimental system, e.g., in a plant leaf cell, and then select the imaging parameters according to the results of the lambda scans. We generally image

GFP using the 488 nm excitation line from an argon laser and collect its emissions between 500 and 530 nm but this detection spectrum is generally narrowed when GFP is being imaged along with a second fluorophore, e.g., 511–530 nm when used in conjunction with aniline blue staining.

It is extremely important to ensure good spectral separation of co-expressed fluorophores. The methods and requirements for this depend on the confocal imaging system used. On the Leica TCS-SP2, for example, where possible we sequentially image combinations such as CFP and YFP, YFP and mRFP, and even GFP and mRFP. Sequential imaging involves alternately turning on and off the excitation lines for the different fluorophores, only collecting the emissions from the fluorophore relating to each excitation line while that line is on.

Sequential imaging is preferred because many fluorescent proteins also have secondary (lower) fluorescence emission peaks that can cause interference when multiple fluorescent proteins are imaged simultaneously. For example, GFP has a secondary fluorescence emission peak within the same spectral region as mRFP and other red fluorescent proteins. Sequential imaging, as opposed to simultaneous imaging, allows only the fluorescence from a single specified wavelength range to be gathered at any one time. The two (or more) images are then overlaid to obtain the merged image. Images can be processed post-acquisition using standard software packages, such as Adobe Photoshop or more specialist suites such as Aviso.

Following *A. tumefaciens* infiltration, plant cells that express the fluorescently tagged effector at different levels are observed; from very high levels to very low levels (see Fig. 1b for example). Proteins expressed to very high levels may result in artifactual localizations, such as protein aggregations, or apparently cytoplasmic localization. Observation of cells with lower levels of expression may reveal more detail about preferential or genuine localization within the cell.

A general consideration for all fluorescent protein tagging is that it is important to determine whether the fluorescent tag remains attached to the protein of interest and that the whole fusion protein is intact. This should be done for all constructs using western blotting.

Effector-interacting protein partners may be identified through either yeast 2 hybrid or co-immunoprecipitation experiments, followed by BiFC to determine if the interaction does indeed occur in vivo, and to localize the interaction within plant cells.

A control for background re-association of YFP could involve co-expression of the two free halves (N-terminal "YN" and C-terminal "YC"), but perhaps a more valid control would involve co-expression of fusions to non-interacting proteins.

If using constitutive overexpression from the *Ham34* promoter, it is first essential to establish when during infection the gene of interest is expressed and likely to be found in the interaction with host plants. For example, genes expressed in the early infection stages of *P. infestans* of potato are most likely to localize at the leading edge of a disease lesion as new cells are contacted and infected. It may also be an advantage to observe localization using the native promoter for the gene of interest and to ascertain the localization of expression in the correct biological context.

Plant epidermal cells do not plasmolyse as readily as mesophyll cells.

Acknowledgments

PCB and SW are funded by the Scottish Government Rural and Environment Research and Analysis Directorate (RERAD) and the Biotechnology and Biological Sciences Research Council (BBSRC), UK. PRJB is funded by the University of Dundee, UK, and the BBSRC.

References

1. Jones, J. D. and Dangl, J. L. (2006) The plant immune system. *Nature* **444**, 323–329.
2. Block, A., Li, G., Fu, Z. Q., and Alfano, J. R. (2008) Phytopathogen type III effector weaponry and their plant targets. *Curr. Opin. Plant Biol.* **11**, 396–403.
3. Shan, W., Cao, M., Leung, D., and Tyler, B. M. (2004) The *Avr1b* locus of *Phytophthora sojae* encodes an elicitor and a regulator required for avirulence on soybean plants carrying resistance gene *Rps1b*. *Mol. Plant Microbe Interact.* **17**, 394–403.
4. Allen, R. L., Bittner-Eddy, P. D., Grenville-Briggs, L. J., Meitz, J. C., Rehmany, A. P., Rose, L. E., and Beynon, J. L. (2004) Host-parasite co-evolutionary conflict between *Arabidopsis* and downy mildew. *Science* **306**, 1957–1960.
5. Rehmany, A. P., Gordon, A., Rose, L. E., Allen, R. L., Armstrong, M. R., Whisson, S. C., et al. (2005) Differential recognition of highly divergent downy mildew avirulence gene alleles by *RPP1* resistance genes from two *Arabidopsis* lines. *Plant Cell* 7, 1839–1850.
6. Armstrong, M. R., Whisson, S. C., Pritchard, L., Bos, J. I., Venter, E., Avrova, A. O., et al. (2005) An ancestral oomycete locus contains late blight avirulence gene *Avr3a*, encoding a protein that is recognised in the host cytoplasm. *Proc. Natl Acad. Sci. U. S. A.* **102**, 7766–7771.
7. van Poppel, P. M. A. J., Guo, J., van de Vondervoort, P. J. I., Jung, M. W. M., Birch, P. R. J., Whisson, S. C., and Govers, F. (2008) The *Phytophthora infestans* avirulence gene *Avr4* encodes an RXLR-dEER effector. *Mol. Plant Microbe Interact.* **21**, 1460–1470.
8. Vleeshouwers, V. G., Rietman, H., Krenek, P., Champouret, N., Young, C., Oh, S. K., et al. (2008) Effector genomics accelerates discovery and functional profiling of potato disease resistance and *Phytophthora infestans* avirulence genes. *PLoS ONE* **3**, e2875. doi:10.1371/journal.pone.0002875.
9. Qutob, D., Tedman-Jones, J., Dong, S., Kuflu, K., Pham, H., Wang, Y., et al. (2009) Copy number variation and transcriptional polymorphisms of *Phytophthora sojae* RXLR effector genes *Avr1a* and *Avr3a*. *PLoS ONE* **4**, e5066. doi:10.1371/journal.pone.0005066.
10. Dong, S., Qutob, D., Tedman-Jones, J., Kuflu, K., Wang, Y., Tyler, B.M., and Gijzen, M. (2009) The *Phytophthora sojae* avirulence

locus *Avr3c* encodes a multi-copy RXLR effector with sequence polymorphisms among pathogen strains. *PLoS ONE* 4, e5556. doi:10.1371/journal.pone.0005556.

11. Whisson, S. C., Boevink, P. C., Moleleki, L., Avrova, A. O., Morales, J. G., Gilroy, E. M., et al. (2007) A translocation signal for delivery of oomycete effector proteins into host plant cells. *Nature* **450**, 115–118.
12. Dou, D., Kale, S. D., Wang, X., Jiang, R. H., Bruce, N. A., Arredondo, F. D., et al. (2008) RXLR-mediated entry of *Phytophthora sojae* effector Avr1b into soybean cells does not require pathogen-encoded machinery. *Plant Cell* **20**, 1930–1947.
13. Win, J., Morgan, W., Bos, J., Krasileva, K. V., Cano, L. M., Chaparro-Garcia, A., et al. (2007) Adaptive evolution has targeted the C-terminal domain of the RXLR effectors of plant pathogenic oomycetes. *Plant Cell* **19**, 2349–2369.
14. Dou, D., Kale, S. D., Wang, X., Chen, Y., Wang, Q., Wang, X., et al. (2008) Conserved C-terminal motifs required for avirulence and suppression of cell death by *Phytophthora sojae* effector Avr1b. *Plant Cell* **20**, 1118–1133.
15. Bos, J. I., Armstrong, M. R., Gilroy, E. M., Boevink, P. C., Hein, I., Taylor, R. M., et al. (2010) *Phytophthora infestans* effector AVR3a is essential for virulence and manipulates plant immunity by stabilizing host E3 ligase CMPG1. *Proc. Natl Acad. Sci. U. S. A.* **107**, 9909–9914.
16. Tyler, B. M., Tripathy, S., Zhang, X., Dehal, P., Jiang, R. H., Aerts, A., et al. (2006) *Phytophthora* genome sequences uncover evolutionary origins and mechanisms of pathogenesis. *Science* **313**, 1261–1266.
17. Haas, B. J., Kamoun, S., Zody, M. C., Jiang, R. H., Handsaker, R. E., Cano, L. M., et al. (2009) Genome sequence and analysis of the Irish potato famine pathogen *Phytophthora infestans*. *Nature* **461**, 393–398.
18. Judelson, H. S., Ah-Fong, A. M., Aux, G., Avrova, A. O., Bruce, C., Cakir, C., et al. (2008) Gene expression profiling during asexual development of the late blight pathogen *Phytophthora infestans* reveals a highly dynamic transcriptome. *Mol. Plant Microbe Interact.* **21**, 433–447.
19. Hein, I, Gilroy E. M., Armstrong, M. R., and Birch, P. R. J. (2009) The zig-zag-zig in oomycete–plant interactions. *Mol. Plant Pathol.* **10**, 547–562. doi:10.1111/j.1364-3703.2009.00547.x.
20. Song, J., Win, J., Tian, M., Schornack, S., Kaschani, F., Ilyas, M., et al. (2009) Apoplastic effectors secreted by two unrelated eukaryotic plant pathogens target the tomato defense protease Rcr3. *Proc. Natl Acad. Sci. U. S. A.* **106**, 1654–1659.
21. Damasceno, C. M., Bishop, J. G., Ripoll, D. R., Win, J., Kamoun, S., and Rose, J. K. (2008) Structure of the glucanase inhibitor protein (GIP) family from *Phytophthora* species suggests coevolution with plant endo-beta-1,3-glucanases. *Mol. Plant Microbe Interact.* **21**, 820–830.
22. Sun, W. X., Jia, Y. J., Feng, B. Z., O'Neill, N. R., Zhu, X. P., Xie, B. Y., and Zhang, X. G. (2009) Functional analysis of Pcipg2 from the straminopilous plant pathogen *Phytophthora capsici*. *Genesis* **47**, 535–544. doi:10.1002/dvg.20530.
23. Liu, Z., Bos, J., Armstrong, M., Whisson, S. C., da Cunha, L., Torto, T., et al. (2005) Patterns of diversifying selection in the phytotoxin-like *scr74* gene family of *Phytophthora infestans*. *Mol. Biol. Evol.* **22**, 659–672.
24. Thieme, F., Szczesny, R., Urban, A., Kirchner, O., Hause, G., and Bonas, U. (2007) New type III effectors from *Xanthomonas campestris* pv. *vesicatoria* trigger plant reactions dependent on a conserved N-myristoylation motif. *Mol. Plant Microbe Interact.* **20**, 1250–1261.
25. Doehlemann, G., van der Linde, K., Assmann, D., Schwammbach, D., Hof, A., Mohanty, A., et al. (2009) Pep1, a secreted effector protein of *Ustilago maydis*, is required for successful invasion of plant cells. *PLoS Pathog.* **5**, e1000290. doi:10.1371/journal.ppat.1000290.
26. Mosquera, G., Giraldo, M. C., Khang, C. H., Coughlan, S., and Valent, B. (2009) Interaction transcriptome analysis identifies *Magnaporthe oryzae* BAS1-4 as biotrophy-associated secreted proteins in rice blast disease. *Plant Cell* **21**, 1273–1290.
27. Chapman, S., Oparka, K. J., and Roberts, A. G. (2005) New tools for in vivo fluorescence tagging. *Curr. Opin. Plant Biol.* **8**, 565–573.
28. Shaner, N. C., Campbell, R. E., Steinbach, P. A., Giepmans, B. N. G., Palmer, A. E., and Tsien, R. Y. (2004) Improved monomeric red, orange and yellow fluorescent proteins derived from *Discosoma* sp. red fluorescent protein. *Nat. Biotechnol.* **22**, 1567–1572.
29. Haseloff, J., Siemering, K. R., Prasher, D. C., and Hodge, S. (1997) Removal of a cryptic intron and subcellular localization of green fluorescent protein are required to mark transgenic *Arabidopsis* plants brightly. *Proc. Natl Acad. Sci. U. S. A.* **94**, 2122–2127.
30. Campbell, R. E., Tour, O., Palmer, A. E., Steinbach, P. A., Baird, G. S., Zacharias, D. A., and Tsien, R. Y. (2002) A monomeric red

fluorescent protein. *Proc. Natl Acad. Sci. U. S. A.* **99**, 7877–7882.

31. Boevink, P., Martin, B., Oparka, K., Cruz, S. S., and Hawes, C. (1999) Transport of virally expressed green fluorescent protein through the secretory pathway in tobacco leaves is inhibited by cold shock and brefeldin A. *Planta* **208**, 392–400.
32. Zheng, H., Kunst, L., Hawes, C., and Moore, I. (2004) A GFP-based assay reveals a role for RHD3 in transport between the endoplasmic reticulum and Golgi apparatus. *Plant J.* **37**, 398–414.
33. Gilroy, E. M., Hein, I., van der Hoorn, R., Boevink, P. C., Venter, E., McLellan, H., et al. (2007) Involvement of cathepsin B in the plant disease resistance hypersensitive response. *Plant J.* **52**, 1–13.
34. Sambrook, J., Fritsch, E. F., and Maniatis, T. (1989) Molecular cloning: a laboratory manual, 2nd edition. Cold Spring Harbor Press, New York.
35. Mattanovich, D., Rüker, F., Machado, A. C., Laimer, M., Regner, F., Steinkellner, H., et al. (1989) Efficient transformation of *Agrobacterium* spp. by electroporation. *Nucleic Acids Res.* **17**, 6747.
36. Mersereau, M., Pazour, G. J., and Das, A. (1990) Efficient transformation of *Agrobacterium tumefaciens* by electroporation. *Gene* **90**, 149–151.
37. Shaw, C. H. (1995) Introduction of cloning plasmids into *Agrobacterium tumefaciens*. pp 33–37. In: Methods in Molecular Biology, vol. 49: Plant gene transfer and expression protocols. H. Jones (ed.). Humana Press, Totowa, NJ.
38. Latijnhouwers, M., Hawes, C., Carvalho, C., Oparka, K., Gillingham, A. K., and Boevink, P. C. (2005) An *Arabidopsis* GRIP domain protein locates to the trans-Golgi and binds the small GTPase ARL1. *Plant J.* **44**, 459–470.
39. Batoko, H., Zheng, H. Q., Hawes, C., and Moore, I. (2000) A rab1 GTPase is required for transport between the endoplasmic reticulum and Golgi apparatus and for normal Golgi movement in plants. *Plant Cell* **12**, 2201–2218.
40. Avrova, A. O., Boevink, P. C., Young, V., Grenville-Briggs, L. J., van West, P., Birch, P. R. J., and Whisson, S. C. (2008) A novel *Phytophthora infestans* haustorium-specific membrane protein is required for infection of potato. *Cell. Microbiol.* **10**, 2271–2284.
41. Judelson, H. S. and Ah Fong, A. M. V. (2009) Progress and challenges in oomycete transformation. pp 435–453. In: Oomycete genetics and genomics: diversity, plant and animal interactions, and toolbox. S. Kamoun and K. Lamour (eds.). John Wiley & sons, Hoboken, NJ.

Chapter 17

Immunolocalization of Pathogen Effectors

Eric Kemen, Kurt Mendgen, and Ralf T. Voegele

Abstract

The use of polyclonal antibodies enables the detection of proteins on a cellular and even subcellular level. Immunolocalization can be used on all pathosystems even if one or both partners of the interaction are unamenable to molecular tools like transformation. This chapter provides detailed information about how to obtain high quality antibodies, how to prepare samples, and finally how to detect the proteins. Methods for light and electron microscopy are presented.

Key words: Immunolocalization, Polyclonal antibodies, Microscopy, Effector

1. Introduction

One of the crucial steps in pathogenicity of bacteria or eukaryotic pathogens is the delivery of effectors into host cells to render the plant susceptible (1). Several approaches like fusion to fluorescent proteins and other tags have been established to localize these important effectors within plant cells to determine their site of action. Tagging methods are useful tools as long as transformation systems are available for the organism of interest. However, such systems have not been established for most obligate biotrophic pathogens yet. To analyze effectors from this class of pathogens, heterologous expression is often the method of choice. Yet, these results should be handled with great caution since expression levels as well as posttranslational modifications carried out by the native organism are known to influence subcellular localizations (2).

In this chapter, we present a methodology for localizing proteins in the native system on a cellular and a subcellular level using immunocytological techniques. This method comprises two

John M. McDowell (ed.), *Plant Immunity: Methods and Protocols*, Methods in Molecular Biology, vol. 712,
DOI 10.1007/978-1-61737-998-7_17, © Springer Science+Business Media, LLC 2011

major steps: generation and purification of high quality polyclonal antibodies and sample preparation for microscopy.

Since concentrations of effector proteins are expected to be low in native systems (3), polyclonal antibodies recognizing different epitopes of the target protein are superior in signal intensity to monoclonal antibodies. Despite this advantage, the mayor criticism toward polyclonal antibodies has long been their potential risk regarding unspecific binding due to cross-reaction with other proteins and polysaccharides. One major problem is, for example, the cross-reaction of antibodies with components of the extrahaustorial matrix surrounding fungal haustoria. A careful choice of the right constructs for overexpression of the antigen of interest prior to immunization and combining different steps of affinity purification greatly reduce these problems.

A meticulous sample preparation is another crucial step in succeeding with immunolocalization. Proteins need to be preserved at their native location but in turn the goal is to reach a high signal to noise ratio. This can only be achieved by unfolding and therefore denaturing the proteins to present enough epitopes to bind antibodies. This struggle between resolution and signal intensity has led to a broad range of embedding protocols and embedding media (for details see ref. 4). Two most frequently used embedding media in light and electron microscopy are acrylic and epoxy resins. Acrylic resins, in general, do not covalently crosslink with either lipids or proteins and are therefore ideal in preserving antigenicity of proteins. However, they are poor in preserving membrane structures. By contrast, epoxy resins show high structural preservations but covalently link to proteins and therefore reduce antigenicity and signal intensity.

2. Materials

2.1. Generation and Purification of Antibodies

1. Suitable overexpression system for antigens (e.g., pET system (Novagen, Merck)), including vectors for overexpression (e.g., pET28a, pET32a (Novagen, Merck)) and overexpression strains (e.g., BL21(DE3), Origami B(DE3), Rosetta B(DE3) (Novagen, Merck)).
2. Gene-specific primers for cloning of the desired portion of the antigen (Fig. 1).
3. Thermal cycler (e.g., Mastercycler Gradient, Eppendorf).
4. DNA polymerase with proofread function (e.g., Expand High Fidelity, Roche).
5. Agarose gel electrophoresis equipment.
6. Standard tools for molecular cloning like restriction endonucleases, buffers, clean up kits for PCR reactions, DNA ligase and buffers, plasmid isolation kits.

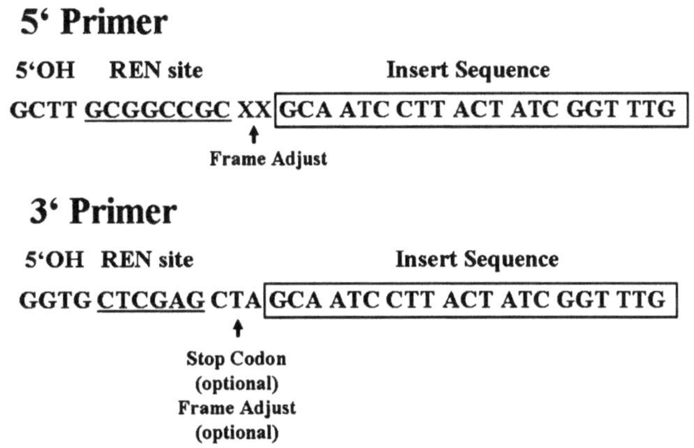

5' Primer

5'OH REN site Insert Sequence

GCTT GCGGCCGC XX | GCA ATC CTT ACT ATC GGT TTG |

Frame Adjust

3' Primer

5'OH REN site Insert Sequence

GGTG CTCGAG CTA | GCA ATC CTT ACT ATC GGT TTG |

Stop Codon (optional)
Frame Adjust (optional)

Fig. 1. Schematic representation of primers to amplify desired target and at the same time insert pertinent restriction endonuclease sites. The 5′-primer is shown with a *Not*I restriction endonuclease site. Random bases (XX) may be inserted after the restriction endonuclease site to adjust the reading frames of tag and antigen. The depicted 3′-primer introduces an *Xho*I site. The 3′-primer may carry an artificial stop codon to prevent read through. Both primers carry 5′ sequence extensions for better performance of the restriction endonucleases.

7. Suitable growth medium (e.g., Luria Bertani medium) containing pertinent antibiotics.
8. Inducer stock solution for overexpression (e.g., 1 M IPTG).
9. Incubation shaker (e.g., Multitron II, Infors) set to 37°C and 300 rpm, 1 l Erlenmeyer flasks for incubation.
10. Refrigerated centrifuge (e.g., RC5B, Sorvall), fixed angle rotor (e.g., GSA), centrifuge bottles.
11. Buffer B, 100 mM NaH_2PO_4, 10 mM Tris base, 8 M urea, pH 8.0 (using HCl) (The QIAexpressionist, Qiagen).
12. Ni-NTA Superflow (Qiagen).
13. Poly-Prep Chromatography column (0.8×4 cm) (BioRad Laboratories).
14. Buffer C, 100 mM NaH_2PO_4, 10 mM Tris base, 8 M urea, pH 6.3 (using HCl) (The QIAexpressionist, Qiagen).
15. Buffer D, 100 mM NaH_2PO_4, 10 mM Tris base, 8 M urea, pH 5.9 (using HCl) (The QIAexpressionist, Qiagen).
16. Buffer E, 100 mM NaH_2PO_4, 10 mM Tris base, 8 M urea, pH 4.5 (using HCl) (The QIAexpressionist, Qiagen).
17. Vertical electrophoresis system for protein separation (SDS-PAGE) (e.g., Mini-PROTEAN Tetra Electrophoresis System, BioRad Laboratories).
18. Protein assay kit (e.g., BioRad Protein-assay, BioRad Laboratories).
19. Adjuvant for immunization (e.g., Freund's Adjuvant, Sigma Aldrich).

20. Western Blot equipment (e.g., Mini Trans-Blot Module of the Mini-PROTEAN Tetra Cell, BioRad Laboratories).
21. CNBr-activated Sepharose 4B (GE Healthcare).
22. 1 mM HCl.
23. Coupling buffer, 0.1 M $NaHCO_3$, 0.5 M NaCl pH 8.3.
24. 15 and 50 ml Falcon tubes.
25. Acetate buffer, 0.1 M Na-acetate, 0.5 M NaCl pH 4.0.
26. Glycine buffer, 0.2 M Glycine pH 8.0.
27. Elution buffer, 0.05 M Glycine pH 2.2.
28. 1 M Tris–HCl pH 8.0 and 10 mM Tris–HCl pH 8.0.

2.2. Light Microscopy

1. Microtome for semithin sections and glass knifes.
2. Light microscope with DIC or phase contrast and epifluorescence.
3. PAP-PEN (Kisko-Biotech). Provides a thin film-like barrier when a circle is drawn around the specimen on a slide. This barrier creates a surface tension to hold a wash, or antibody solution within the target area on the slide.
4. Fixing medium should be prepared fresh shortly before use: mix glacial acetic acid with ethanol (1:3).
5. Embedding medium (all chemicals from Sigma): Butyl-methacrylate, Methyl-methacrylate, Benzoyl-peroxide 75% (stabilized with 25% H_2O), Dithiothreitol (DTT), *N,N*- dimethylaniline.

   The acrylic resin mix should be stored at 4°C and kept dry all the time. Once turning slightly milky it should be discarded.

   Mix 80 ml Butyl-methacrylate, 20 ml Methyl-methacrylate and 1.4 g Benzoyl-peroxide 75% in a separating funnel. Invert until Benzoyl-peroxide is completely dissolved and leave funnel until water can be easily removed. Add DTT and invert until dissolved. Purge the mixture with nitrogen for 20 min to remove traces of water and oxygen. Keep resin in an airtight container until use.
6. TBS, 10 mM Tris–HCl and 150 mM NaCl. Adjust pH to 7.5.
7. Incubation buffer: prepare a 1% BSA-C (Aurion) (v/v) dilution in TBS. Add up to 3 mM sodium azide and store at 4°C.
8. Block buffer: dissolve 5% BSA (Albumin bovine Fraction V) (w/v) in TBS. When dissolved add 5% Serotec Goat Serum (Serotec GmbH) (see Note 1) and supplement with 3 mM sodium azide. Store at 4°C.
9. Buffer for pre-absorption: grind leaves of your host plant in liquid nitrogen until pulverized. Add TBS 0.6:1 (w/v) and

sonicate for 10 min. Spin down for 15 min and 16,000 rpm in an SS34-rotor and resuspend the pellet in half the start volume. Sonicate for 5 min to homogenize the pellet and autoclave. Store the buffer at –20°C.

10. Mounting medium DABCO pH 8.5 (5): prepare a solution of 82 mM 1,4-Diazabicyclo[2.2.2]octan in 50% glycerol in water. Store mounting medium at –20°C and bring to room temperature before use.

2.3. Additional Material for Electron Microscopy

1. Microtome for ultrathin sections and diamond knifes.
2. Ni-slot grits.
3. Substitution medium: precool a mixture of 50% acetone and 50% methanol on dry ice. Make sure that your acetone and methanol are free of water. Add 0.2% (w/v) uranyl acetate, 1.0% osmium tetroxide and 2.0% Glutaraldehyde. It is extremely important to make sure that you prepare the mixture at less than –60°C, otherwise, the chemicals start to react.
4. Epon/Araldit (6): mix 20.7% (v/v) Epoxy embedding medium (Fluka GmbH), 17.0% (v/v) Durcupan® ACM (Fluka GmbH), 2.3% (v/v) dibutylphthalate, and 50.0% Epoxy embedding medium (Fluka GmbH). Aliquot resin and store at –20°C.
5. Glycine buffer: 0.1 M Glycine in TBS.
6. Uranyl acetate: prepare a 3% uranyl acetate solution in double distilled water. Use 0.2 nm filter each time before staining.
7. Lead citrate: (adapted from (7)) Mix 1.33 g $Pb(NO_3)_2$ and 1.76 g $Na_3(C_6H_5O_7)\times 2H_2O$ in boiled water. Mix well for 30 min (!) to allow complete conversion of lead nitrate into lead citrate. Add 8 ml decarbonated 1 N NaOH (use 0.1 N Fixanal® buffer concentrate (Sigma) and add 1:10th of the volume recommended). Top up to 50 ml and mix until solution is clear. Leave at least 2 days at 4°C before using it for the first time. Do not shake before use and discard once it turns milky.

3. Methods

3.1. Generation and Purification of Antibodies

3.1.1. Generation of Expression Vector

1. Select a suitable portion of your target gene based on in silico prediction of hydrophilicity and antigenicity (see Note 2).
2. Design pertinent primers to amplify the desired region and at the same time introducing suitable restriction sites (Fig. 1, see Note 3).
3. Amplify your target using DNA polymerase exhibiting proofreading function to minimize nucleotide exchanges during amplification.

4. Check PCR amplification on an agarose gel and clean up PCR product with a suitable kit.
5. Digest PCR product and vector with the pertinent restriction endonucleases (see Note 4).
6. Clean up digests using plasmid clean up kits, and ligate the insert into the vector. Following clean up of the ligation mix, a pertinent *Escherichia coli* strain is transformed (see Note 5).
7. Re-isolate plasmid and verify by sequencing using vector-specific primers covering the insert sites and the complete insert.
8. Transform a suitable expression strain (see Note 6).

3.1.2. Overexpression

1. Check the best duration for induction on a test scale overexpression (see Note 7).
2. For large-scale overexpression start with a 5 ml over night (ON) culture in LB medium. Subculture into 250 ml fresh LB containing antibiotic in a 1 l Erlenmeyer flask to a starting OD_{600} of around 0.1. You can do four flasks in parallel.
3. Grow cells to an OD_{600} of 0.5 and induce the system by adding IPTG to a final concentration of 1 mM.
4. After the predetermined time point for maximum overexpression harvest cells by centrifugation (4°C, 10 min, 10,000 × *g*).
5. Resuspend each pellet in 5 ml buffer B and let stir for 1 h at room temperature (RT).
6. Pellet insoluble material by centrifugation (RT, 15 min, 14,000 × *g*) and keep the supernatant (see Note 8).

3.1.3. Purification of Antigen

1. Resuspend Ni-NTA column material by carefully shaking. Remove 1 ml suspension and fill into Poly-Prep column. Let the material settle and equilibrate with 5 ml buffer B.
2. Apply the supernatant (step 6 in Subheading 3.1.2) to the column while collecting the flow through. Apply the flow through again to the column 2–3 times.
3. Wash column twice using 5 ml buffer C (collect two fractions).
4. Elute proteins using 4 × 0.5 ml buffer D (collect four fractions).
5. Elute tagged proteins using 4 × 0.5 ml buffer E (collect four fractions).
6. Finish the elution process by adding 2 × 0.5 ml buffer B (columns may be stored at 4°C for up to 3 months for reuse if trace amounts of Na-azide are added to the buffer).
7. Check the purification protocol. At each purification step (lysed cells, column flow throughs, elution fractions) take 4.5 μl aliquots and mix with sample buffer for SDS-PAGE.

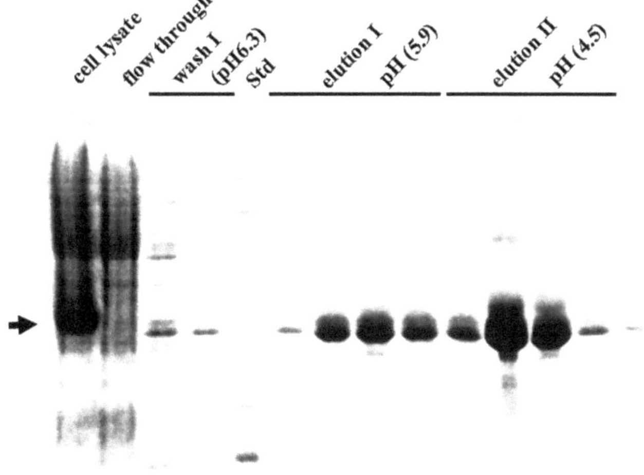

Fig. 2. Purification procedure for the antigen. Samples from left to right represent aliquots of: cell lysate, flow through, wash I (buffer C, two fractions), protein size marker (Std), elution I (buffer D, four fractions), elution II (buffer E, four fractions), re-equilibration of column into buffer B. The arrow indicates the position of the tagged antigen (size 35 kDa).

Incubate at 95°C for 5 min, spin down samples and apply the supernatant to a 12% SDS-PAGE gel (Fig. 2).

8. Combine fractions containing the desired antigen in high quantities (Fig. 2, elution I (fractions 1 and 2) and elution II (fraction 2 and 3)).
9. Do a protein determination to quantify the purified antigen (see Note 9).
10. Store antigen until use at –20°C.

3.1.4. Immunization

1. Aliquot antigen to 250 µg (smaller animals like guinea pigs) or 500 µg (larger animals like rabbits).
2. Mix one aliquot of antigen with an equal volume of suitable adjuvant (see Note 10).
3. Take a sample of the serum prior to first immunization of the animal. Inject the mixture of antigen and adjuvant subcutaneously at least on four different sites.
4. Repeat injections bi-weekly. At least two boosts are necessary for efficient IgG production.
5. Take a test serum 1 week after the second boost and analyze using a Western Blot together with the null-serum taken prior to immunization. As samples use aliquots of antigen and dummy (see Fig. 3). After running the gel and transfer, cut blot into strips (see Fig. 4) and analyze one using the serum and the other using the null-serum as primary antibody.
6. If immune response is sufficient, collect serum from the test animal (see Note 11) by a low speed spin of total blood (maximum 1,000 × *g*).

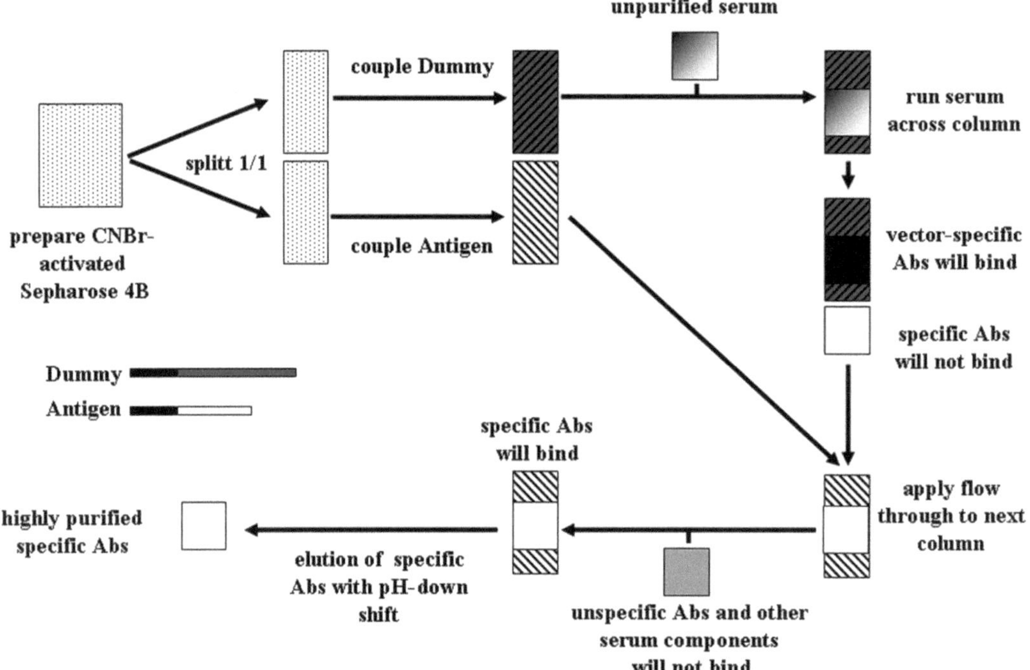

Fig. 3. Schematic representation of the steps required for antibody purification.

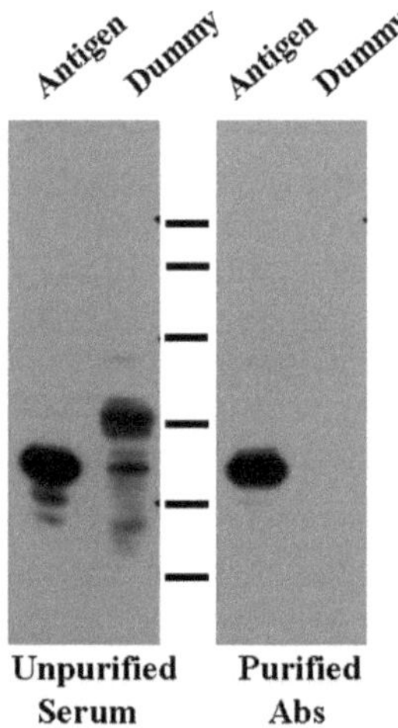

Fig. 4. Western blot control of the antibody purification procedure. The left half of the blot is probed with unpurified serum as primary antibody, the right half is probed using purified antibody. Purified antibody only produces one antigen-specific immune response.

3.1.5. Antibody Purification

1. Hydrate and swell 1 g of CNBr-activated Sepharose 4B (GE Healthcare) in 1 mM HCl for 1 h at RT.
2. Wash column material with 250 ml 1 mM HCl (use a sintered glass funnel, pore size 3).

3. Mix at least 1.5 mg purified antigen (or dummy) in a maximum volume of 2 ml with 4 ml coupling buffer (see Note 12).
4. Wash column material with 10 ml coupling buffer (see Note 13).
5. Mix antigen (or dummy) with column material and transfer to 15 ml Falcon tube.
6. Incubate mixture for 1 h at RT or ON at 4°C.
7. Remove liquid by filtration and resuspend material in 40 ml glycine buffer to block reactive groups which have not bound protein.
8. Incubate for 2 h at RT.
9. Perform washing steps to remove unbound protein using 50 ml coupling buffer, followed by 50 ml Acetate buffer. Repeat these alternating washing steps three times.
10. Resuspend the column material and store at 4°C until further use.
11. Thaw serum on ice and take a 5 ml aliquot.
12. Fill the column material into fresh Poly-Prep Chromatography columns (you should have two columns ready now, one with the bound antigen and a second with the bound dummy). Let excess buffer run off. Do not let the columns run completely dry.
13. Apply serum aliquot to the dummy column. Record the void volume (usually 1–2 ml) (see Note 14).
14. Once the buffer in the column has been displaced by the serum, collect the flow through in a fresh 15 ml Falcon tube.
15. Apply and collect the flow through four more times to ensure binding of antibodies directed against the tag portion of the fusion protein.
16. After the last cycling, add one void volume of coupling buffer to the column to collect all serum in the flow through.
17. Apply the complete flow through (should be around 5 ml again) to the second column (the one with the antigen coupled).
18. Proceed with this second column as you have done with the first. Only in this case desired antibodies should bind to the antigen coupled to the matrix. The flow through contains serum proteins and other antibodies (see Note 15).
19. Wash the column twice with 5 ml 10 mM Tris–HCl pH 8.0.
20. Elute specific antibodies from the column using 4.5 ml elution buffer. The tube used for collection should contain 500 µl 1 M Tris–HCl pH 8.0 to immediately neutralize the pH.

21. Apply one void volume 10 mM Tris–HCl to the column in order to collect all of the applied elution buffer (see Note 16).
22. Do a Western Blot probed with purified antibody and with unpurified serum to ensure proper purification of the antibody.
23. Antibodies may be stored at −20°C until further use. Short-time storage may be at 4°C.

3.2. Light Microscopy

1. To reduce background due to unspecific binding of antibodies to nonprotein plant components (e.g., starch, cell wall), it is good practice to add an additional step we call "pre-absorption": mix in a 2 ml Eppendorf tube 40% (v/v) incubation buffer, 10% (v/v) block buffer, 49% (v/v) buffer for pre-absorption, and 1% (v/v) of your antibody. Invert mixture for 10 min and spin 10 min at 10,000 × *g* in a table top centrifuge. Keep supernatant and discard pellet.
2. Use a cork borer (diameter: 6 mm) to punch out pieces of plant material you are interested in.
3. Transfer samples immediately to ~10 ml fixing medium. Avoid drying out.
4. Ethanol-wash two times (or until you do not smell acetic acid any more). Between washing steps incubate for at least 30 min.
5. Gradually change ethanol against resin:

   10% for 1 h, 25% for 1 h, 50% for 2 h, 75% for 3 h, 100% for 30 min, 100% ON, 100% for 3 h, 100% for 3 h
6. To activate polymerization of your resin add 20 µl *N,N*-dimethylaniline to 1 ml of embedding medium. Mix by inversion and place on ice while preparing your samples. Using a spatula, carefully transfer your samples to 0.2 ml PCR tubes and top up with your activated embedding medium.
7. For polymerization leave your samples on ice for 2 days (place your ice box in a fridge during polymerization or change ice after 1 day).
8. Polymerize at least two additional days at RT before you start sectioning.
9. Sectioning: for high resolution and still sufficient signal, 0.5 µm serial sections are a good compromise. Glass knifes are ideal for this purpose (for further reading (8)).
10. Immunostaining: sections should be mounted on Polysine slides and surrounded by two circles of PAP-PEN, leaving enough space for 100 µl liquid. It is important to wait for the PAP-PEN to be dry. If possible leave ON at 4°C.

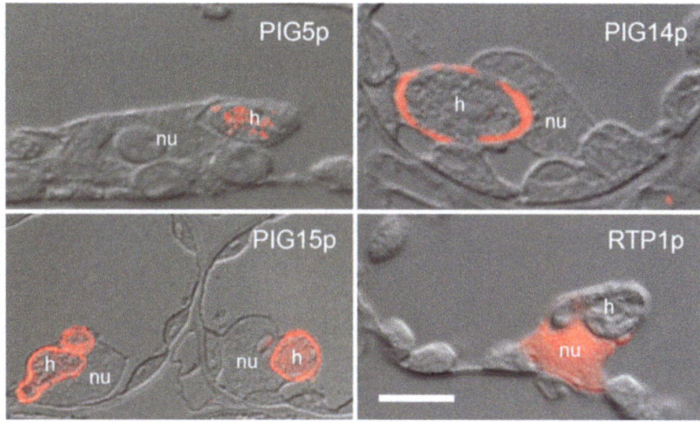

Fig. 5. Light microscopy. Distribution of PIG5p within the haustorial cytoplasm, PIG14p within the extrahaustorial matrix, PIG15p within the haustorial cytoplasm and the extra-haustorial matrix and RTP1p within the host nucleus (h = haustorium, nu = nucleus of the host plant, bar = 5 μm).

11. For deep etching incubate slides for 2.5 min in 100% acetone.
12. Transfer slides immediately to TBS to rehydrate samples and wash off acetone.
13. To reduce background due to unspecific binding of secondary antibody, incubate sections for 10 min in block buffer.
14. Wash sections once for 10 min using incubation buffer.
15. Replace incubation buffer by pretreated antibody solution and incubate for 2 h on an orbital shaker at RT.
16. Using incubation buffer, wash six times 5 min each.
17. Use incubation buffer to dilute your secondary antibody (see Note 17) and incubate samples for 1 h.
18. After three final TBS washes, mount your sample in DABCO, seal using nail polish and start examining your samples (see Fig. 5) after 10 min (see Note 18).

3.3. Electron Microscopy

1. To achieve best results high pressure freezing followed by freeze substitution is one of our recommended methods (for details see ref. 9). Of course, other cryo methods can be used as well.
2. Before transferring your samples into the substitution medium, make sure that the temperature is below -80°C. Incubate samples for at least 80 h at -80°C before gradually rising to -15°C (6 h at -60°C, 3 h at -35°C). Incubate 1 h at -15°C and start washing three times using 1:1 (v/v) acetone:methanol for 1 h at -15°C.
3. Gradually, transfer your samples into propylenoxide incubating at -20°C for 1 h each (20, 40, 60, 80%, and three times 100%).

4. Increase temperature to RT and wash once more to avoid traces of methanol interfering with resin polymerization.
5. Carefully infiltrate Epon/Araldit into samples by gradually increasing concentration (10, 30, 50, 60, 70, 80, 90, and 100%; 12 h each). Once reaching 70% add 50 µl per ml polymerization starter and mix thoroughly.
6. Repeat 100% three times.
7. Polymerize samples for 24 h at 60°C and leave samples at RT for at least 48 h before sectioning.
8. Best results are achieved by using 50–80 nm serial sections mounted on nickel slot supported by a pioloform film (10).
9. Immunostaining: to increase the otherwise poor antigenicity of epoxy embedded samples it is indispensable to etch. Therefore:
10. Incubate sections for 8 min in 10% H_2O_2 before transferring to Glycine buffer.
11. After 5 min incubation follow steps 13–16 described in Subheading 3.2 for primary antibody binding.
12. Secondary gold-labeled antibodies (between 2 and 15 nm) should be used according to the descriptions given by individual producers (see Note 19).
13. Following 1 h of secondary-antibody incubation, wash three times 10 min in TBS.
14. To covalently link antigen – antibody complexes incubate 5 min in 2% Glutaraldehyde and rinse thoroughly using double distilled water.
15. Best contrast (see Fig. 6) can be achieved by combining lead citrate and uranyl acetate staining (11).

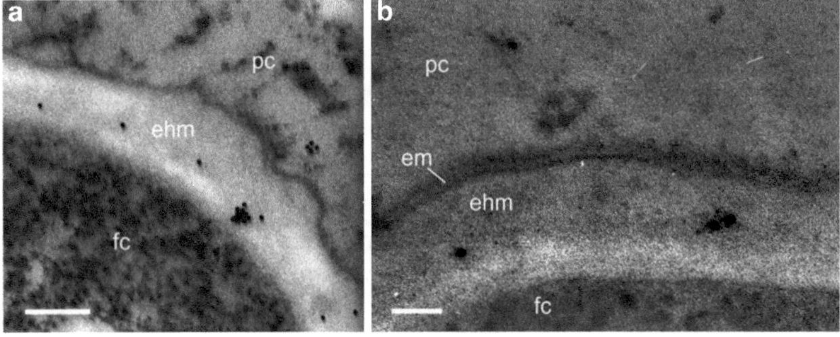

Fig. 6. Electron microscopy. Two sections of the host–parasite interface showing the distribution of RTP1p within the extrahaustorial matrix and the host cytoplasm. The close up (**b**) demonstrates the advantage of this method in preserving membrane structures, an important tool to study effector transfer (ehm = extrahaustorial matrix, em = extrahaustorial membrane, pc = plant cytoplasm, fc = fungal cytoplasm; bar in (**a**) = 200 nm, bar in (**b**) = 50 nm).

16. Incubate 20 min in uranyl acetate, wash and finally incubate 2 min in lead citrate. Wash again thoroughly and leave samples to dry at RT.

4. Notes

1. Serum for block buffer should be adapted to the source of secondary antibody. For example: In case, rabbit is used as primary and goat – anti-rabbit as secondary antibody use goat serum to block.
2. The fusion protein generated by the portion of your gene of interest and the tag introduced through the vector should have a final size of around 30–40 kDa. Smaller and larger antigens usually have reduced overexpression rates. Sizes can be adjusted by using a different vector (e.g., pET32 only carries a small tag, 6× His-tag with some additional amino acids), while pET28 also introduces around 100 amino acids (thioredoxin, a protein with low antigenicity and therefore is well suited for smaller antigens).
3. Cloning primers should contain around 20 nucleotides specific for the target sequence, pertinent restriction sites and 5′-overhang sequences to facilitate the task of restriction endonucleases. 5′-end primer may contain some nucleotides past the restriction site to adjust the open reading frame and 3′-primer may introduce a stop codon.
4. Digests should be done over night since most enzymes show reduced activity when used on DNA ends.
5. For initial transformation use *E. coli* K12 derivatives (e.g., DH5α) because of the higher transformation efficiency compared to the *E. coli* B strains which usually show better overexpression.
6. For overexpression of antigens using a T7-based system, *E. coli* B strains are superior to *E. coli* K12 derivatives. Origami strains assist cytoplasmic disulfide bond formation, while Rosetta strains provide rare tRNAs which might be helpful for expression of eukaryotic proteins.
7. Do a trial with a small culture volume to determine the best time point to harvest the cells. Different antigens show a remarkable difference. Time frames may vary between 30 min and 4 h of induction. Make sure you remove the same number of cells for each sample point (i.e., the equivalent to 1 ml culture of an OD_{600} of 0.5).
8. Supernatant may be used either directly or may be stored at 4°C overnight.

9. From the purification procedure you should get at least 2 ml protein in solution with a concentration of at least 1 mg/ml. If concentration should be lower, but enough antigen is available, samples may be concentrated.
10. Total volumes should range between 0.5 and 1.0 ml. Complete Freund's adjuvant has proven very effective for first immunization of rabbits. Subsequent boosts are carried out using incomplete Freund's adjuvant. However, other adjuvants have to be chosen for smaller animals.
11. Serum can be collected by complete exsanguination or by repeated collection of smaller volumes of serum. Antibody titer may drop if boosts are discontinued.
12. Do a coupling of the desired antigen and a different tagged protein (dummy antigen) in parallel. Successive chromatographic steps as depicted in Fig. 3 efficiently remove unwanted components of the serum.
13. This step has to be done very quickly.
14. It is important to take the void volume of the column into account in order not to dilute or lose your antibody.
15. Up to now the void volume could be easily determined by the slight reddish color of the serum. In this step these components will be removed, so you have to take great care not to lose any of your antibodies.
16. If everything worked out, you should have your purified antibody in the same volume (5 ml) as you started out with (5 ml serum).
17. Secondary antibody: dilution 1:400 Cy3-labeled goat – anti-rabbit (Biotrend Chemikalien GmbH).
18. Depending on signal intensity samples are usually stable for at least 24 h. After that the signal gradually decreases. If samples need to be kept for longer time, store at 4 C.
19. Although there is a range of possibilities to detect primary antibodies in electron microscopy (for details see ref. 4), gold particles, adjusted to the expected resolution show the best signal to noise ratio. For instance: Goat – anti-rabbit 6 or 10 nm secondary antibodies from Aurion were diluted 1:50 (v/v) in incubation buffer and spun for 1 min at 1,300 × *g* using a table top centrifuge.

Acknowledgments

We are grateful to Heinz Vahlenkamp for technical assistance. This work was supported by various grants provided by the Deutsche Forschungsgemeinschaft.

References

1. Greenshields, D. L., and Jones, J. D. (2008) Plant pathogen effectors: getting mixed messages. *Curr. Biol.* **18**, R128–R130.
2. Stulemeijer, I. J., and Joosten, M. H. (2008) Post-translational modification of host proteins in pathogen-triggered defence signalling in plants. *Mol. Plant Pathol.* **9**, 545–560.
3. Echtenkamp, F., Deng, W., Wickham, M. E., Vazquez, A., Puente, J. L., Thanabalasuriar, A., Gruenheid, S., Finlay, B. B., and Hardwidge, P. R. (2008) Characterization of the NleF effector protein from attaching and effacing bacterial pathogens. *FEMS Microbiol. Lett.* **281**, 98–107.
4. Newman, G. R., and Hobot, J. A. (2001) Resin Microscopy and On-Section Immunocytochemistry, 2nd Edition. Springer, Berlin, ISBN: 978-3-540-67277-7.
5. Longin, A., Souchier, C., French, M., and Bryon, P. A. (1993) Comparison of anti-fading agents used in fluorescence microscopy: image analysis and laser confocal microscopy study. *J. Histochem. Cytochem.* **41**, 1833–1840.
6. Graham, L. L., and Beveridge, T. J. (1990) Evaluation of freeze-substitution and conventional embedding protocols for routine electron microscopic processing of eubacteria. *J. Bacteriol.* **172**, 2141–2149.
7. Reynolds, E. S. (1963) The use of lead citrate at high pH as an electron-opaque stain in electron microscopy. *J. Cell Biol.* **17**, 208–212.
8. Dashek, W. V. (2000) Methods in Plant Electron Microscopy and Cytochemistry. Springer, Berlin, ISBN 978-0-89603-809-7.
9. Stark-Urnau, M., and Mendgen, K. (1995) Sequential deposition of plant glycoproteins and polysaccharides at the host–parasite interface of *Uromyces vignae* and *Vigna sinensis*. *Protoplasma* **186**, 1–11.
10. Strobel, G., and Stuermer, C. A. (1994) Growth cones of regenerating retinal axons contact a variety of cellular profiles in the transected goldfish optic nerve. *J. Comp. Neurol.* **346**, 435–448.
11. Knauf, G. M., Welter, K., Müller, M., and Mendgen, K. (1989) The haustorial host–parasite interface in rust-infected bean leaves after high-pressure freezing. *Physiol. Mol. Plant Pathol.* **34**, 519–530.

Chapter 18

Laser Capture Microdissection of Nematode Feeding Cells

Nagabhushana Ithal and Melissa G. Mitchum

Abstract

Obligate plant-parasitic nematodes, such as cyst nematodes (*Heterodera* and *Globodera* spp.) and root-knot nematodes (*Meloidogyne* spp.), form specialized feeding cells in host plant roots. These feeding cells provide the sole source of nutrition for the growth and reproduction of the nematode to complete its life cycle. Feeding cell formation involves complex physiological and morphological changes to normal root cells and is accompanied by dramatic changes in plant gene expression. The distinct features of feeding cells suggest that their formation entails a unique gene expression profile, a better understanding of which will assist in building models to explain signaling pathways that modulate transcriptional changes in response to nematodes. Ultimately, this knowledge can be used to design strategies to develop resistance against nematodes in crop plants.

Feeding cells comprise a small fraction of the total root cell population, and identification of plant gene expression changes specific to these cells is difficult. Until recently, the specific isolation of nematode feeding cells could be accomplished only by manual dissection or microaspiration. These approaches are limited in that only mature feeding cells can be isolated. These limitations in tissue accessibility for macromolecule isolation at different stages of feeding cell development can be overcome through the use of laser microdissection (LM), a technique that enables the specific isolation of feeding cells from early to late stages for RNA isolation, amplification, and downstream analysis.

Key words: Soybean cyst nematode, Syncytia, Laser capture microdissection (LCM), RNA extraction, RNA amplification, Root-knot nematode, Giant-cell, *Heterodera*, *Meloidogyne*

1. Introduction

Laser microdissection (LM) is a technique that allows for the isolation of a single cell or highly pure population of cells from heterogeneous tissue sections for molecular or biochemical analysis. The technique, which was originally developed for the isolation of individual cells in mammalian systems (1, 2), has been shown to be effective for gene expression analysis of microdissected plant cells (3–5). More recently, this technique has been employed for

John M. McDowell (ed.), *Plant Immunity: Methods and Protocols*, Methods in Molecular Biology, vol. 712,
DOI 10.1007/978-1-61737-998-7_18,

capturing feeding cells induced by root-knot and cyst nematodes in host plant roots for RT-PCR and cDNA library construction, respectively (6–8). The objective of our own research has been to obtain pure samples of soybean cyst nematode-induced feeding cells, called syncytia, from soybean roots at different stages of development, including very early stages, for RNA extraction, amplification, and subsequent gene expression analyses (Fig. 1; (9)). This is an alternate approach to, for example, manual dissection or microaspiration of nematode feeding cells, techniques that have been employed to study gene expression changes in mature feeding cells (10–13).

There are two general classes of LM systems. One class is the infrared (IR) laser capture microdissection (LCM) system originally developed by Arcturus Bioscience and now available from MDS Analytical Technologies (www.moleculardevices.com). The other class is the ultraviolet (UV) laser cutting system provided by Zeiss P.A.L.M (http://www.zeiss.com), Leica (www.leica-microsystems.com), and Molecular Machines and Industries (www.molecular-machines.com). MDS Analytical Technologies also provides combined IR/UV models (Arcturus^{XT} and Arcturus Veritas). Both IR and UV systems are being used for isolation of nematode feeding cells (6–9, 14, 15). Although we discuss here a protocol for use with the Arcturus PixCell IIe IR LCM system, which we have used extensively, the protocol can be easily adapted for use with other laser microdissection systems and applied to study other plant–nematode interactions.

To conduct LCM, slides containing tissue sections with cells of interest are placed on the microscope stage. The tissue sections are covered with an isolation cap that harbors a thermolabile polymer, which is activated by an IR laser above the region containing the cells of interest. This activation leads to the formation of a polymer–cell complex. When the cap is lifted, the cells adhered to the polymer are removed from the heterogeneous tissue section (Fig. 1). This process can be subsequently repeated on new tissue sections to increase the numbers of cells harvested.

LCM has been used extensively in animal systems, and protocols for tissue fixation, sectioning, and macromolecule extraction have been optimized (16). The structural and compositional differences between animal and plant cells and the presence of the cell wall and vacuole in plant cells, however, required the development of optimal methodologies for plants. Protocols have now been developed for several plant species, including rice, maize, and Arabidopsis, and for different tissues, including developing embryo, pollen, phloem, and leaf (3–5, 17–21). The method described below is a modification of some of these protocols and adopted for LCM of nematode feeding cells in soybean.

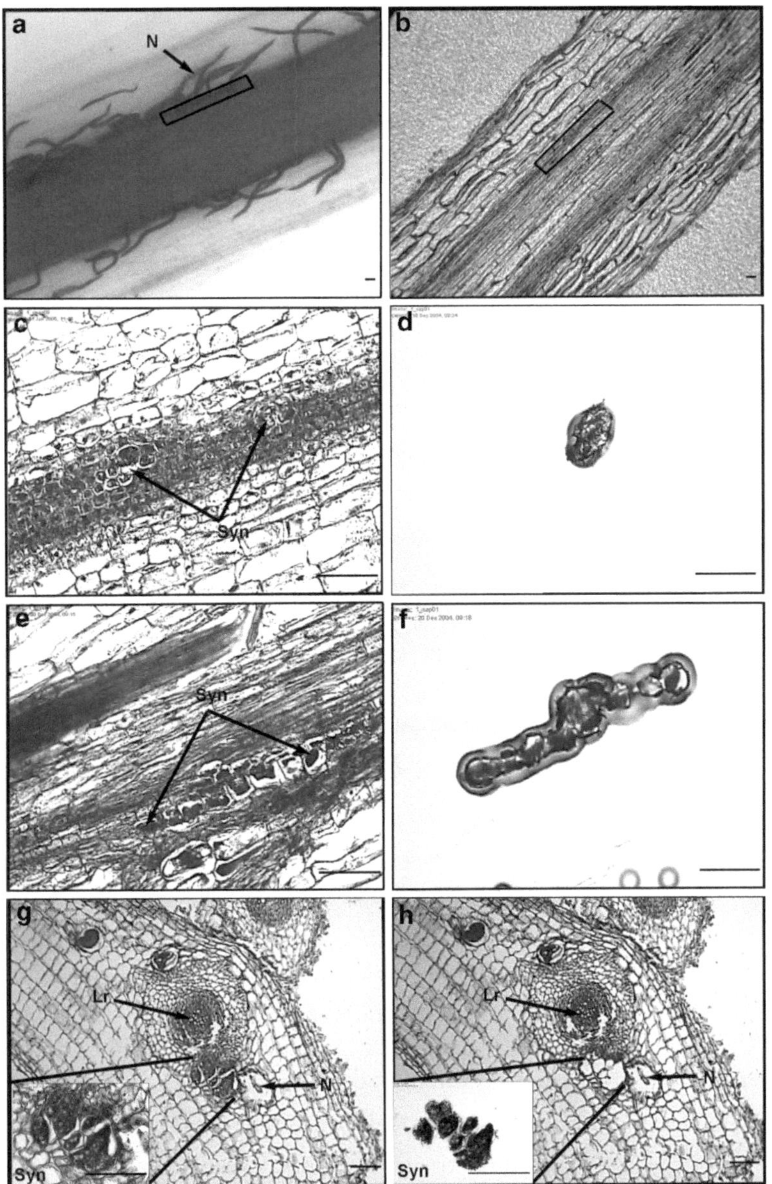

Fig. 1. Laser capture microdissection of soybean cyst nematode (SCN)-induced syncytia in soybean roots. (**a**) Light micrograph of an acid fuchsin stained soybean root 2 days postinfection (dpi) with second-stage juveniles (J2). (**b**) Longitudinal section of an uninfected soybean root; boxed area indicates the region within the pericycle from which the J2s select a cell to induce a syncytium. Longitudinal sections of SCN-infected soybean roots showing syncytia at (**c**) 2- and (**e**) 5-dpi. Syncytial cells captured on a Capsure HS LCM cap from (**d**) 2- and (**f**) 5-dpi sections. Longitudinal section of an SCN-infected soybean root at 10-dpi (**g**) before and (**h**) after LCM. Inset in (**g**) shows the 10-dpi syncytium at a higher magnification and the inset in (**h**) is the syncytium captured on a LCM cap. *N* nematode, *Syn* syncytium, *Lr* lateral root. Bars = 50 μm (Reproduced from ref. (9) with permission from *Molecular Plant-Microbe Interactions*).

2. Materials

2.1. Nematode Egg Hatching

1. Purified nematode eggs.
2. Egg sterilization solution: 2% sodium azide. (Sodium azide is a toxic chemical. Use gloves when handling.)
3. No. 500-mesh sieve (25 μm).
4. Egg hatching chamber (see Note 1).
5. Gentamycin: 22.5 mg/ml gentamycin sulfate (see Note 2).
6. Nystatin: 1.5 mg/ml nystatin (see Note 3).
7. Aluminum foil.
8. Siliconized micropipette tips and microfuge tubes (Denville) (see Note 4).
9. 27°C incubator.

2.2. Plant Culture

1. Soybean seeds.
2. Nontoxic germination paper (12" × 18", regular weight) (Seedburo).
3. 3MM chromatography paper (Whatman).
4. Plastic cafeteria trays (18" × 14").
5. Indelible black marker, thick point (e.g., Sharpie).
6. Saran wrap.
7. 0.01% agarose solution. Autoclaved and stored at room temperature.
8. Commercial bleach (5% sodium hypochlorite (v/v)).
9. Hoagland's solution (Sigma).

2.3. Tissue Preparation and Embedding

1. Fixative: ethanol:glacial acetic acid (3:1).
2. Disposable or stainless steel base molds.
3. Unisette tissue processing/embedding cassettes (Simport Plastics).
4. Biopsy foam pads (30.2 × 25.4 × 2 mm) (Fisher Scientific).
5. Ethanol solutions.
6. Xylene (use in fume hood).
7. HM 325 rotary microtome (Richard Allen).
8. Superfrost Plus glass slides (Fisher Scientific).
9. Dri-Can Reusable Desiccating Canisters (Multisorb Technologies).
10. Automated tissue processor (Tissue-Tek VIP2000; Miles Scientific) (see Note 5).

11. Clear-Rite 3 (formerly Richard-Allen Scientific, now Thermo Scientific).
12. Paraffin (Surgipath).
13. Slide warmer (Fisher Scientific).

2.4. Laser Capture Microdissection

1. PixCell IIe system (formerly Arcturus, now MDS Analytical Technologies) (see Note 6).
2. CapSure HS LCM caps (MDS Analytical Technologies).
3. ExtraSure Sample Extraction Device (MDS Analytical Technologies).
4. Prep-strips (MDS Analytical Technologies).
5. 0.5 ml thin-walled reaction tubes (Applied Biosystems – Cat#N8010611).
6. CapSure HS alignment tray and incubation block (MDS Analytical Technologies).
7. 42°C incubator.

2.5. RNA Extraction

1. PicoPure RNA Isolation Kit (MDS Analytical Technologies).
2. RNase-Free DNase Set (Qiagen).

2.6. RNA Amplification

1. RiboAmp HS^{Plus} RNA Amplification Kit (MDS Analytical Technologies).
2. GeneChip IVT Labeling Kit (Affymetrix).

3. Methods

3.1. Nematode Egg Hatching

1. Collect nematode eggs onto a 500-mesh sieve (25 μm). Rinse well with tap water. Using a water bottle, rinse eggs into a clean 100 ml glass beaker (see Note 7).
2. Add 1 ml of 2% sodium azide solution to beaker containing nematode eggs and make up to 100 ml with tap water. The final egg suspension should have a final concentration of 0.02% sodium azide.
3. Using a stir bar, stir the egg suspension for 20 min on a stir plate.
4. Pour egg suspension back onto a 500-mesh sieve (25 μm). Rinse eggs with tap water extensively to remove all sodium azide. Using a squirt bottle, rinse eggs into a clean 50 ml polypropylene tube with about 10 ml of tap water.
5. Thaw aliquots of antibiotics from –20°C freezer. Add 1 ml of 22.5 mg/ml gentamycin sulfate and 1 ml of 1.5 mg/ml nystatin

to egg suspension. Increase volume to 15 ml with tap water and mix (see Note 2).

6. Pour egg suspension onto an autoclaved hatching chamber (see Note 1) and then cover with aluminum foil.
7. Put hatching chamber into 27°C incubator.
8. Allow eggs to hatch for 2–3 days.
9. Pour hatched preparasitic second-stage juveniles (ppJ2) into a small glass beaker.
10. Check the ppJ2 suspension under a dissecting scope. The juveniles should be active and devoid of any signs of heavy bacterial or fungal contamination.

3.2. Plant Culture and Nematode Infection

1. Count the number of soybean seeds required for the experiment (see Note 8).
2. Place the seeds in a 1-L beaker and then put the beaker under running tap water for 30 min. Adjust the water flow so that the seeds are in constant swirling motion.
3. Prepare 100 ml of 10% commercial bleach (0.5% sodium hypochlorite) by measuring 10 ml of bleach in a 100 ml graduated cylinder and make the volume up to 100 ml by adding 90 ml tap water.
4. After 30 min, pour off water and add 10% bleach solution to seeds. Incubate for 10 min. Mix the contents of the beaker at 2–3 min intervals.
5. Pour off bleach from the beaker and rinse thoroughly with tap water. Run the seeds under tap water for 30 min to remove residual bleach.
6. Cut the required number of pieces of germination paper (18" × 7.5") to roll ragdolls. Fold the germination paper in half and label the fold (with name of seed and date) so the label appears on the outside when rolled. Wet the germination paper with sterile water. Open the folded paper (like you would a book) and arrange about ten seeds on the right side of the paper, in a row 1" from the top edge of the germination paper with hilum facing downward. Fold the left side over as you would close a book.
7. Roll the germination paper from open ends tight enough to hold the seeds in place (approximate diameter will be 1 in.). The label should be seen on the outside of the ragdoll.
8. Place the ragdolls upright in a 1-L beaker with a few milliliters of deionized water. Cover the beaker with saran (plastic) wrap and then poke a few small holes in the wrap with a dissecting needle.
9. Place the beaker of seeds into a 27°C chamber in the dark for 48 h.

10. After 48 h, remove germinated seedlings from 27°C incubator.
11. Select seedlings with a healthy white radicle at least 3–4 cm in length (see Note 9).
12. Prepare ppJ2 inoculum. Use only freshly hatched ppJ2s (2–3 days old) from antibiotic solution. Rinse ppJ2s with sterile water. Make a dilution of ppJ2s to estimate the concentration. For example, dilute the ppJ2 stock by adding 1–9 ml of water in a small glass beaker. (Dilution depends on the concentration of ppJ2 stock.)
13. Mix inoculum well. (The ppJ2 will sink in water, so you must mix before dispensing each time). Aliquot five drops of 25 μl onto a regular glass slide using siliconized tips. Count the number of ppJ2s/drop using a dissecting microscope. Take the average of five drops to get an estimate of the number of ppJ2s in 25 μl of diluted solution. Multiply this number by 40 to get the total number of ppJ2s in 1 ml of dilute solution.
14. Depending on the concentration of inoculum and the number of seedlings to be inoculated (we typically use ~200 ppJ2/seedling for LCM work), calculate the total number of ppJ2s needed and the volume of ppJ2 stock solution required.
15. Aliquot the required volume of ppJ2 stock into siliconized tubes and give a short spin of about 15 s to pellet the ppJ2s. Calculate the volume of 0.01% agarose needed (at the rate of 100 μl per seedling). Aliquot this volume into a 50 ml polypropylene tube. Resuspend the ppJ2 pellet in 0.01% agarose and transfer to 50 ml tube. Vortex the tube to resuspend the ppJ2s evenly in the solution.
16. Arrange a single layer of thick 3MM chromatography paper (16" × 12") on a plastic cafeteria tray. Place two layers of germination paper (16" × 12") on top of the chromatography paper. Wet the paper with sterile deionized water and remove air bubbles by rolling a glass tube across the tray while laying each layer (see Note 10).
17. Arrange seedlings in rows of ten across the width of the tray. Keep about a 1-cm distance between each seedling in a row. (This allows four rows of seedlings/tray). Using a thick-end black indelible marker, gently dot each seedling about 1 cm above the root tip (see Note 11).
18. Inoculate each seedling with 100 μl of ppJ2 suspension (using siliconized tips) at required concentration by placing the inoculum exactly at the center of the black dot. For root-knot nematodes, the root tip should be inoculated.
19. Cover the root tip and the inoculated area with a moist (not too wet) strip of germination paper (12" × 2").
20. Wrap the tray with two layers of saran wrap. Poke several holes in saran wrap with dissecting needle.

21. Use a second, inverted cafeteria tray as a cover.
22. Place in growth chamber at 26°C for 24 h.
23. For synchronized infection, wash each seedling root tip under tap water 24 h after inoculation and reroll in ragdolls (germination paper) as described in Subheading 3.2, steps 6–7.
24. Place ragdolls upright in a 1 L beaker with 300 ml of 10× diluted sterile Hoagland's solution.
25. Place in a growth chamber under long-day growth conditions (16 h light/8 h dark) at 26°C. Add glass Pasteur pipette bubblers (connected to an aquarium pump) to each beaker to aerate roots.

3.3. Tissue Sampling, Processing, and Embedding

1. Using a scalpel, excise a piece of root tissue (about 0.5 cm) from around the dot at a given time point after inoculation (see Note 12).
2. Transfer the tissue immediately to 50 ml fixative (3:1-ethanol: glacial acetic acid) in 100 ml glass beaker and incubate for 19 h at 4°C.
3. Transfer the tissue to 70% ethanol at 4°C until ready to process (up to a maximum of 24 h).
4. Place excised root piece between two biopsy pads and mount in a tissue processing cassette.
5. Process the tissue using an automated tissue processor with the following program: dehydration at room temperature in a graded series of ethanol (1 h each [v/v], 65, 80, 80, 95, 95, 100, 100, and 100%), two changes of Clear-Rite 3 for 1 h each (see Note 13).
6. After the second change of Clear-Rite 3, transfer the root tissue to paraffin at 58°C.
7. Replace the paraffin four times at 50 min intervals.
8. Embed the tissue sample in paraffin and allow blocks to set at 4°C (see Note 14).

3.4. Sectioning

1. Cut 10-μm tissue sections from the paraffin blocks using a rotary microtome.
2. Float the sections in diethyl pyrocarbonate (DEPC)-treated water at 50°C.
3. Mount the sections on positively charged Superfrost Plus glass slides (see Note 15).
4. Dry the sections at 42°C for 24 h on a slide warmer.

3.5. Laser Capture Microdissection

1. Just prior to LCM, deparaffinize the slides twice in xylene for 5 min each, dry at room temperature, and then transfer the slides to an airtight slide box containing a dessicating canister (see Note 16).

2. Prepare the slides for LCM using a Prep-strip to remove any loosely adhered tissue.
3. Place individual slides into position on the LCM microscope stage.
4. Lower a CapSure HS LCM cap onto the section using the placement arm.
5. Set the laser beam to 7.5 μm. Focus the syncytial cell of interest using the 10× magnification lens, capture the cells with laser power of 70–80 mW and laser pulse duration of 700 μs (see Note 17).
6. Remove the captured cells from the parent tissue section by rapidly lifting the LCM cap using the placement arm. Nonspecific material adhered to the cap can be removed using a sticky note (see Note 18).
7. Attach an ExtraSure Sample Extraction Device to the LCM cap in preparation for RNA extraction. Place the CapSure-ExtraSure assembly in a CapSure HS alignment tray.

3.6. RNA Extraction

1. Extract RNA from the LCM cells using the PicoPure RNA isolation kit (see Note 19). In brief, add 10 μl of extraction buffer (XB) into the buffer well and place a new 0.5 ml microfuge tube onto the CapSure-ExtraSure assembly. Cover with an incubation block preheated to 42°C and incubate the assembly for 30 min at 42°C.
2. Centrifuge the microfuge tube with the CapSure-ExtraSure assembly at 800 × *g* for 2 min to collect cell extract into the microcentrifuge tube.
3. Remove the microfuge tube from the CapSure-ExtraSure assembly and proceed with RNA isolation or freeze the cell extract at −80°C.
4. Precondition the RNA purification column by adding 250 μl of conditioning buffer (CB) onto the column filter membrane, incubate for 5 min at room temperature and centrifuge the column with the collection tube at 16,000 × *g* for 1 min.
5. Add 10 μl of 10% ethanol to the cell extract, mix well, and pipette the mixture into the preconditioned purification column. Centrifuge the column with the collection tube for 2 min at 100 × *g* to bind RNA immediately followed by a centrifugation at 16,000 × *g* for 30 s to remove flow through.
6. Add 100 μl wash buffer 1 (W1) to the purification column and centrifuge for 1 min at 8,000 × *g*.
7. Treat with on column DNase (Qiagen) to remove contaminating genomic DNA.
8. Pipette 100 μl of wash buffer 2 (W2) into the column and spin for 1 min at 8,000 × *g*. Add another 100 μl of W2 and spin for 1 min at 16,000 × *g*.

9. Transfer the purification column to a new 0.5 ml microfuge tube. Pipette 11 μl of elution buffer (EB) directly onto the membrane of the column. Incubate for 1 min at room temperature.
10. Centrifuge the column for 1 min at 1,000 × g to distribute EB in the column. Spin for 1 min at 16,000 × g to elute RNA.

3.7. RNA Amplification

1. Perform cDNA synthesis and RNA amplification using Riboamp HS RNA Amplification Kit to amplify picogram or low nanogram amounts of starting RNA. Reactions are carried out in a thermal cycler.
2. For first strand synthesis, mix 1 μl of primer 1 with 11 μl of the eluted RNA and incubate at 65°C for 5 min. Chill the samples at 4–8°C for 1 min. Spin the tubes briefly to collect the samples at the bottom of the tube. Add 2 μl of HS enhancer, 5 μl of first strand master mix, and 2 μl of first strand enzyme mix. Incubate at 42°C for 1 h and then chill the samples to 4–8°C for 1 min. Immediately add 2 μl of first strand nuclease mix to the samples and incubate at 37°C for 30 min followed by 95°C for 5 min. Chill the samples to 4–8°C for 1 min.
3. For second strand synthesis, mix 1 μl of primer 2 to the first strand synthesis mix, incubate at 95°C for 2 min, and then chill and maintain the samples at 4–8°C for 2 min. Add 29 μl of second strand master mix, and 1 μl of second strand enzyme mix. Incubate at 25°C for 10 min, then at 37°C for 30 min, and then 70°C for 5 min. Hold at 4–8°C up to a maximum of 30 min until you are ready to proceed with cDNA purification.
4. Purify cDNA using the binding columns provided and according to the manufacturer's instructions. Elute cDNA in 11 μl of DNA elution buffer.
5. For in vitro transcription, mix 11 μl cDNA with 2 μl HS enhancer, 2 μl IVT buffer, 6 μl IVT master mix, and 2 μl IVT enzyme mix. Incubate at 42°C for 6 h and then chill the samples to 4–8°C. Treat the amplified RNA (aRNA) with 1 μl of the DNase mix and incubate at 37°C for 15 min.
6. Purify aRNA using the binding columns provided and according to the manufacturer's instructions. Elute aRNA into 12 μl of RNA elution buffer.
7. A second round of amplification can be performed using the same kit. Alternatively, biotinylated aRNA for microarray hybridization can be prepared using the GeneChip IVT Labeling Kit (see Note 20). Typically, two rounds of RNA amplifications are required to produce microgram quantities of aRNA (see Note 21).

4. Notes

1. Nematode egg hatching chambers can be constructed using two 400 ml tricornered polypropylene beakers (Fisher Scientific). Using a hack saw, cut one beaker evenly at the 300 ml mark and the other at the 240 ml mark. Place a piece of mesh (30 μm opening/21% open area mesh, Sefar) over the bottom of the smaller piece of cut beaker (top half) and slide the larger piece of cut beaker (top half) over it so that the mesh is stretched tautly. Place onto a polypropylene dish made from the base of a 1 L tri-cornered beaker. Layer the egg suspension on top of the mesh. The hatched juveniles will migrate through the mesh into the collecting base.
2. Gentamycin is an antibiotic to control bacterial contamination in nematode cultures. To make 25 ml of 22.5 mg/ml gentamycin sulfate, weigh out 562.5 mg of gentamycin on an analytical balance, pour into a 50 ml polypropylene tube, make up to 25 ml with double distilled water, mix well, and filter through a 0.22-μm syringe filter in laminar flow hood. Dispense 1 ml aliquots into sterile 1.5 ml tubes. Store in –20°C freezer. Use at a final concentration of 1.5 mg/ml.
3. Nystatin is an antifungal agent used to control fungal contamination in nematode cultures. To make 25 ml of 1.5 mg/ml nystatin, weigh out 37.5 mg (.038 g) of nystatin on an analytical balance, pour into a 50 ml polypropylene tube, make up to 25 ml with double distilled water, mix well (this antibiotic will not go into solution completely; do not filter sterilize), and dispense 1 ml aliquots into sterile 1.5 ml tubes. Store in –20°C freezer. Use nystatin at a final concentration of 0.05 mg/ml.
4. We recommend using commercially available (e.g., Denville) siliconized micropipette tips and microfuge tubes to handle nematodes. This avoids loss of worms due to adhering to the walls of ordinary tips and microfuge tubes.
5. To maintain consistency in tissue quality and to reduce processing time, we recommend processing the tissues using an automated tissue processor using RNase-free solutions. However, all the tissue processing steps described can be conducted manually with minor modifications.
6. The method described here was optimized for the Arcturus PixCell IIe system; however, it is also compatible with upgraded models (i.e., Arcturus^{XT} and Arcturus Veritas). In addition, several other laser microdissection platforms have been used for nematode feeding cell isolation (6–8, 14, 15).

7. The nematode egg hatching method described here is suitable for both soybean cyst nematode and root-knot nematode.
8. Prior to setting seeds for germination, check the germination percentage of the seed stock because this varies based on age of the seed.
9. Never allow the seedlings to dry out. Always cover the seedlings with moist germination paper.
10. Paper should not be too wet. Drain off excess water from the tray.
11. Marking the inoculation spot with a dot using a black marker enables the precise location of the infection zone to facilitate the excision of nematode infection sites at later time points.
12. A subset of samples can be stained with acid fuchsin (22) to confirm adequate infection in the sampling zone (e.g., Fig. 1a).
13. Prepare all solutions in RNase-free water to maintain RNA integrity.
14. We position the root tissue horizontally for longitudinal sections. Blocks can be stored in an airtight box containing a desiccating canister at 4°C for up to 14 days. We recommend quick processing of the tissues for LCM to obtain high quality RNA.
15. Several sections can be collected per slide, especially if you can cut ribbons. At this point, sections can be screened under a light microscope for the presence of feeding cells. Select only the slides with sections containing feeding cells for further processing.
16. Complete dehydration of the sections is required for successful LCM. Following xylene treatment, you must ensure that the sections do not absorb any atmospheric moisture. Although, slides can be stored for up to a week, we recommend quick processing to obtain high-quality RNA.
17. Focus the microscope on the black ring of the LCM HS cap and not on the specimen. Reduce the illumination so the beam appears as a white spot. Focus until you have an intense, tight spot with a faint halo. After setting the initial parameters, test fire the laser making sure the laser spot is targeting an area of the slide where there are no cells to pick up. The melted spot on the polymer should show proper "wetting" (i.e., there should be a visible thin dark ring surrounding a clear area where the laser was fired). Once you are satisfied with the settings, proceed to capturing the tissue. One cap can be used to capture cells from several sections. If the laser does not cut after a capture, make sure that the cap is sitting correctly in the placement arm.

18. Sometimes, cell capturing results in unwanted cells adhering to the cap. These cells are loosely adhered to the cap and are not securely attached as the LCM-targeted cells. These cells can be easily removed using the adhesive on a sticky note. Gently place adhesive surface of the sticky note on the cap and slowly peel it off. Examine the cap to see if the unwanted cells are removed and repeat if required. This approach does not affect the quality of the RNA. The amount of cells required to obtain sufficient aRNA depends on the time point postinoculation. More samples may be required at early time points and relatively less at later time points. We estimate approximately 7 μg amplified RNA could be obtained per square millimeter of captured cell area.
19. RNA extraction and amplification kits specifically designed for application with LCM tissue are available from different manufacturers. Due to the small quantities of RNA isolated from LCM tissues, we recommend the use of one of these kits.
20. Note that the orientation of the aRNA is reversed after first round of cDNA synthesis and in vitro transcription. Use appropriate primers for subsequent amplification or other downstream applications.
21. You can monitor the quality and quantity of aRNA after each round of amplification by using standard spectroscopy, a Nanodrop ND-1000 spectrophotometer, an Agilent Bioanalyzer, or gel electrophoresis.

Acknowledgments

The authors gratefully acknowledge research support from the Missouri Soybean Merchandising Council and the National Research Initiative of the United States Department of Agriculture (Award #2005-35604-15434). We also thank Melody Kroll for proofreading the manuscript.

References

1. Bonner, R. F., Emmert-Buck, M. R., Cole, K., Pohida, T., Chuaqui, R. F., Goldstein, S. R. and Liotta, L. A. (1997) Cell sampling – laser capture microdissection: molecular analysis of tissue. *Science* **278**, 1481–1483.
2. Emmert-Buck, M. R., Bonner, R. F., Smith, P. D., Chuaqui, R. F., Zhuang, Z. P., Goldstein, S. R., Weiss, R. A. and Liotta, L. A. (1996) Laser capture microdissection. *Science* **274**, 998–1001.
3. Asano, T., Matsumura, T., Kusano, H., Kikuchi, S., Shimada, H. and Kadowaki, K.-I. (2002) Construction of a specialized cDNA library from plant cells isolated by laser capture microdissection: toward comprehensive analysis of the genes expressed in the rice phloem. *Plant J.* **32**, 401–408.
4. Casson, S., Spencer, M., Walker, K. and Lindsey, K. (2005) Laser capture microdissection for the analysis of gene expression during

embryogenesis of *Arabidopsis. Plant J.* **42**, 111–123.

5. Spencer, M. W. B., Casson, S. A. and Lindsey, K. (2007) Transcriptional profiling of the *Arabidopsis thaliana* embryo. *Plant Physiol.* **143**, 924–940.
6. Ramsay, K., Wang, Z. H. and Jones, M. G. K. (2004) Using laser capture microdissection to study gene expression in early stages of giant cells induced by root-knot nematodes. *Mol. Plant Pathol.* **5**, 587–592.
7. He, B., Magill, C. and Starr, J. L. (2005) Laser capture microdissection and real-time PCR for measuring mRNA in giant-cells induced by *Meloidogyne javanica. J. Nematol.* **37**, 308–312.
8. Klink, V. P., Alkharouf, N., MacDonald, M. and Matthews, B. (2005) Laser capture microdissection (LCM) and expression analyses of *Glycine max* (soybean) syncytium containing root regions formed by the plant pathogen *Heterodera glycines* (soybean cyst nematode). *Plant Mol. Biol.* **59**, 965–979.
9. Ithal, N., Recknor, J., Nettleton, D., Maier, T., Baum, T. J. and Mitchum, M. G. (2007) Developmental transcript profiling of cyst nematode feeding cells in soybean roots. *Mol. Plant Microbe Interact.* **5**, 510–525.
10. Bird, D. M. and Wilson, M. A. (1994) DNA sequence and expression analysis of root-knot nematode-elicited giant-cell transcripts. *Mol. Plant Microbe Interact.* 7, 419–424.
11. Juergensen, K., Scholz-Starke, J., Sauer, N., Hess, P., Van-Bel, A. J. E. and Grundler, F. M. W. (2003) The companion cell-specific Arabidopsis disaccharide carrier *AtSUC2* is expressed in nematode-induced syncytia. *Plant Physiol.* **131**, 61–69.
12. Wang, Z. H., Potter, R. H. and Jones, M. G. K. (2003) Differential display analysis of gene expression in the cytoplasm of giant cells induced in tomato roots by *Meloidogyne javanica. Mol. Plant Pathol.* **4**, 361–371.
13. Szakasits, D., Heinen, P., Wieczorek, K., Hofmann, J., Wagner, F., Kreil, D. P., Sykacek, P., Grundler, F. M. W. and Bohlmann, H. (2009) The transcriptome of syncytia induced by the cyst nematode *Heterodera schachtii* in Arabidopsis roots. *Plant J.* **57**, 771–784.
14. Fosu-Nyarko, J., Jones, M. G. K. and Wang, Z. (2009) Functional characterization of transcripts expressed in early-stage *Meloidogyne javanica*-induced giant cells isolated by laser microdissection. *Mol. Plant Pathol.* **10**, 237–248.
15. Klink, V. P., Overall, C. C., Alkharouf, N. W., MacDonald, M. H. and Matthews, B. F. (2007) Laser capture microdissection (LCM) and comparative microarray expression analysis of syncytial cells isolated from incompatible and compatible soybean (*Glycine max*) roots infected by the soybean cyst nematode (*Heterodera glycines*). *Planta.* **226**, 1389–1409.
16. Goldsworthy, S. M., Stockton, P. S., Trempus, C. S., Foley, J. F. and Maronpot, R. R. (1999) Effects of fixation on RNA extraction and amplification from laser capture microdissected tissue *Mol. Carcinog.* **25**, 86–91.
17. Matsunaga, S., Schutze, K., Donnison, I. S., Grant, S. R., Kuroiwa, T. and Kawano, S. (1999) Single pollen typing combined with laser-mediated manipulation. *Plant J.* **20**, 371–378.
18. Kerk, N. M., Ceserani, T., Tausta, S. L., Sussex, I. M. and Nelson, T. M. (2003) Laser capture microdissection of cells from plant tissues. *Plant Physiol.* **132**, 27–35.
19. Nakazono, M., Qui, F., Borsuk, L. A. and Schnable, P. A. (2003) Laser-capture microdissection, a tool for the global analysis of gene expression in specific plant cell types: identification of genes expressed differentially in epidermal cells or vascular tissues of maize. *Plant Cell* **15**, 583–596.
20. Liu, X., Wang, H., Li, Y., Tang, Y., Liu, Y., Hu, X., Jia, P., Ying, K., Feng, Q., Guan, J., Jin, C., Zhang, L., Lou, L., Zhou, Z. and Han, B. (2004) Preparation of a single rice chromosome for construction of a DNA library using a laser microbeam trap. *J. Biotechnol.* **109**, 217–226.
21. Inada, N. and Wildermuth, M. C. (2005) Novel tissue preparation method and cell-specific marker for laser microdissection of Arabidopsis mature leaf. *Planta* **221**, 9–16.
22. Daykin, M. E. and Hussey, R. S. (1985) Staining and histopathological techniques in nematology, in *An Advanced Treatise on Meloidogyne Methodology* (Barker, K. R., Carter, C. C., Sasser, J. N., eds.), North Carolina State University Graphics, Raleigh, pp. 39–48.

Chapter 19

Laser Microdissection of Plant–Fungus Interaction Sites and Isolation of RNA for Downstream Expression Profiling

Divya Chandran, Noriko Inada, and Mary C. Wildermuth

Abstract

The molecular mechanisms that mediate the intimate interaction of an adapted obligate biotroph, such as the powdery mildew *Golovinomyces orontii*, on its host plant are spatially and temporally distinct. As *G. orontii* exclusively infects epidermal cells with a dominant host response occurring in the underlying mesophyll cells, we sought to develop a method to accurately and reproducibly perform global expression profiling on *Arabidopsis thaliana* leaf epidermal and mesophyll cells at the site of infection. Specific stages of *G. orontii* disease progression on *Arabidopsis* are visible by microscopy thus allowing distinct phases of the interaction to be studied. Tissue preparation, laser microdissection, and RNA isolation protocols that allow for temporally and spatially defined global expression profiling are described. By using these procedures to examine the growth and reproduction phase (5 days postinfection) of *G. orontii* on *Arabidopsis*, we identified known and novel processes, process components, and putative regulators of these processes that mediate the sustained growth and reproduction of this adapted obligate biotroph.

Key words: *Arabidopsis thaliana*, *Golovinomyces orontii*, Powdery mildew, Laser microdissection, Modified microwave method, Tissue preparation, RNA isolation, Biotroph

1. Introduction

Investigating molecular responses mediating plant–pathogen interactions at the cellular level is complicated by the temporal and spatial nature of the infection process. Therefore, it is essential to harvest specific cell populations of interest (e.g., cells at the site of infection) from heterogeneous plant tissue and extract RNA of sufficient quantity and quality for use in gene expression analysis. We have used the *Golovinomyces orontii* (powdery mildew)–*Arabidopsis thaliana* pathosystem to elucidate processes associated with the sustained growth and reproduction of an

John M. McDowell (ed.), *Plant Immunity: Methods and Protocols*, Methods in Molecular Biology, vol. 712,
DOI 10.1007/978-1-61737-998-7_19,

adapted fungal biotroph at the site of infection (1). *G. orontii* was selected because (1) the structural development and progression of disease on *Arabidopsis* is well defined, limited to leaf epidermal cells and visible by light microscopy, (2) the extent of *G. orontii* infection can be controlled by the dose and method of application of the conidia, and (3) infected epidermal and underlying mesophyll cells, the site of significant host response, can be directly observed and isolated using laser microdissection.

In order to prepare mature infected, or parallel uninfected, *Arabidopsis* leaf tissue for laser microdissection, we evaluated a number of tissue preparation methods and determined that a modified microwave paraffin technique, employing phosphate buffer instead of a chemical fixative, resulted in the highest preservation of leaf internal structure and nucleic acids (2). This method facilitates even and rapid cell permeation in a total preparation time of ~5 h compared with conventional paraffin tissue preparation methods which require ~3 days. To isolate specific cell populations, we used laser microdissection (LMD), which provides a rapid and reproducible means of precise, contamination-free isolation of specific cell groups or single cells from heterogeneous tissue sections (3–5). With LMD, prepared tissue sections are viewed under a microscope, dissected via UV-laser excision and collected into a reaction tube by gravity. LMD has several critical advantages over other cell-isolation methods, such as microcapillary or protoplasting with cell sorting, for the isolation of plant cells responding to pathogens. First, it allows for the highly automated dissection and collection of cells and is not limited to cell layers near the leaf surface as are microcapillary techniques. Second, contamination from neighboring nonselected cells is minimal. Third, cells of interest can be distinguished by their morphological traits with no requirement for a molecular marker as is typically the case for protoplasting with cell sorting and/or in situ or immunohistological labeling/staining. Finally, the use of microwave-prepared sections limits the induction of isolation-associated responses. For example, as altered cell wall integrity plays a role in powdery mildew resistance and host defense, protoplasting was not a suitable approach. Isolated cells then provide RNA for the profiling of gene expression from individual cell types or targeted groups of cells.

Our focus was on the development of methods of cell isolation for the downstream application of global expression profiling. The pooling of reasonable numbers of LMD-isolated single cells or groups of cells typically yields ng quantities of RNA thus preventing the direct use of this RNA for downstream microarray analysis which typically requires μg quantities of RNA. To overcome this limitation, isolated RNA is amplified in a linear fashion to obtain sufficient quantities for use in microarray analysis (e.g., Affymetrix ATH1 GeneChip®). As tissue preparation, laser microdissection, and/or RNA amplification could impact mRNA quality, distribution, and/or microarray processing and output, it is very important to include quality control assessments at every step of the protocol to ensure that the final data is of high quality with minimal impact due to experimental procedures. As part of this quality control assessment, we highly recommend the collection of parallel leaves from the same experiment to assess the impact of each of these steps on the data output.

This chapter describes (1) the modified microwave method and paraffin embedding of *G. orontii*-infected and uninfected *Arabidopsis* leaf tissue, (2) the isolation of specific groups of cells by laser microdissection, and (3) the extraction of RNA from laser microdissected cells. In addition, we discuss the collection and analysis of additional samples to assess the impact of these procedures on data output. The subsequent chapter describes the (1) two-cycle RNA amplification, (2) ATH1 GeneChip® hybridization, and associated quality control assessments.

2. Materials

2.1. Tissue specimen preparation

1. Uninfected and *G. orontii* (powdery mildew) infected 4-week-old *Arabidopsis* plants grown in environmentally controlled (pH, T, RH) plant growth chambers. Sufficient plants should be grown and treated to allow for quality control samples to be collected and processed (see Subheading 3.5).
2. Diethylpyrocarbonate (DEPC, ≥97%, Sigma-Aldrich, St. Louis, MO). Stored at 2–8°C.
3. RNaseZap® (Applied Biosystems/Ambion, Austin, TX). RNaseZap® is slightly corrosive and should be handled with gloves.
4. RNase-free water (0.1% DEPC-treated) or nuclease-free water (Applied Biosystems/Ambion). DEPC-treated water is prepared by adding 1 ml DEPC (see Subheading 2.1) to 1 L of water and stirring well until the DEPC is completely dissolved. The solution is then incubated overnight and autoclaved. DEPC is toxic and solutions containing DEPC must be prepared in a fume hood. Autoclaving decomposes DEPC into ethanol and carbon dioxide rendering it inactive and nontoxic.
5. Clear glass threaded vials, capacity 15 ml (Fisher Scientific, Pittsburgh, PA).
6. Pelco 3440 MAX lab microwave oven (Ted Pella, Redding, CA).
7. Sørensen's phosphate buffer solution is prepared by mixing appropriate volumes of 0.2 M sodium dihydrogen phosphate

and 0.2 M disodium hydrogen phosphate stock solutions with DEPC-treated water to make a final 10 mM Sørensen's phosphate buffer solution, pH 7.2. 10 mM solution is prepared fresh and chilled and stored at 4°C until used that day.

8. Trypan blue (Sigma-Aldrich) is dissolved at 1 mg/ml in freshly prepared 10 mM Sørensen's phosphate buffer solution, pH 7.2 and chilled and stored at 4°C until used that day. Trypan blue is toxic and must be handled with gloves in a fume hood.
9. Alcohol series: 30, 50, 70, 95, and 100% ethanol (200 proof); ethanol:isopropanol (1:1); and 100% isopropanol (high purity solvent).
10. Safranin-O (Sigma-Aldrich) is dissolved at 1% (w/v) in 100% ethanol.
11. Paraffin wax (Paraplast X-TRA®, Fischer Scientific).
12. Aluminum weighing dish with handle (Fisher Scientific).
13. Rotary microtome.
14. Paint brush (camel hair brush, brush width 1.3 cm).
15. Xylene, histological/cytological grade (VWR International, West Chester, PA). Xylene is a hazardous solvent and must be handled in a fume hood with impervious gloves (e.g., nitrile gloves) to avoid skin contact.
16. Polyethylene naphthalate (PEN)-slides, 2.0 μm, RNase-free (JH Technologies Inc., San Jose, CA) required for use with the Leica AS-LMD system.

2.2. Laser Microdissection

1. Leica AS LMD (Leica Microsystems GmbH, Wetzlar, Germany).
2. 0.2 ml PCR tubes, RNase-free (Biozym Scientific GmbH, Germany). Specific PCR tubes are required that are compatible with the Leica AS LMD.
3. PicoPure™ RNA extraction buffer (Molecular Devices, Sunnyvale, CA).

2.3. RNA Isolation

1. PicoPure™ RNA isolation kit (Molecular Devices). Use within 6 months of purchase.
2. RNase-free DNase set (Qiagen, Valencia, CA). RNase-free DNase I is stored at –20°C and RNase-free buffer RDD at 4°C.

2.4. RNA Quality and Yield Assessment

1. Nanodrop 1000 (Thermo Scientific, Wilmington, DE).
2. Agilent 2100 bioanalyzer with equipment (Agilent Technologies, Santa Clara, CA).
3. RNA 6000 Pico Kit (Agilent Technologies). RNA 6000 ladder is stored at –20°C. All other reagents and reagent mixes are

stored at 4°C when not in use. Dye and dye mixtures must be protected from light. Gel mix is stable at 4°C up to 1 month after preparation.

4. RNase-free water (Ambion) or DEPC-treated water (see Subheading 2.1).
5. RNaseZap® (Ambion).
6. Gene-specific PCR primers.
7. Reagents for PCR.

3. Methods

The methods described below outline (1) preparation of the leaf tissue specimen, (2) the process of LMD using Leica AS LMD, (3) the isolation of high quality RNA from laser microdissected cells, and (4) RNA yield and quality assessment including the collection of samples for use in assessing the impact of each experimental step on data output. The subsequent chapter includes methods describing the subsequent RNA amplification, microarray hybridization, and associated quality control procedures.

RNA integrity is a major factor that influences the quality of data obtained using this method. To avoid RNase-mediated RNA degradation and to avoid the introduction of contaminating RNA, it is recommended to work in an RNase-free environment and observe the following general precautions while performing the entire protocol.

1. All work spaces and labware, including pipetteman plungers, must be decontaminated by rinsing or wiping with RNase-removal agents, such as RNaseZap®. Only sterile, nuclease-free filter-barrier tips and reaction tubes should be used from dedicated, covered containers.
2. If possible, RNA-only dedicated pipetteman should be utilized and an RNA-only dedicated lab bench space should be used.
3. All solutions must be made with DEPC-treated water or RNase-free water.
4. Gloves must be worn at all times and changed frequently. Lab coats should be worn and hair pulled back to avoid contamination.

3.1. Tissue Specimen Preparation

The preparation of *Arabidopsis* leaf tissue for LMD and RNA isolation includes (1) microwave fixation and dehydration, (2) embedding in paraffin wax, and (3) sectioning and slide preparation. The goal is to obtain tissue sections with excellent histology and preservation of nucleic acids.

3.1.1. Microwave Fixation and Dehydration

1. Mature *Arabidopsis* leaves (*G. orontii*-infected and uninfected) are dissected into small transverse sections (~1 cm × 5 mm) in a drop of cold 10 mM Sørenson's phosphate buffer on an RNase-free glass plate using a new razor blade (see Note 1).
2. Leaf tissue pieces are immediately transferred into clear glass vials filled with 10 ml cold 10 mM Sørenson's buffer with 1 mg/ml trypan blue, and placed on ice. Trypan blue stains fungal haustoria and thus aids in the identification of infected epidermal cells during the LMD step. Addition of trypan blue to uninfected samples is optional (see Note 2).
3. Samples are labeled by inserting a small piece of paper with experimental information written in pencil into each vial (e.g., plant genotype, treatment type, experiment date, etc.).
4. The microwave power setting is adjusted to 450 W from the standard 650 W.
5. Uncapped vials with infected or uninfected leaf pieces are placed in a plastic tube rack and the rack is placed in a water bath [plastic container filled with tap water (see Note 3)]. The microwave temperature probe is completely immersed in a separate vial containing cold 10 ml 10 mM Sørenson's buffer placed in the water bath (see Note 4).
6. Samples are microwaved at the 37°C setting for 15 min. This step is repeated three times with the vials replaced with Sørenson's buffer each time, and the water bath replaced with fresh tap water after each step (see Note 5).
7. Samples are then dehydrated in an ethanol series of ~10 ml of 30, 50, 70, 95, and 100% ethanol (100% ethanol step repeated twice) with each step performed in the microwave at 67°C setting for 1 min 15 s (see Note 6). Approximately five drops of Safranin-O is added to each vial in the last 100% ethanol step to stain transparent leaf tissue (see Note 7).
8. The 100% ethanol–Safranin-O step is followed by ethanol:isopropanol (1:1) and 100% isopropanol steps (~10 ml each), with each step performed in the microwave at the 77°C setting for 1.5 min. Care is taken to remove all isopropanol during the last 100% isopropanol step (see Note 8).
9. 5 ml 100% isopropanol is immediately added to each vial.
10. Just before embedding in paraffin wax (see below), each vial is microwaved at the 77°C setting for 1 min 30 s.

3.1.2. Paraffin Wax Embedding

1. An appropriate amount of paraffin wax is melted at 60°C in a wax dispenser (see Note 9).
2. The microwave temperature probe is immersed directly into the water bath for the paraffin wax embedding steps.

3. Approximately 5 ml of melted paraffin wax is dispensed into each vial containing 5 ml 100% isopropanol yielding a final concentration of 1:1. Vials are swirled immediately and microwaved at 77°C setting for 10 min.
4. The 100% isopropanol:paraffin wax mixture is replaced with ~10 ml of 100% melted paraffin wax and microwaved at 67°C setting for 10 min (see Note 10).
5. The 100% melted paraffin wax is replaced five times, microwaving at 67°C setting for ~30 min after each replacement (see Note 11).
6. After the final paraffin wax change, vials are incubated overnight at 62°C to completely remove any traces of isopropanol from the tissue.
7. A hot plate is turned on ~20 min prior to tissue embedding.
8. Each paraffin wax sample is poured into an embedding mold (aluminum dish with handle) placed on top of the hot plate (see Note 12). Using a straight needle, the sample label is oriented to one side of the embedding mold and tissue pieces are oriented in rows (see Note 13). The wax is allowed to harden slowly, at room temperature, to minimize air bubble formation.
9. Embedded tissues are stored at 4°C with a desiccant (see Note 14).

3.1.3. Sectioning and Slide Preparation

1. RNase-free PEN slides are prepared as follows (see Note 15):
   (a) PEN slides are placed in a slide holder and immersed in a glass container filled with 0.01% DEPC solution.
   (b) Slides are incubated at room temperature for 30 min to 1 h.
   (c) Slides are then allowed to air dry (~1 h) and baked at 100°C for 1 h in an oven.
2. Paraffin-embedded samples are cut into small cubes of ~1 × 0.5 × 0.8 cm, with a razor blade, with one leaf tissue piece centered in each cube, and mounted on wooden blocks. To mount on the wooden block, heat the spatula in a flame, place the paraffin cube on top of the spatula, and transfer to the wooden block. Then, heat the spatula again and press against the sides of the paraffin cube to help attach it to the block.
3. The wooden block is clamped in a microtome and sections of 10 μm thickness are cut (see Note 16).
4. Section strips are picked up carefully using a dedicated paint brush and floated on drops of RNase-free water on the membrane side of the PEN slide (see Note 17).
5. The PEN slide is placed on a heating plate and incubated at 42°C until the sections appear stretched (more taut and transparent; ~5–10 min).

6. Excess RNase-free water is removed carefully using a KimWipe and slides are allowed to dry overnight at 42°C in an oven.
7. PEN slides with paraffin sections are then stored at 4°C with a desiccant until LMD (see Note 18).

3.2. Laser Microdissection

The process of LMD involves deparaffinization of sections, followed by laser cutting and collection of cut cells by gravity using the Leica AS LMD system. Prior to laser cutting, one must confirm that the tissue preparation method results in excellent preservation of leaf internal structure as evidenced by well-defined vascular bundles, phloem, and chloroplasts, and expanded and rounded epidermal cells as shown in Fig. 1 (2). For laser microdissection, it is important to define your cells of interest [e.g., using a reporter construct for a gene of interest (Fig. 1e)] and to standardize the cell types and locale of the cells to be isolated prior to cutting. As shown in Fig. 1, we isolated ~20 cells per infection site. As described below, batches of 1,250 cells (infected) or 2,500 cells (uninfected) were harvested at one time into a single PCR tube cap. 7,500 total cells were then pooled for each sample to result in sufficient amplified RNA for microarray analysis as described in the following chapter.

3.2.1. Deparaffinization of Sections

1. All steps are performed in a fume hood.
2. Fresh xylene is poured into a glass container fitted with a glass slide holder.

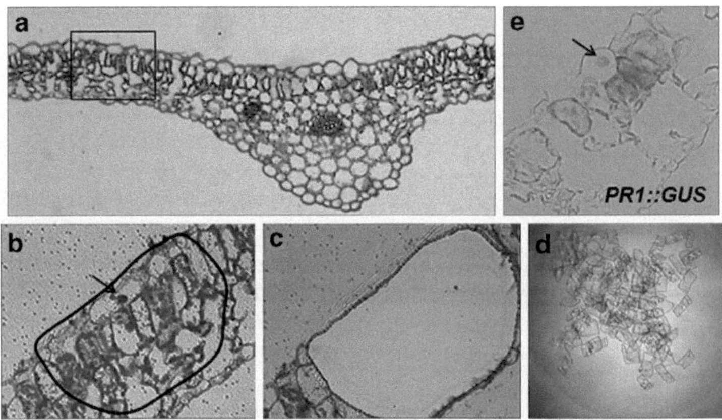

Fig. 1. Laser microdissection of *Arabidopsis* leaf epidermal and mesophyll cells. (**a**) Area targeted for LMD in box. (**b**) Before and (**c**) after LMD of group of (~20) epidermal and mesophyll cells at PM infection site. (**d**) Captured laser microdissected cells. (**e**) PM-induced expression of *PR1::GUS* is observed in the mesophyll cells neighboring the infected epidermal cell. *Arrow* indicates *Golovinomyces orontii* haustorium in infected epidermal cell (reproduced from ref. 1 with permission from the National Academy of Sciences of the USA).

3. PEN slides with paraffin sections are placed on a hot plate (set to low) and heated for ~15 s until the paraffin begins to melt (see Note 19).
4. Slides are transferred immediately to the slide holder immersed in xylene and incubated at room temperature for 2 min.
5. Slides are air-dried in the fume hood for 15–30 min until they are completely dry. Slides are further dried at 42°C in an oven for 15 min and used immediately for LMD (see Note 20).

3.2.2. Laser Microdissection and Cell Collection

1. The Leica AS LMD instrument is turned on, the AS LMD operating software initialized, and the laser calibrated following the manufacturer's instructions (see Note 21).
2. The PEN slide with leaf sections is mounted on the slide holder in an inverted position and inserted into the slide holder groove on the motorized microscope stage.
3. A 0.2 ml RNase-free PCR tube is inserted into the tube holder and filled with 40 μl PicoPure RNA extraction buffer. The tube holder is inserted below the slide holder on the microscope stage with the tube cap positioned directly beneath the slide.
4. For infected leaf tissue samples, sections are quickly scanned at a lower objective (20×) to identify *G. orontii* haustorium-containing cells. Once an infected epidermal cell is identified, the 40× XT objective is used to target and cut epidermal and mesophyll cells surrounding the infected cell (see Note 22). Tissues are visualized on a computer monitor through a video camera. Cells to be isolated are encircled on the computer screen using a mouse and then cut by UV laser. Direct the laser to ablate the cells surrounding the group of cells of interest to preserve the integrity of the cells of interest. Laser settings are adjusted manually using the AS LMD operating software to obtain reproducible, clean, excision of cells with a minimum cutting diameter. For *Arabidopsis* mature leaf 10 μm sections, the following laser conditions are optimal when cutting with the 40× XT objective: aperture 6, intensity 35, and speed 5 (see Note 23). An example of a *G. orontii*-infected transverse leaf section before and after LMD is shown in Fig. 1 with ~20 cells at the site of infection collected per infection site.
5. For uninfected leaf tissue sections, epidermal and mesophyll cells are targeted and cut with the 40× XT objective using the same laser settings as for infected sections. Leaf sections and targeted cells are chosen to be similar to those used for infected leaves.
6. Sample collection time per batch is limited to 2 h to avoid evaporation of the RNA extraction buffer and limit potential

RNA degradation. Batch size was chosen based on required collection time and reproducibility of efficient extraction of RNA from that number of cells (see Note 24). For infected samples, collection of 2,500 cells (groups of ~20 per infection site) requires ~35–40 cross sections (1–2 slides). For uninfected samples, collection of 2,500 cells requires ~5 sections (1 slide).

7. After each batch of cells is captured, the sample is collected at the bottom of the tube by spinning for 1 min at full speed using a microcentrifuge. To ensure that all collected cells are transferred to the bottom of the tube, an additional 10 μl of RNA extraction buffer is pipetted into the tube cap and contents collected at the bottom by microcentrifuging for 1 min at full speed.
8. Samples are then incubated at 42°C for 30 min and vortexed vigorously for 3 min. The cell extract is collected at the bottom of the tube by microcentrifuging for 2 min at full speed.
9. Cell extract samples may be immediately used for RNA isolation or stored at −80°C up to 3 months without any impact on RNA integrity.

3.3. RNA Isolation from Laser Microdissected Cells

The PicoPure™ RNA isolation kit (Arcturus) is employed for RNA isolation from laser microdissected cells according to the manufacturer's instructions as outlined below (see Note 25). RNA was isolated from batches of 2,500 LMD-isolated cells with 7,500 cells total collected for each GeneChip sample to be processed. As our amplification efficiency is reproducibly higher with 2,500 cells than with 7,500 cells, the RNA from batches of 2,500 cells are amplified separately and then pooled to obtain sufficient amplified RNA for GeneChip analysis (detailed in subsequent chapter).

1. The RNA purification column is preconditioned by pipetting 250 μl Conditioning Buffer (CB) onto the purification column filter membrane, incubating the column for 5 min at room temperature, and centrifuging at 16,000 × *g* for 1 min.
2. 50 μl of 70% ethanol is added to the cell extract from step 10 (Subheading 3.2.2) and mixed well by pipetting up and down.
3. The cell extract and ethanol mixture (combined volume of 100 μl) is pipetted into the preconditioned purification column. The tube is centrifuged for 2 min at 100 × *g*, immediately followed by centrifugation at 16,000 × *g* for 30 s to remove flow through.
4. 100 μl Wash Buffer (W1) is pipetted into the purification column and centrifuged for 1 min at 8,000 × *g*.

5. DNase treatment is performed at this stage to reduce risk of DNA interference during RNA amplification and microarray hybridization.
   (a) 5 μl DNase I stock solution is added to 35 μl Buffer RDD and mixed gently by inverting.
   (b) 40 μl DNase incubation mix is pipetted directly into the purification column membrane and incubated at room temperature for 15 min.
   (c) 40 μl PicoPure RNA kit Wash Buffer (W1) is pipetted into the purification column membrane and centrifuged at 8,000 × *g* for 15 s.
6. 100 μl Wash Buffer (W2) is pipetted into the purification column membrane and centrifuged at 8,000 × *g* for 1 min.
7. Another 100 μl Wash Buffer (W2) is pipetted into the purification column membrane and centrifuged at 16,000 × *g* for 2 min.
8. The purification column is recentrifuged at 16,000 × *g* for 1 min to completely remove any residual wash buffer (see Note 26).
9. The purification column is transferred to a new 0.5 ml microcentrifuge tube.
10. 11 μl Elution Buffer (EB) is pipetted directly onto the membrane of the purification column and the column incubated for 1 min at room temperature (see Note 27).
11. The column is centrifuged for 1 min at 1,000 × *g* to distribute the EB in the column, then for 1 min at 16,000 × *g* to elute RNA.
12. 0.3 μl RNase inhibitor is added to protect samples from RNase-mediated RNA degradation during storage or downstream applications. The isolated RNA sample may be used immediately or stored at –80°C until use.

3.4. RNA Yield and Quality Assessment

Total RNA is quantitated using the NanoDrop 1000 and its purity assessed by A_{260}/A_{280} ratio and microcapillary electrophoresis using the RNA 6000 PicoLabChip® on the Agilent 2100 bioanalyzer. The A_{260}/A_{280} ratio is an indicator of purity as RNA absorbs at A_{260} whereas proteins absorb more strongly at A_{280}. Microcapillary electrophoresis provides a quantitative readout of RNA by size and a visual assessment of RNA degradation. RT-PCR may also be used to assess degradation of mRNA. Gene-specific primers designed to amplify 5′- , middle-, and 3′- regions of the mRNA transcript can allow one to assess RNA degradation by RT-PCR, which typically occurs at the 5′ end due to the action of RNases.

RT-PCR with gene-specific primers may also be used to verify cell type specificity and contamination from adjacent cell types using cell type specific markers. For example, to evaluate the specificity of RNA harvested from *Arabidopsis* mature leaf epidermal and mesophyll cells, molecular markers specific for these cell types were identified and RT-PCR performed on these specific genes. PCR products for the epidermal and mesophyll cell specific markers CUT1 and chloroplastic carbonic anhydrase, respectively, were detected only in samples from the appropriate cell type (Fig. 2, (2)). In addition, RT-PCR may be used to verify that

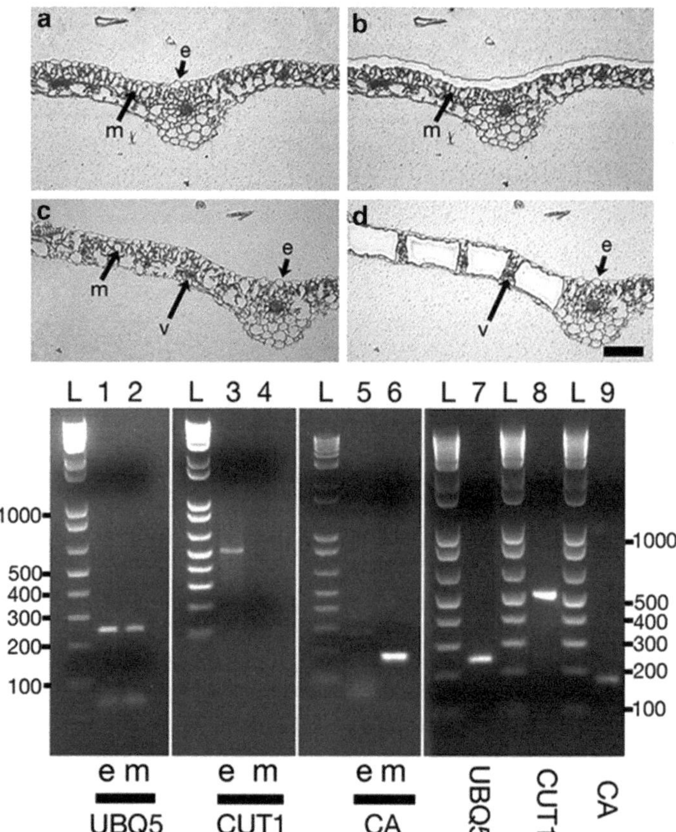

Fig. 2. Isolation of epidermal and mesophyll cells from *Arabidopsis* leaves and cell type specific RT-PCR. *Upper panel*: Sectioned tissue before (**a**, **c**) and after (**b**, **d**) LMD. Epidermal cells (**a**, **b**) and mesophyll cells (**c**, **d**) were collected separately from serial sections. Mature *Arabidopsis* rosette leaves were prepared using the modified microwave paraffin method with Sørenson's buffer. *e* epidermal cells, *m* mesophyll cells, *v* vascular bundles. *Bar* = 100 μm. *Lower panel*: RNA was extracted from LMD-harvested epidermal (e; lane 1, 3, and 5) and mesophyll (m; 2, 4, and 6) cells using the Qiagen RNeasy micro kit and subjected to RT-PCR. Ubiquitin5-specific primers give a 250 bp product (*UBQ5*: lanes 1, 2, and 7). The epidermal specific marker CUT1 primers yield a 501 bp product (*CUT1*: 3, 4, and 8), and the mesophyll specific marker plastidic carbonic anhydrase primers result in a 180 bp product (*CA*: 5, 6, and 9). Lanes 7–9 show control reactions performed with RNA isolated from whole leaves. L, 1 kB plus ladder (Invitrogen) (reproduced from ref. 2 with permission from Springer-Verlag).

a known gene enriched in the LMD-isolated cells (e.g., at the site of powdery mildew infection) is enriched in the LMD-isolated cell sample (e.g., *PR1* for 5 dpi of *A. thaliana* with *G. orontii* as in (1)). Though the RT-PCR procedures are not detailed here, we strongly advise their use for verifying that genes expected to be enriched in the collected sample set are enriched prior to amplification, target preparation, and microarray hybridization as presented in the following chapter.

3.4.1. RNA Quantification Using the NanoDrop

The NanoDrop allows for the analysis of 0.5–2.0 μl samples, without the need for cuvettes or capillaries. Outlined below is the procedure to quantify RNA using the NanoDrop 1000.

1. The NanoDrop software is initialized and the *Nucleic Acids* module is used to select RNA-40 as the constant for measuring total RNA.
2. The pedestal is cleaned with a KimWipe and RNase-free water.
3. In general, all measurements are performed by pipetting 1–2 μl of the appropriate solution directly onto the pedestal with the NanoDrop arm open, and closing the arm during measurement. The surface is wiped with a lint-free tissue between each sample.
4. A test measurement is performed with 1–2 μl RNase-free water.
5. The instrument is then blanked with 1–2 μl RNase-free water.
6. Sample RNA measurements are made by pipetting 1 μl sample onto the pedestal (see Note 28).
7. RNA samples with A_{260}/A_{280} ratios ~2.0, as determined by NanoDrop 1000, are suitable for downstream applications. We typically observe A_{260}/A_{280} ratios of 2.0–2.1.
8. RNA is automatically quantitated based on the absorbance at A_{260}, wavelength-dependent extinction coefficient (40 ng cm/μl for RNA) and path length of 0.2–1.0 mm, and displayed as ng/μl. We typically obtain ~1 ng RNA from 2500 LMD-isolated cells.

3.4.2. RNA Quality Assessment Using the Bioanalyzer

The Agilent Bioanalyzer 2100 is employed for total RNA quality assessment according to the manufacturer's instructions.

1. All reagents and reagent mixes (excluding RNA ladder) are kept at 4°C. Reagents are warmed to room temperature for at least 30 min before use. The RNA samples and ladder are placed on ice. The dye and gel-dye mix are wrapped in aluminum foil to protect from light.
2. The heating block is turned on and set to 70°C.

3. The RNA samples and RNA ladder are incubated at 70°C for 2 min and placed on ice for 5 min. The tubes are briefly centrifuged to clear any condensate from the tube's walls and cap.
4. The bioanalyzer electrodes are cleaned before use as follows. An electrode cleaner chip filled with 350 μl RNaseZap® is placed in the machine for 5 min. It is replaced with a second chip filled with 350 μl RNase-free water for an additional 5 min. The electrode is then allowed to air-dry for 1 min.
5. The gel-mix is prepared as follows. 550 μl of prewarmed gel matrix is added to a spin filter and centrifuged for 10 min at 1,500 × *g* (4,000 rpm) at room temperature. The filtered gel is aliquoted into 0.5 ml RNase-free tubes in 65 μl amounts and stored at 4°C. The gel-mix should be used within 1 month.
6. The gel-dye mix is prepared as follows. The dye is vortexed for 10 s and briefly spun down. 1 μl of the dye is added to 65 μl of the filtered gel-mix, mixed well by vortexing and centrifuged at 13,000 × *g* for 10 min at room temperature.
7. The seal of the chip priming station is checked before use as follows. A syringe is screwed onto the chip priming station and pulled to the 1 ml position. An empty chip is placed in the priming station and the lid closed. The plunger is pressed down until it is held by the syringe clip and released after 5 s. If sealed well, the plunger will release to about 0.7 ml within 2 s. If not, this step is repeated.
8. The gel-dye mix is loaded onto the chip as follows. A new chip is placed in the priming station and 9 μl of gel–dye mix is carefully loaded onto the bottom of the well marked G, avoiding the formation of air bubbles. A timer is set to 30 s. The plunger is positioned at 1 ml and the chip priming station closed. The latch will click when locked. The plunger is pressed down until it is held by the syringe clip. After 30 s, the clip is released. After another 5 s the plunger is slowly pulled back up to the 1 ml position and the chip priming station opened.
9. 9 μl of the gel dye matrix is loaded into two additional wells marked G.
10. 9 μl of Conditioning Solution is loaded into the well marked CS.
11. 5 μl of marker is loaded into the well marked ladder and each of the 12 sample wells (see Note 29).
12. 1 μl ladder is loaded into the well marked "ladder" and 1 μl RNA is loaded into each of the 12 wells. Unused RNA sample wells, if any, are loaded with 1 μl RNase-free water each to ensure proper running of samples in the other wells (see Note 30).

13. The sample is vortexed by placing the chip in a vortexer and mixed for 1 min at 612 × *g* (see Note 31).
14. The loaded chip is placed in the bioanalyzer and the lid closed.
15. In the instrument context, Assay > RNA > picochip is selected and the assay started.
16. The *End of Run* message appears when the chip run is finished and files are automatically saved. The run time is ~30 min.
17. The electrodes are cleaned after use as described in step 4.
18. An example of RNA profiles for laser microdissected cells is shown in Fig. 3 (see Note 32).

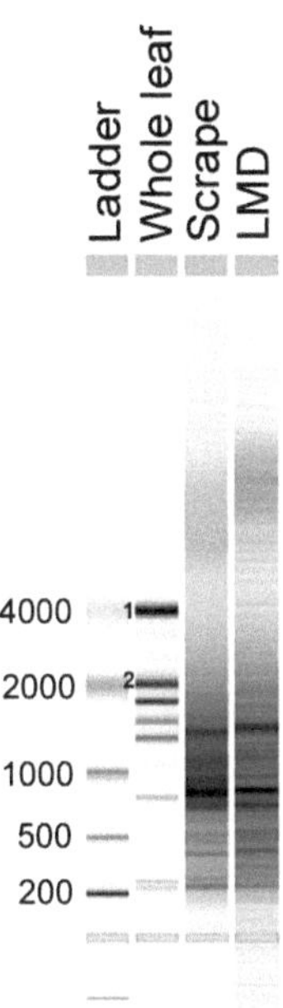

Fig. 3. Bioanalyzer electrophoretic gel image showing total RNA isolated from whole leaf (fresh tissue), whole leaf scrape (prepared tissue), and laser microdissected cells (prepared tissue, laser microdissected). Distinct cytoplasmic ribosomal RNA bands marked as 1 (25S) and 2 (18S). Chloroplastic and mitochondrial ribosomal RNA bands are also visible in the whole leaf sample RNA. It is also good to see bands in the 500–2,500 nt range consistent with the bulk of mRNA transcripts. RNA profiles of prepared tissue samples (LMD and whole leaf scrape) are similar indicating no significant impact from LMD on RNA quality. However, prepared tissue RNA profiles differ from that of fresh tissue (whole leaf) indicating that tissue preparation does have an impact on RNA quality. We do observe distinct bands in the mRNA size range in the prepared tissue samples and subsequent analysis found that RNA degradation associated with the tissue preparation process did not significantly impact GeneChip microarray output.

3.4.3. Other Assessments of RNA Quality

PCR-based assessment of RNA quality. The extent of RNA degradation may also be assessed by performing RT-PCR on RNA extracted from laser microdissected cells using gene-specific primers designed to amplify 5′-, middle-, and 3′- regions of the transcript. Absence of a PCR product with the 5′-specific primers would indicate RNA degradation at the 5′ end. However, even though degradation may be observed at the 5′ end of the transcript, this may not impact Affymetrix ATH1 GeneChip results as the probesets are designed to be 3′ biased (within 600 nt of the 3′ end). For this reason, gene-specific primers that amplify a ~200 bp product that is 600–400 nt from the 3′ end may be of particular interest. Housekeeping genes that are moderately expressed and not rapidly turned over may be preferable for this assessment.

Affymetrix ATH1 GeneChip assessment of RNA quality. As the goal is to use the LMD-isolated RNA for ATH1 expression profiling, the ultimate assessment of RNA quality uses the output from the ATH1 arrays. Methods for assessing RNA quality using the ATH1 arrays are presented in the following chapter.

3.5. Assessment of Impact of Tissue Specimen Preparation, Laser Microdissection, and RNA Amplification on RNA Quality and Microarray Results

Tissue preparation, LMD, and/or RNA amplification can have an impact on microarray gene expression data. RNA degradation or truncation can occur during tissue preparation, laser microdissection, and/or RNA amplification where mRNA may not have been fully converted to cDNA. Therefore, it is extremely important to evaluate RNA quality and the impact of RNA degradation on ATH1 gene expression data before proceeding with the microarray analysis. Assessment of RNA quality by NanoDrop, bioanalyzer and RT-PCR is described in Subheading 3.4. In addition to impacting RNA quality, these procedures could alter the distribution of mRNA and therefore the microarray output. To investigate the impact of tissue preparation, laser microdissection, and RNA amplification (see next chapter) on RNA quality (below) and microarray output (see next chapter), parallel samples should be collected as described below. Using the below controls, we determined that tissue preparation does result in some RNA degradation as evaluated by Bioanalyzer and RT-PCR, but that this degradation does not significantly impact the ATH1GeneChip results, which uses probesets that are 3′ biased (1). We further found that only the tissue preparation method altered the ATH1 output and that this alteration was minimal with altered expression of ~107 probesets (<0.5% of probesets on the ATH1 array) independent of infection (1).

The following control samples in addition to the LMD samples (prepared tissue, laser microdissected, and amplified) discussed above are included to allow one to assess the impact of tissue preparation, laser microdissection, and RNA amplification (see following chapter) on RNA quality and ATH1 expression profiling output. These should be collected in parallel from the same experiment used for the LMD samples.

1. Whole leaf (WL) sample (fresh, unamplified, and not laser microdissected). Fresh mature whole leaves (~100 mg sample) are harvested and immediately frozen in liquid Nitrogen. RNA isolation and microarray hybridization is performed as described above and in the following chapter.
2. Whole leaf amplified (WLA) sample (fresh, amplified and not laser microdissected). A 10 ng aliquot of RNA isolated from each of the WL samples above is subject to two-round amplification and microarray hybridization as described in the next chapter.
3. Whole leaf scrape (Scrape) sample (prepared tissue, amplified and not laser microdissected). Tissue prepared and sectioned whole leaf samples are scraped from PEN slides prior to laser microdissection. RNA isolation, two-round amplification, and microarray hybridization are performed as described above and in the following chapter.

To assess the impact of tissue preparation on RNA quality and ATH1 output, samples 2 (WLA) and 3 (Scrape) are compared. To assess the impact of LMD isolation, LMD samples are compared with Scrape samples, and to assess the impact of RNA amplification, WL samples are compared with WLA samples. Assessments include:

1. Impact on RNA quantity and quality (see Subheading 3.4 and Fig. 3).
2. Impact on ATH1 gene expression output, detailed in the following chapter.

4. Notes

1. During dissection, each freshly harvested leaf is placed on a drop of cold Sørenson's buffer to avoid drying of tissue. Razor blades are changed as often as required (approximately once every three leaves) to avoid any physical damage to the leaf tissue from dull edges. Mature Arabidopsis leaves of similar age are chosen and the leaf is first cut to remove edge regions resulting in a rectangle. Transverse sections (~1 cm across the leaf and 5 mm width) are then cut.
2. Forceps are predipped in Sørenson's buffer to transfer leaf pieces into vials without puncturing them. In general, pieces from three to four leaves can be accommodated in one vial. It is important to keep samples cold at this stage to minimize RNA degradation.

3. The level of water in the water bath should be adjusted so that the vials do not float.
4. The probe should NOT be cleaned with RNase-free water or RNaseZap® as this damages the temperature probe.
5. Vials are placed on ice during each solution change step. It is important not to remove all solution from vials while changing solutions to avoid drying of tissue. Note that trypan blue is added to Sørensen's buffer only in the initial step, not with subsequent solution changes. Remember to also change solution in the temperature probe vial.
6. The microwave temperature probe is immersed in a separate vial containing the appropriate ethanol solution during each dehydration step. The water in the water bath should not be replaced following the 30% ethanol change.
7. Addition of Safranin-O gives the solution a deep fruit-punch color.
8. This is the only step where all solution is removed from the vial and has to be performed quickly to avoid drying of tissue.
9. Turn on the paraffin wax dispenser ~30 min to 1 h prior to use to premelt the wax. Take care not to set the melting temperature too high (>62°C) as this reduces the plasticizing ability of the wax and impacts smooth cutting during tissue sectioning.
10. A handful of ice is added to the water bath to bring the temperature down quickly from 77 to 67°C.
11. Alternatively, samples can be incubated overnight at 62°C in an incubator, with paraffin wax changes performed the following day.
12. A hockey needle is used to gently scoop out any tissue that is remaining in the vial.
13. Leaf pieces should be well spaced to avoid sample loss when making paraffin blocks.
14. It is important to store samples with a desiccant to prevent tissue rehydration and potential activation of cellular RNases, which could degrade RNA. For example, we place the aluminum dishes in a white freezer box with a humidity sponge. Under these storage conditions, paraffin-embedded samples may be stored for up to 6 months.
15. PEN slides are glass slides covered with a special UV-absorbing membrane (polyethylene naphthalate). The membrane allows the attached tissue to drop into the cap of a microcentrifuge tube upon laser microdissection. Membranes mounted on steel frames (PET, polyethylene terephthalate) are also available and depending on tissue type and section thickness,

a particular slide type may be optimal. It is advisable to test different combinations of membranes and slide supports for optimal results.

16. Thickness of sections required for LMD may vary depending on the type of tissue used and the targeted cells. Note that section thickness impacts LMD parameters.
17. It is important to align the section strips within the edges of the PEN membrane to ensure that all sections are available for laser cutting.
18. Prepared paraffin slides can be stored for several weeks at 4°C with a desiccant without any impact on RNA integrity or yield.
19. It is critical to completely remove paraffin wax from sections because paraffin has a negative impact on laser cutting. However, the paraffin sections should not be overheated (on the hot plate) as this melts the PEN membrane.
20. It is important to completely dry the slides prior to LMD since moisture can not only interfere with optimal laser cutting, but also rehydrate samples resulting in potential activation of cellular RNases and degradation of RNA.
21. LMD using the Leica AS LMD allows for rapid, contamination-free isolation of cells using a UV laser beam for cutting. Alternative laser-based systems, namely, laser microdissection and pressure catapulting (LMPC) and laser capture microdissection (LCM) have also been employed to isolate specific cell populations from embedded plant tissue sections (6–9). LMPC is similar to LMD except that the cut cells are catapulted into the collection tube using an additional laser pulse. In LCM, a CapSure Cap is placed over the target area and an infrared laser beam is pulsed through the cap, which causes the thermoplastic film to form a thin protrusion that bridges the gap between the cap and tissue. Lifting of the cap removes the target cell(s) attached to the cap. This process could lead to contamination from cells nonspecifically adhering to the plastic membrane. For our studies, the use of the LCM system was unsuitable as we could not reproducibly isolate selected cells at the site of infection without obtaining neighboring unwanted cells. To ensure cells targeted for isolation using the AS LMD reproducibly fell and were collected into the underlying tube cap, we assessed the efficiency of collection and found 98% efficiency for our 20 cell group at the site of powdery mildew infection.
22. *G. orontii* haustoria are easily visible since the leaves were stained with trypan blue during the tissue preparation step. A number of parameters were taken into account in the selection

of infected sites for cell collection (1) an isolated *G. orontii* colony, (2) infected epidermal cell not directly adjacent to stomata or trichome, (3) underlying cells did not include the vasculature, and (4) not near edge of section.

23. Laser settings should be adjusted based on the laser objective and type of tissue used. We found that the use of the 40× XT objective was critical, as the high (extended) UV transmission of this objective allowed for narrower, more precise excision of the leaf material. Depending upon the tissue and laser settings, a cutting line of 2 μm can be obtained with the 40× XT objective.

24. The laser microdissected cells are protected from RNA degradation once they are in the RNA extraction buffer. We found that 2,500 cells were required for the reproducible, efficient isolation of RNA. 2,500 cells can be collected in 2 h (uninfected samples) or 4 h (infected samples). To limit RNA degradation and buffer evaporation, infected samples are collected in two sets of 1,250 cells each and pooled prior to RNA isolation. To compensate for extraction buffer evaporation during the 2 h collection time, ~10 μl of RNase-free water is added to the tube cap every 30 min. For each sample to be analyzed by microarray, we collected 7,500 cells total consisting of three batches of 2,500 cells; this yielded sufficient and reproducible amplified RNA for downstream global expression profiling using the GeneChip. In the future, fewer cells may be required as mRNA isolation and/or quantification technologies improve. However, at least 25 groups of cells at the infection site (or ~500 cells) would likely be required to minimize stochastic effects.

25. We evaluated an optimized phenol extraction method TRIzol (Invitrogen), three RNA-binding resin spin column kits (Qiagen RNeasy® micro kit, Arcturus PicoPure™ kit, and Ambion RNAqueous®-micro kit) and beta-tested a not yet released magnetic bead-based RNA isolation kit (Agencourt® Chloropure™ kit). RNA isolation kits need to reproducibly and efficiently extract small amounts of RNA (~1 ng) without the addition of contaminants or components (e.g., carrier RNA) that can interfere with downstream applications (e.g., microarray hybridization). In addition, the RNA buffer should exhibit minimal evaporation during the LMD collection time (2 h). The PicoPure™ RNA isolation kit (Arcturus) was optimal for our purposes. However, the Ambion RNAqueous®-micro kit was also suitable.

26. Any residual wash buffer in the column may interfere with downstream processes.

27. Due to small sample volumes, the elution buffer must be directly pipetted onto the column membrane to avoid sample loss.

28. For optimal results, the RNA sample is thoroughly mixed and briefly spun down prior to removing 1 μl from the top of the solution.
29. Marker must be loaded into each unused sample for the chip to run properly.
30. Use chip within 5 min of loading.
31. Make sure that the chip is tightly fastened onto the vortexer (with tape) to prevent it from excessive movement of the chip. If properly taped, no liquid will spill during vortexing.
32. The 2100 bioanalyzer outputs an electrophoretic gel image with RNA profiles for individual samples. Distinct cytoplasmic ribosomal RNA bands (25S, 18S) and chloroplastic ribosomal RNA bands (23S, 16S) are visible in the whole leaf sample RNA as shown in Fig. 3. It is also good to see bands in the 500–2,500 nt range consistent with the bulk of mRNA transcripts. We observe some degradation of the RNA in the Scrape and LMD samples due to the tissue preparation method (Fig. 3). However, the observed banding in these samples in the mRNA size range is a good sign. For example, you do not want to just observe a smear in the lower size range (e.g., <500 nt). RT-PCR with gene-specific primers designed to amplify 5′- and 3′-regions indicated the tissue specific degradation occurred at the 5′ end of the transcript. Subsequent microarray analysis indicated that this degradation did not significantly impact Genechip output as the probesets are 3′ biased. Using less optimal tissue preparation and/or RNA isolation methods, we observed fewer distinct bands of higher molecular weight and/or "smearing." If problems with RNA quantity and quality are observed, it is recommended to evaluate tissue processing, LMD, and RNA extraction procedures before proceeding, and repeat steps as necessary. As shown in Fig. 3, we found the dominant impact on RNA quality was due to tissue preparation.

References

1. Chandran, D., Inada, N., and Wildermuth, M. C. (2010) Lasermicrodissection at the site of powdery mildew infection identifies novel regulators and processes mediating the *Golovinomyces orontii* interaction. *Proc Natl Acad Sci USA 107*, 460–5.
2. Inada, N., and Wildermuth, M. C. (2005) Novel tissue preparation method and cell-specific marker for laser microdissection of Arabidopsis mature leaf. *Planta 221*, 9–16.
3. Nelson, T., Tausta, S. L., Gandotra, N., and Liu, T. (2006) Laser microdissection of plant tissue: what you see is what you get. *Annu Rev Plant Biol 57*, 181–201.
4. Murray, G. I. (2007) An overview of laser microdisection technologies. *Acta Histochem* ***109***, 171–6.
5. Day, R. C., Grossniklaus, U., and Macknight, R. C. (2005) Be more specific! Laser-assisted microdissection of plant cells. *Trends Plant Sci* ***10***, 397–406.
6. Fosu-Nyarko, J., Jones, M. G., and Wang, Z. (2009) Functional characterization of transcripts expressed in early-stage *Meloidogyne*

javanica-induced giant cells isolated by laser microdissection. *Mol Plant Pathol* **10**, 237–48.

7. Ithal, N., Recknor, J., Nettleton, D., Maier, T., Baum, T. J., and Mitchum, M. G. (2007) Developmental transcript profiling of cyst nematode feeding cells in soybean roots. *Mol Plant Microbe Interact* **20**, 510–25.
8. Klink, V. P., Overall, C. C., Alkharouf, N. W., MacDonald, M. H., and Matthews, B. F. (2007) Laser capture microdissection (LCM) and comparative microarray expression analysis of syncytial cells isolated from incompatible and compatible soybean (Glycine max) roots infected by the soybean cyst nematode (*Heterodera glycines*). *Planta* **226**, 1389–409.
9. Ramsay, K., Wang, Z., and Jones, M. G. K. (2004) Using laser capture microdissection to study gene expression in early stages of giant cells induced by root-knot nematodes. *Mol Plant Pathol* **5**, 587–92.

Chapter 20

Global Expression Profiling of RNA from Laser Microdissected Cells at Fungal–Plant Interaction Sites

Divya Chandran, Greg Hather, and Mary C. Wildermuth

Abstract

Global expression profiling of RNA isolated from laser microdissected cells allows one to profile a specific set of cells allowing for enhanced sensitivity and for cell- or site-specific patterns of expression to emerge. In Chapter 19, we detail our optimized methods of tissue preparation, laser microdissection (LMD), and RNA isolation of cells at the site of *Golovinomyces orontii* infection of mature *Arabidopsis* leaves. Here, we describe (1) amplification of the RNA to obtain sufficient starting material for microarray analysis, (2) microarray hybridization and associated quality control assessments. As tissue preparation, LMD, and/or RNA amplification could impact mRNA quality, distribution, and/or microarray processing and output, it is important to include quality control assessments at every step of the protocol to ensure that the final data is a reproducible and accurate readout of the biological source material. The collection of parallel samples to evaluate these components of the experimental protocol allows one to determine their impact on mRNA quality and distribution (described in Chapter 19) and on microarray output (discussed here). In addition, one likely wants to compare similarly processed whole leaf samples to LMD-isolated samples in order to identify genes and processes specifically impacted or more highly impacted at the infection site compared with the whole leaf.

Using the procedures described herein to profile cells specifically at the site of powdery mildew infection of *Arabidopsis* (Chandran et al., Proc Natl Acad Sci U S A 107(1):460–465, 2010), we determined that our site-specific global expression data was a highly reproducible, sensitive, and accurate readout of the infection site. Furthermore, this site-specific analysis allowed us to identify novel processes (e.g., endoreduplication), regulators (e.g., MYB3R4), and process components associated with the sustained growth and reproduction of the powdery mildew *G. orontii* on *Arabidopsis thaliana* at 5 days postinfection that were hidden in whole leaf analyses (Chandran et al., Proc Natl Acad Sci U S A 107(1):460–465, 2010).

Key words: *Arabidopsis thaliana*, RNA amplification, ATH1 GeneChip® profiling, quality control assessment, RNA degradation plots

1. Introduction

Global expression profiling of plant–pathogen interaction sites using laser microdissection (LMD) allows for spatial resolution of host molecular responses to pathogens. In Chapter 19, we

John M. McDowell (ed.), *Plant Immunity: Methods and Protocols*, Methods in Molecular Biology, vol. 712,
DOI 10.1007/978-1-61737-998-7_20,

described the LMD of *Arabidopsis*–powdery mildew interaction sites and isolation of RNA for downstream expression profiling. Very small amounts of RNA (~1 ng) are obtained from laser microdissected cells, limiting the direct use of this technology in downstream microarray applications, which typically require microgram quantities of RNA. To increase yield, RNA can be amplified using a linear or PCR-based approach (1). For expression profiling of *Arabidopsis*–powdery mildew interaction sites, we found that the two-cycle target linear amplification method reproducibly produced sufficient RNA of high quality for gene expression profiling studies. In the first cycle, total RNA extracted from LMD-isolated cells is reverse transcribed into double-stranded cDNA using a T7-oligo(dT) primer. The double-stranded cDNA is converted to complementary RNA (cRNA) using T7 RNA Polymerase and unlabeled ribonucleotide mix. In the second cycle, the unlabeled cRNA is reverse transcribed using random primers and the T7-oligo(dT) primer. The resulting double-stranded cDNA is amplified and labeled using a biotinylated nucleotide analog/ribonucleotide mix. The labeled cRNA is then cleaned up, fragmented, and hybridized to Affymetrix GeneChip® *Arabidopsis* ATH1 expression arrays.

When generating gene expression data from a small number of cells, artifacts or bias could be introduced at any part of the process. Therefore, it is very important to evaluate the impact of the tissue preparation, LMD, and amplification procedures on RNA quality (see Chapter 19) and microarray gene expression output (discussed here). As detailed below, this chapter describes (1) two-cycle RNA amplification and (2) microarray hybridization and associated quality control procedures. As part of the quality control analyses, we discuss both novel and standard quality assessments of Affymetrix GeneChip data (reviewed in ref. 2).

2. Materials

2.1. Two-Round RNA Amplification

1. GeneChip® two-cycle target labeling and control reagents (Affymetrix, Inc., Santa Clara, CA) containing IVT labeling kit, two-cycle cDNA synthesis kit, Poly-A RNA control kit, hybridization controls, and sample cleanup module. Sample cleanup module is stored at room temperature. All other reagents are stored at –20°C (see Note 1).
2. 3 M sodium acetate, pH 5.0, prepared using RNase-free water.
3. Alcohol series: 96–100% ethanol (200 proof) and 80% ethanol prepared using RNase free water.

4. Diethylpyrocarbonate (DEPC, ≥97%, Sigma-Aldrich, St. Louis, MO). Stored at 2–8°C.
5. RNase-free water (0.1% DEPC-treated water) or nuclease-free water (Applied Biosystems/Ambion, Austin, TX). DEPC-treated water is prepared by adding 1 mL DEPC to 1 L of water and stirring well until the DEPC is completely dissolved. The solution is then incubated overnight and autoclaved. DEPC is toxic and solutions containing DEPC must be prepared in a fume hood. Autoclaving decomposes DEPC into ethanol and carbon dioxide rendering it inactive and nontoxic.
6. Thermocycler.
7. Speed Vac.

2.2. Amplified RNA Quality and Yield Assessment

1. Nanodrop 1000 (Thermo Scientific, Wilmington, DE).
2. Agilent 2100 bioanalyzer with equipment (Agilent Technologies, Santa Clara, CA).
3. RNA 6000 Pico Kit (Agilent Technologies). RNA 6000 ladder is stored at −20°C. All other reagents and reagent mixes are stored at 4°C when not in use. Dye and dye mixtures must be protected from light. Gel mix is stable at 4°C up to 1 month after preparation.
4. Diethylpyrocarbonate (DEPC, ≥97%, Sigma-Aldrich, St. Louis, MO). Stored at 2–8°C.
5. RNase-free water (0.1% DEPC-treated water) or nuclease-free water (Applied Biosystems/Ambion, Austin, TX). DEPC-treated water is prepared by adding 1 mL DEPC to 1 L of water and stirring well until the DEPC is completely dissolved. The solution is then incubated overnight and autoclaved. DEPC is toxic and solutions containing DEPC must be prepared in a fume hood. Autoclaving decomposes DEPC into ethanol and carbon dioxide rendering it inactive and nontoxic.
6. RNaseZap® (Applied Biosystems/Ambion). RNaseZap® is slightly corrosive and must be handled with gloves.

2.3. Microarray Hybridization

1. 5× Fragmentation buffer (Affymetrix, Inc.). Stored at room temperature.
2. GeneChip® *Arabidopsis* ATH1 Genome Array (Affymetrix, Inc.). Stored at 4°C.
3. GeneChip® Fluidic Station 450 (Affymetrix, Inc.).
4. GeneChip® Scanner 3000 7G (Affymetrix, Inc.) with an autoloader.

3. Methods

The methods described below outline (1) two-cycle RNA amplification, amplified RNA yield and quality assessment, (2) microarray hybridization, and (3) quality control assessments of microarray output. RNA integrity is a major factor that influences the quality of data obtained using the described protocols. To avoid RNase-mediated RNA degradation and to avoid the introduction of contaminating RNA, an RNase-free work environment is strongly recommended as are the below general precautions.

1. All work spaces and labware, including pipetteman plungers, must be decontaminated by rinsing or wiping with RNase-removal agents, such as RNaseZap®. Only sterile, nuclease-free filter-barrier tips and reaction tubes should be used from dedicated, covered containers.
2. If possible, RNA-only dedicated pipetteman should be utilized and an RNA-only dedicated lab bench space should be used.
3. All solutions must be made with DEPC-treated water or RNase-free water.
4. Gloves must be worn at all times and changed frequently. Lab coats should be worn and hair pulled back to avoid contamination.

In addition, for enhanced reproducibility and accuracy in pipetting, reaction mixes should not be made individually, but as a master mix for all samples. When making the master mix, include 1–2 extra samples worth of reaction mixture to ensure sufficient volume for all samples. To minimize shearing of RNA when mixing, all samples are gently flicked and not vortexed.

3.1. RNA Amplification

The Affymetrix two-cycle target labeling kit is employed for two-cycle RNA amplification according to the manufacturer's instructions as outlined below. As described in Chapter 19, total RNA was isolated from batches of 2,500 LMD-isolated cells with 7,500 cells total collected for each GeneChip sample to be processed. We find amplification efficiency is reproducibly higher with RNA from 2,500 cells than with 7,500 cells, with efficiencies of 1.0×10^4 vs. 3.0×10^3, respectively. Therefore, RNA from batches of 2,500 cells are amplified separately and then pooled to obtain sufficient amplified RNA for GeneChip analysis.

3.1.1. Preparation of Poly-A Controls and T7-Oligo(dT) Primer/Poly-A Controls Mix

The Poly-A controls are spiked into the RNA samples and amplified and labeled together with the samples. These controls allow for the assessment of the labeling process, independent of the sample RNA and are for genes not found in eukaryotic tissue.

1. We typically obtain ~1 ng RNA from 2,500 LMD-isolated cells. For 1 ng RNA starting amount, the Poly-A RNA serial dilutions are prepared as follows to obtain a final dilution of 1:500,000 (see Note 2).
   (a) For the first dilution (1:20), 2 μL Poly-A control stock is added to 38 μL Poly-A control dilution buffer, mixed thoroughly by pipetting and spun down briefly to collect contents at the bottom of the tube (see Note 3).
   (b) For the second dilution (1:50), 2 μL first dilution is added to 98 μL Poly-A control dilution buffer, mixed thoroughly and spun down.
   (c) For the third dilution (1:50), 2 μL second dilution is added to 98 μL Poly-A control dilution buffer, mixed thoroughly and spun down.
   (d) For the fourth dilution (1:10), 2 μL third dilution is added to 18 μL Poly-A control dilution buffer, mixed thoroughly and spun down. The fourth dilution (final dilution of 1:500,000) is used to prepare the solution described in the next step.
2. The T7-oligo(dT) primer/Poly-A control mix is prepared by mixing 2 μL 50 μM T7-oligo(dT) primer, 2 μL diluted Poly-A control (fourth dilution; step 1d), and 16 μL RNase-free water. This recipe is sufficient for ten samples.

3.1.2. First-Cycle, First-Strand cDNA Synthesis

1. Total RNA/T7-oligo(dT) primer/Poly-A control mix is prepared as follows. The total RNA sample (1–10 ng) in 1–3 μL volume is placed in a 0.2 mL RNase-free PCR tube (see Note 4). 2 μL of the T7-oligo(dT) primer/Poly-A controls mix and RNase-free water is added to a final volume of 5 μL. The tube is gently flicked to mix contents and briefly spun down to collect the contents at the bottom of the tube. The sample is then incubated for 6 min at 70°C, cooled at 4°C for at least 2 min and spun down (see Note 5).
2. The First-Cycle, First-Strand Master Mix is assembled in a separate tube. Sufficient First-Cycle, First-Strand Master Mix is prepared for all RNA samples. The following recipe is for a single reaction. 2 μL 5× First-Strand reaction mix, 1 μL 0.1 M DTT, 0.5 μL 10 mM dNTP, 0.5 μL RNase Inhibitor, and 1.0 μL SuperScript II RT enzyme are added to a 0.2 mL PCR tube in the order mentioned, mixed well by gentle flicking and spun down.
3. 5 μL of First-Cycle, First-Strand Master Mix is added to each total RNA sample/T7-oligo(dT) primer/Poly-A controls mix (from step 1) for a final volume of 10 μL. The contents are mixed thoroughly by gently flicking, spun down, and immediately placed at 42°C for 1 h.

4. Samples are then heated at 70°C for 10 min to inactivate the RT enzyme, cooled for at least 2 min at 4°C and spun down (see Note 6).

3.1.3. First-Cycle, Second-Strand cDNA Synthesis

1. The First-Cycle, Second-Strand Master Mix is assembled in a separate tube. Sufficient First-Cycle, Second-Strand Master Mix is prepared for all samples, fresh, before use. The following recipe is for a single reaction. 4.8 μL RNase-free water, 4 μL freshly diluted 17.5 mM magnesium chloride, 0.4 μL 10 mM dNTP, 0.6 μL *Escherichia coli* DNA Polymerase I and 0.2 μL RNase H enzyme are added to a 0.2 mL PCR tube in the order mentioned, mixed well by gentle flicking and spun down to collect contents at the bottom of the tube.
2. 10 μL of the First-Cycle, Second-Strand Master Mix is added to each sample from step 4 (Subheading 3.1.2) for a total volume of 20 μL. The contents are mixed by gently flicking the tube a few times and spun down.
3. The Second-Strand reaction is incubated for 2 h at 16°C, 10 min at 75°C, cooled at least 2 min at 4°C and spun down (see Note 7).

3.1.4. First-Cycle, IVT Amplification of cRNA

The MEGAscript® T7 Kit (Ambion, Inc.) is used for this step.

1. The First-Cycle, IVT Master Mix is assembled in a separate tube at room temperature. Sufficient First-Cycle, IVT Master Mix is prepared for all samples. The following recipe is for a single reaction. 5 μL each of 10× Reaction Buffer, ATP, CTP, UTP, GTP solution, and enzyme mix is mixed in a tube for a total volume of 30 μL. The contents are mixed well by gently flicking the tube a few times and spun down.
2. 30 μL of the First-Cycle, IVT Master Mix is added at room temperature to each 20 μL of cDNA sample from step 3 (Subheading 3.1.3) for a final volume of 50 μL, mixed by gently flicking the tube a few times and spun down.
3. The IVT Master reaction is incubated for 16 h at 37°C and spun down. The sample is then purified as described in the following section (see Note 8).

3.1.5. First-Cycle, Cleanup of cRNA

The Sample Cleanup Module is used for this step (see Note 9).

1. 50 μL of RNase-free water is added to the IVT reaction from step 3 (Subheading 3.1.4) and mixed by vortexing for 3 s.
2. 350 μL IVT cRNA Binding Buffer is added to the sample and mixed by vortexing for 3 s.
3. 250 μL ethanol (96–100%) is added to the lysate and mixed well by pipetting.
4. 700 μL of the sample is applied to the IVT cRNA Cleanup Spin Column placed in a 2 mL collection tube. The sample is

centrifuged for 15 s at ≥8,000 × *g*. The flow-through and collection tube is discarded.

5. The spin column is transferred into a new 2 mL collection tube (supplied with kit) and 500 μL IVT cRNA Wash Buffer is pipetted onto the spin column. The sample is centrifuged for 15 s at ≥8,000 × *g* and flow-through discarded.
6. 500 μL 80% (v/v) ethanol is pipetted onto the spin column and centrifuged for 15 s at ≥8,000 × *g*. The flow-through is discarded.
7. The cap of the spin column is opened and the tube centrifuged for 5 min at maximum speed (~25,000 × *g*). The flow-through and collection tube is discarded (see Note 10).
8. The spin column is transferred into a new 1.5 mL collection tube, and 13 μL of RNase-free water is pipetted directly onto the spin column membrane. The tube is centrifuged for 1 min at maximum speed (~25,000 × *g*) to elute.
9. The entire cRNA eluate is used in the following steps – Subheading 3.1.6 (see Note 11).

3.1.6. Second-Cycle, First-Strand cDNA Synthesis

1. A fresh dilution of Random Primers (final concentration 0.2 μg/μL) is prepared by mixing 2 μL of Random Primers (3 μg/μL) with 28 μL RNase-free water.
2. 2 μL of diluted random primers is added to purified cRNA from step 8 (Subheading 3.1.5) and RNase-free water is added for a final volume of 11 μL.
3. The sample is incubated for 10 min at 70°C, cooled at 4°C for at least 2 min and spun down.
4. The Second-Cycle, First-Strand Master Mix is assembled in a separate tube. Sufficient Second-Cycle, First-Strand Master Mix is prepared for all samples. The following recipe is for a single reaction. 4 μL 5× First-Strand reaction mix, 2 μL 0.1 M DTT, 1.0 μL RNase Inhibitor, 1.0 μL 10 mM dNTP, and 1.0 μL SuperScript II RT enzyme are added to a 0.2 mL PCR tube in the order mentioned, mixed well by gentle flicking and spun down.
5. 9 μL of Second-Cycle, First-Strand Master Mix is added to each cRNA/random primer sample from step 4, for a final volume of 20 μL. The contents are mixed thoroughly by gently flicking, spun down, and immediately placed at 42°C.
6. The First-Strand reaction is incubated for 1 h at 42°C, cooled for at least 2 min at 4°C and spun down.
7. 1 μL of RNase H is added to each sample for a final volume of 21 μL, mixed thoroughly by gently flicking the tube a few times, spun down and incubated for 20 min at 37°C.
8. The sample is heated at 95°C for 5 min and cooled for at least 2 min at 4°C.

3.1.7. Second-Cycle, Second-Strand cDNA Synthesis

1. A fresh dilution of the T7-oligo(dT) primer (final concentration 5 μM) is prepared by mixing 2 μL of T7-oligo(dT) primer (50 μM) with 18 μL of RNase-free water.
2. 4 μL of diluted T7-oligo(dT) primer is added to the sample from step 8 (Subheading 3.1.6) for a final volume of 25 μL. The tube is gently flicked a few times to mix and contents spun down.
3. The sample is incubated for 6 min at 70°C, cooled at 4°C for at least 2 min and spun down.
4. The Second-Cycle, Second-Strand Master Mix is assembled in a separate tube. Sufficient Second-Cycle, Second-Strand Master Mix is prepared for all samples. The following recipe is for a single reaction: 99 μL RNase-free water, 30 μL 5× Second-Strand reaction mix, 3.0 μL 10 mM dNTP, and 4.0 μL *E. coli* DNA Polymerase I are added to a 0.2 mL PCR tube in the order mentioned for a total volume of 125 μL, mixed well by gentle flicking and spun down.
5. 125 μL of the Second-Cycle, Second-Strand Master Mix is added to each sample from step 3 for a total volume of 150 μL. The contents are mixed by gently flicking the tube a few times and spun down.
6. The Second-Strand reaction is incubated for 2 h at 16°C (see Note 7).
7. 2 μL of T4 DNA Polymerase is added to each sample for a final volume of 152 μL. The contents are mixed by gently flicking the tube a few times and spun down.
8. The Second-Strand reaction is incubated for 10 min at 16°C, cooled at 4°C for at least 2 min and spun down (see Note 12).

3.1.8. Cleanup of Double-Stranded cDNA

Sample Cleanup Module is used for cleaning up the double-stranded cDNA.

1. 600 μL of cDNA Binding Buffer is added to the double-stranded cDNA synthesis preparation from step 8 (Subheading 3.1.7) and mixed by vortexing for 3 s (see Note 13).
2. 500 μL of the sample is applied to the cDNA Cleanup Spin Column placed in a 2 mL collection tube, and centrifuged for 1 min at ≥8,000 × *g*. Flow-through is discarded.
3. The spin column is reloaded with the remaining mixture and centrifuged as above. The flow-through and collection tube is discarded.
4. The spin column is transferred into a new 2 mL collection tube. 750 μL of the cDNA Wash Buffer is applied to the spin column and centrifuged for 1 min at ≥8,000 × *g*. Flow-through is discarded.

5. The cap of the spin column is opened and the tube centrifuged for 5 min at maximum speed (~25,000 × *g*). The flow-through and collection tube is discarded.
6. The spin column is transferred into a 1.5 mL collection tube, and 14 μL of cDNA elution Buffer is pipetted directly onto the spin column membrane. The sample is incubated for 1 min at room temperature and centrifuged for 1 min at maximum speed (~25,000 × *g*) to elute.

3.1.9. Synthesis of Biotin-Labeled cRNA

GeneChip IVT Labeling Kit is used for this step.

1. All of the template cDNA from step 6 (Subheading 3.1.8) (~12 μL) is transferred to a tube and the reaction components are added in the following order (see Note 14). 4 μL 10× IVT Labeling Buffer, 12 μL IVT Labeling NTP Mix, 4 μL IVT Labeling Enzyme Mix, and 8 μL RNase-free water is added to 12 μL of template cDNA for a final volume of 40 μL. This recipe is for a single reaction.
2. The reagents are mixed carefully and spun down and incubated at 37°C for 16 h.
3. The labeled cRNA may be stored at –20°C or –70°C if not purified immediately.

3.1.10. Cleanup of Biotin-Labeled cRNA

Sample Cleanup Module is used for cleaning up the biotin-labeled cRNA.

1. The initial cleanup steps are identical to steps 1–7 outlined in Subheading 3.1.5.
2. The final elution using RNase-free water is carried out in two steps:
   (a) In the first step, the spin column is transferred into a new 1.5 mL collection tube, and 11 μL of RNase-free water is pipetted directly onto the spin column membrane. The tube is centrifuged for 1 min at maximum speed (~25,000 × *g*) to elute.
   (b) In the second step, 10 μL of RNase-free water is pipetted directly onto the spin column membrane and the tube is centrifuged for 1 min at maximum speed (~25,000 × *g*) to elute.
3. The cRNA eluate is stored at –20°C or –70°C if not quantitated immediately.

3.2. Complementary RNA Yield and Quality Assessment

cRNA is quantitated using the NanoDrop 1000 and its purity assessed by A_{260}/A_{280} ratio and microcapillary electrophoresis using the RNA 6000 PicoLabChip® on the Agilent 2100 Bioanalyzer as outlined in Chapter 19. We reproducibly achieve RNA amplification rates of 1.0×10^4 from 2,500 cells, similar to

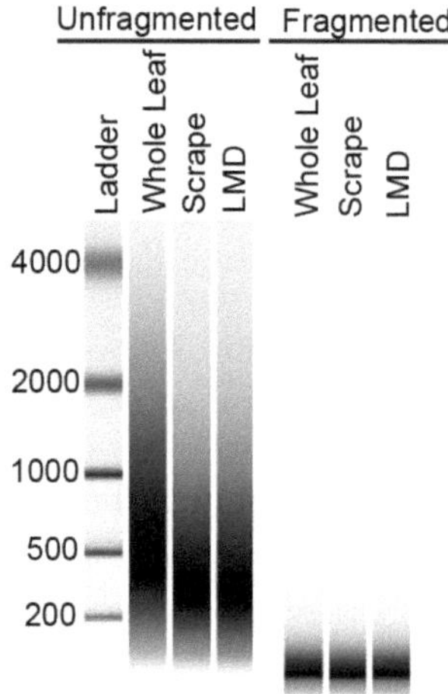

Fig. 1. Bioanalyzer (Agilent 2100 Bioanalyzer) electrophoretic gel image showing unfragmented and fragmented two-cycle amplified cRNA from whole leaf (fresh tissue), whole leaf scrape (prepared tissue), and laser microdissected cells (prepared tissue, laser microdissected).

rates observed in other plant LMD studies (1, 3). Figure 1 shows typical profiles from two-round amplified cRNA, from 2,500 laser microdissected cells. The size distribution for two-round cRNA is typically between 200 and 1,000 bp. cRNA samples with A_{260}/A_{280} ratios of ≥1.9, as determined by NanoDrop 1000, are suitable for GeneChip® hybridization. We typically observe A_{260}/A_{280} ratios of 2.0–2.1.

3.3. Microarray Hybridization

Microarray experiments require a minimum of two independent biological replicates per sample type, and we recommend three replicates and the collection of additional replicates in the event that a given sample does not meet quality control standards. Technical replicates for which the same fragmented cRNA is hybridized to independent GeneChips can also be performed. Technical replicates are recommended if you are inexperienced with target preparation for GeneChip microarrays and/or you are utilizing a new microarray facility for hybridization of your samples. We routinely provide a microarray facility with fragmented cRNA. Typically, 20 µg of fragmented cRNA/per gene GeneChip hybridization is required by a facility. The Affymetrix two-cycle target labeling kit is employed for cRNA fragmentation according to the manufacturer's instructions as outlined below. Fragmentation provides labeled target of optimal and uniform size (50–200 bp). The microarray hybridization protocols, performed according to Affymetrix GeneChip Expression Analysis Technical Manual, are not provided here. It is optimal if all samples for a given experiment are processed at the array facility in parallel.

3.3.1. cRNA Fragmentation

1. 8 µL of 5× Fragmentation buffer is added to each 20 µg cRNA sample (1–21 µL).
2. RNase-free water is added to a final volume of 40 µL.
3. Samples are incubated at 94°C for 35 min and then placed on ice. An aliquot of the sample is saved for analysis on the Bioanalyzer. The final concentration of the fragmented cRNA should be ~0.5 µg/µL. A typical fragmented cRNA sample is shown in Fig. 1.
4. Fragmented cRNA samples and GeneChips® are transported to the microarray facility on ice.

3.4. Quality Control Assessments of Microarray Output

In Chapter 19, we described the assessment of the impact of tissue preparation and LMD on RNA quantity and quality using the Nanodrop, Bioanalyzer and qPCR. We also discussed the collection of control samples (1–3) in addition to the LMD samples (4):

1. Whole leaf (WL) sample (fresh, not laser microdissected, unamplified)
2. Whole leaf amplified (WLA) sample (fresh, not laser microdissected, amplified)
3. Whole leaf scrape (Scrape) sample (prepared tissue, not laser microdissected but scraped from slide, amplified)
4. LMD samples (prepared tissue, laser microdissected, amplified)

In addition, one likely wants to compare similarly processed whole leaf samples to LMD-isolated samples in order to evaluate genes and processes specifically impacted at the infection site or with a greater difference in expression in infected vs. uninfected samples at the infection site compared with the whole leaf as in (4).

Here, we briefly describe both standard (reviewed in ref. 2) and novel quality control assessments used to evaluate the effect of our experimental procedures on Affymetrix GeneChip output. To assess the potential impact of tissue preparation on ATH1 output, WLA, Scrape, and LMD samples are compared. To assess the impact of LMD, Scrape and LMD samples are contrasted. To assess the impact of RNA amplification, WL samples are compared with WLA samples. The final goal is to obtain highly reproducible ATH1 GeneChip data from LMD samples that is an accurate readout of the transcriptional status at the site of infection.

3.4.1. GeneChip Data Processing and Quality Control Assessments

GeneChip data processing and standard quality control assessments were performed using the *gcrma* and *affyPLM* packages in *Bioconductor* (5) and are reviewed elsewhere (e.g., (2)). In brief, extract expression values ($\log_2$) using robust multiarray analysis

(6, 7) with background correction performed using perfect match only. Employ quantile normalization to make the distribution of signal intensities uniform across chips (8). This is then followed by standard quality diagnostics (2), including residual, relative log expression (RLE), normalized unscaled standard error (NUSE), and RNA degradation plots performed using *affyPLM*. Residual plots (not shown) are used to ascertain the presence of large artifacts that indicate results from a particular chip are of poor quality; this is unusual (<5% of GeneChips in our experience). RLE and NUSE plots are used to assess reproducibility and identify sample outliers within samples of one class (e.g., LMD) as well as to compare data from different sample classes (e.g., LMD vs. WL vs. WLA vs. Scrape). Samples should have similar metrics of high quality (see ref. 2). For example, as shown in Fig. 2a, our RLE boxes have relatively small and similar spread and are centered near 0 (indicating that the majority of genes are not changing in expression between experimental conditions/samples). Similarly, NUSE boxes should be centered near 1 with limited spread, as observed for all samples in Fig. 2b. A problem array would have an elevated or more spread out NUSE box (e.g., centered on 1.2 (2)).

In Chapter 19, we ascertained that the tissue preparation method had a negative impact on RNA quality with some associated 5′ mRNA degradation. However, ATH1 array probesets were designed to be 3′ biased (within 600 bp of the 3′ end of the transcript) to minimize the impact of processive degradation of mRNA from the 5′ end. Therefore, the observed RNA degradation might not have a significant impact on ATH1 output. Standard RNA degradation plots, generated using AffyRNAdeg in *Bioconductor*, plot expression values vs. probes presented in order from 5′ to 3′ for each transcript. This degradation plot showed steeper slopes for amplified samples (WLA, Scrape, and LMD) compared with the unamplified whole leaf samples (not shown). Steeper slopes are indicative of less of a signal from 5′ probes compared to 3′ probes and can reflect processive RNA degradation from the 5′ end and/or incomplete reverse transcription. As WLA samples were more similar in profile to Scrape and LMD samples than to unamplified whole leaf samples, this suggests amplification (likely via incomplete reverse transcription) is the major contributor to the observed differences in the standard RNA degradation plots. For this reason, when comparing LMD and whole leaf sample data, amplified samples must be used in the comparison (e.g., see ref. 4). However, these RNA degradation plots are not effective at assessing RNA integrity. For example, Archer et al. found RNA degradation plots for renal cell samples of high and poor RNA quality exhibited similar slopes (9). Indeed, we established that our tissue preparation procedure resulted in some RNA degradation (see Chapter 19); however, this impact is

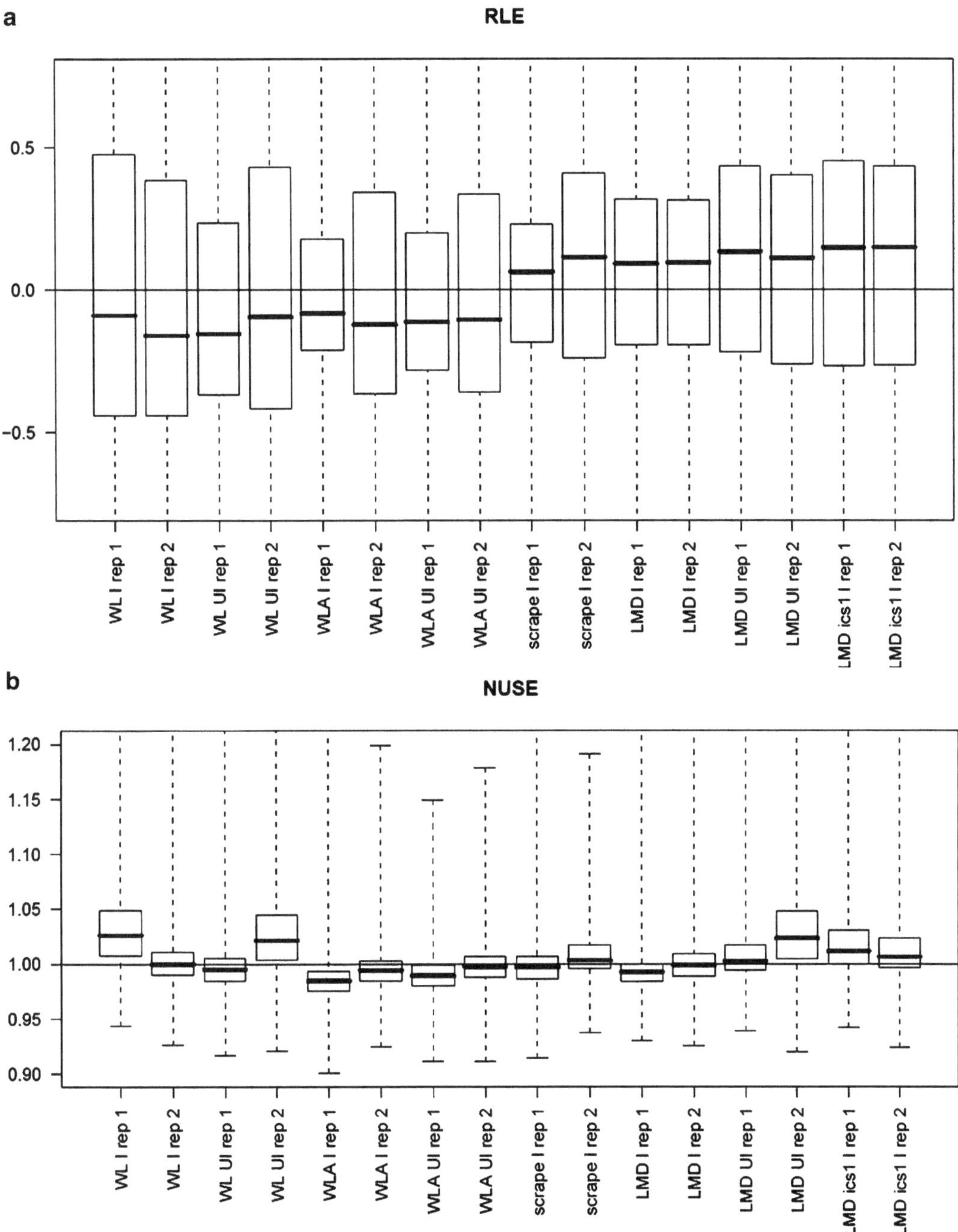

Fig. 2. RLE (**a**) and NUSE (**b**) plots for 16 ATH1 arrays of site-specific powdery mildew profiling project. Reps are independent biological replicates. Included are two arrays performed in the *ics-2* mutant *Arabidopsis* background.

not reflected in the RNA degradation plot as WLA samples exhibit similar or steeper slopes than prepared tissue samples (Scrape and LMD). Archer et al. used mixed-effect models to better represent the nested structure and relationship of probes within probesets to provide a summary estimate of quality with an associated confidence limit that allowed them to differentiate expression output from samples of high vs. low RNA quality (9). However, neither the standard RNA degradation plot nor the Archer mixed-effect model assessment takes the position of the probes in the transcript with respect to their physical distance from the 5′ end into account. For example, for shorter transcripts, individual probes might reside very near the 5′ end, while for longer transcripts even the 5′ most probe might be >1,000 bp from the 5′ end. In addition, updated annotations (e.g., TAIR9) indicate some of the ATH1 probesets are not within 600 bp of the 3′ end. Therefore, we recommend the use of an assessment that takes into account the physical positions of the probes on the transcript, such as the one described below. Probesets for which all probes are "far" from the 5′ end (i.e., ≥0.70 fractional distance from the 5′ end) using current annotation are selected and their expression is compared with that of all genes (probesets) on the array. Using this approach, we found ATH1 $\log_2$ expression correlation values for fresh (WLA) vs. prepared tissue (Scrape) samples using either all probesets or just the 3′ biased probesets were not significantly different (Fig. 3). This indicates that RNA

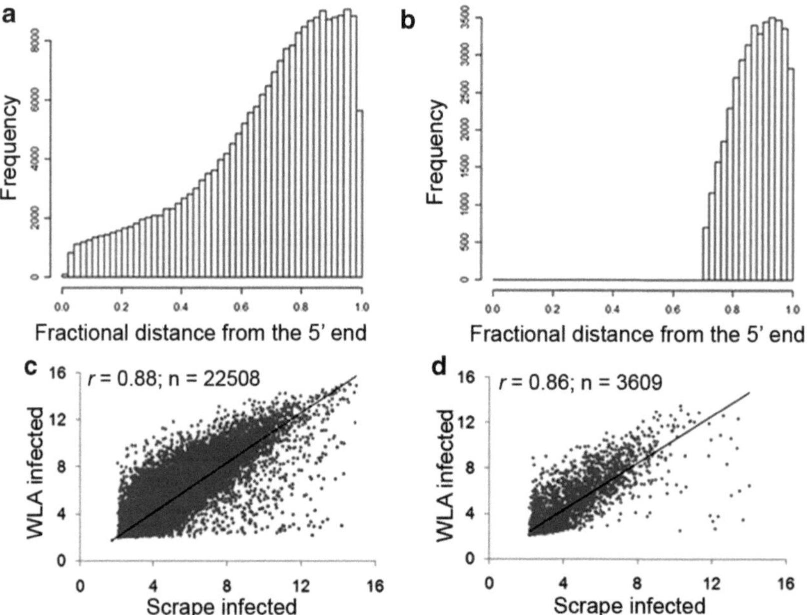

Fig. 3. RNA degradation associated with tissue preparation does not significantly impact ATH1 expression data. Distribution of probesets based on fractional distance from the 5′ end for all genes in (**a**) or for 3′ biased probesets (probesets where the distance between each probe and the 5′ end of the transcript is at least 70% of the coding sequence) in (**b**). ATH1 $\log_2$ expression correlation plots for WLA (fresh tissue, amplified) vs. Scrape (prepared tissue, amplified) infected samples for all genes in (**c**) or for 3′ biased probesets in (**d**). r = Pearson's correlation. Reproduced from Supporting Information of ref. 4 with permission from the National Academy of Sciences of the USA.

degradation associated with tissue preparation does not negatively impact ATH1 expression data.

Correlation analysis of expression data from independent biological replicates of one sample type or different sample types is extremely useful in ascertaining reproducibility of independent biological replicates and identifying whether expression is impacted by an experimental protocol. As appropriate, we recommend comparing both expression of (a) all probesets and (b) probesets associated with the process of interest. In our case, the process of interest is infection and we selected genes with ≥2-fold change in infected vs. uninfected samples computed using one-way ANOVA ($p < 0.05$) with Partek Genomics suite; selected genes also had a false discovery rate (q-value) ≤5%. ATH1 $\log_2$ expression plots and correlation values (Pearson's r) can be easily performed using standard software, such as Excel in Microsoft Word. In addition to comparing the absolute expression levels (shown here), it can be useful to compare the relative levels of expression for infected vs. uninfected samples for both biological replicates and for different sample types (not shown). This allows one to determine whether processes that impact absolute expression also influence relative expression (infected vs. uninfected).

In Fig. 4, we provide correlation plots used to examine the role of tissue preparation on ATH1 output as an example. To assess the impact of tissue preparation on ATH1 output, correlation analysis is performed on WLA (no tissue preparation) vs. Scrape and LMD samples, both of which use prepared tissue. We observed a small subset of 107 probesets (0.5% of those on the ATH1 array) were not well correlated between prepared tissue samples (Scrape and LMD) and fresh tissue samples (WLA), exhibiting a ≥10-fold change in expression in Scrape vs. WLA samples (Fig. 4a; (4)). These same 107 genes also exhibited a ≥10-fold change in expression when LMD samples were compared with WLA samples (Fig. 4b) as would be expected if their expression difference was due to the tissue preparation method. This was true when uninfected samples were compared or when infected samples were compared (Fig. 4c), again indicating that the altered expression of these genes was not associated with infection but with the preparation method. A majority of these tissue preparation-associated genes are known targets of early induced heat shock transcription factors and are likely induced during the initial step of tissue preparation using our modified microwave method (4). These genes are removed from all subsequent microarray analysis because the large differences in expression of these genes as a result of tissue preparation could either

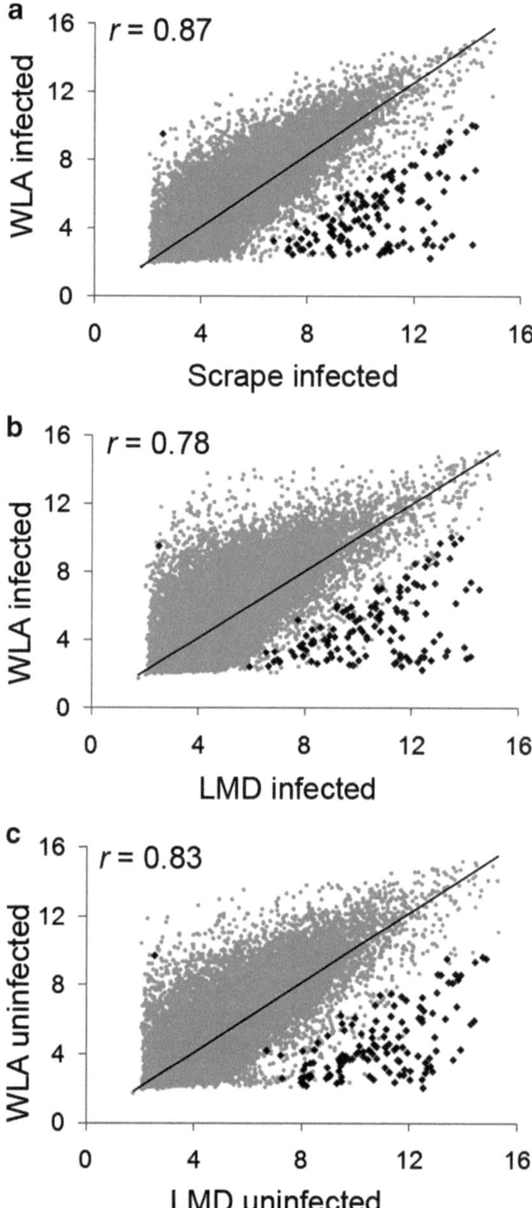

Fig. 4. Correlation analysis indicates tissue preparation impacts expression of <0.5% of probesets on ATH1 array. ATH1 $\log_2$ expression correlation plots for (**a**) Scrape vs. whole leaf amplified (WLA), infected samples, (**b**) LMD vs. WLA, infected samples, and (**c**) LMD vs. WLA, uninfected samples. r = Pearson's correlation. A small subset of 107 genes (shown in *black*) was not correlated between prepared tissue samples (Scrape and LMD amplified) and fresh tissue samples (WLA). Reproduced from Supporting Information of ref. 4 with permission from the National Academy of Sciences of the USA.

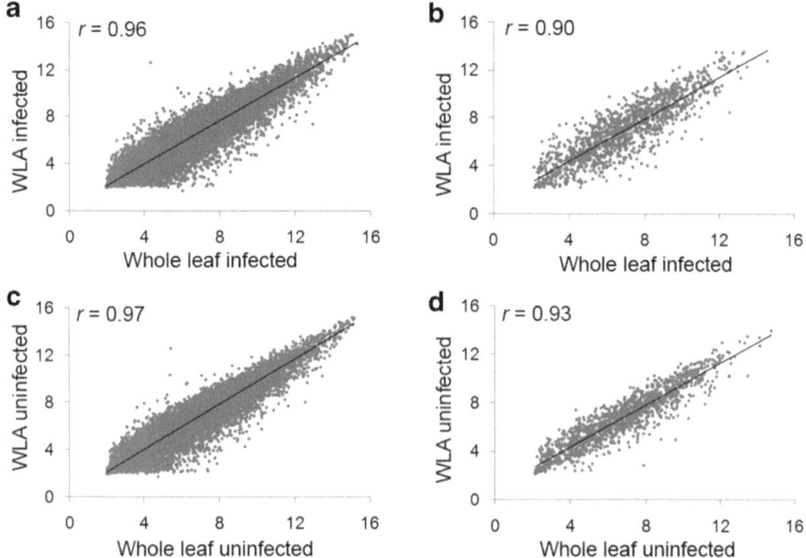

Fig. 5. Correlation analysis indicates two-round amplification does not significantly impact ATH1 expression data. ATH1 log_2 expression correlation plots for whole leaf (WL) vs. WLA, infected and uninfected (UI) replicates for all genes in (**a**) and (**c**), and for selected (twofold change, $p \leq 0.05$) genes in (**b**) and (**d**). r = Pearson's correlation. Reproduced from Supporting Information of ref. 4 with permission from the National Academy of Sciences of the USA.

(a) mask small differences in expression associated with infection or (b) result in false positives, e.g., if variation in expression associated with tissue preparation is greater than the difference in expression between infected and uninfected samples. In contrast to the tissue preparation method, other components of the experimental procedure did not impact ATH1 output as analyzed by correlation analysis. For example, expression of WLA vs. whole leaf (unamplified) samples was highly correlated ($r \geq 0.90$) indicating amplification did not significantly impact ATH1 output (Fig. 5).

To assess reproducibility of all sample types, expression correlation plots for independent biological replicates of the same sample type are compared. As shown for our LMD samples (Fig. 6), these correlations were very high ($r \geq 0.98$) both for all probesets and for those whose expression changed in response to infection. Further validation of our LMD microarray data by qPCR, reporter constructs, and phenotypic analyses confirmed that our site-specific global expression profiling provided a sensitive and accurate readout of expression at the infection site (4).

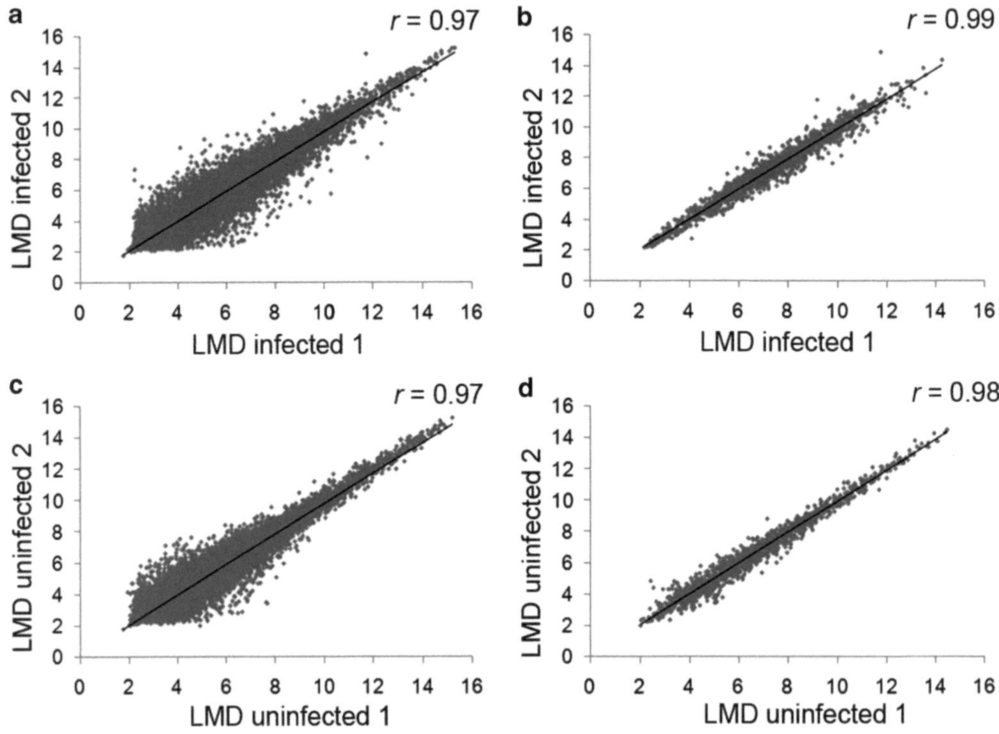

Fig. 6. Independent biological replicates of LMD samples are highly correlated. Expression correlation plots ($\log_2$) for LMD infected and uninfected samples for all genes on array in (**a**) and (**c**), and for selected (twofold change, $p \leq 0.05$) genes, in (**b**) and (**d**). *r*-Values denote Pearson's correlations. Reproduced from Supporting Information of ref. 4 with permission from the National Academy of Sciences of the USA.

4. Notes

1. In Fall 2009, Affymetrix two-cycle target labeling kit was replaced with the GeneChip® 3 IVT Express Kit; however, two-cycle kits may still be available upon request for a limited time.
2. Thin-walled PCR tubes are used to perform all reactions to ensure even temperature distribution.
3. First dilution can be stored at –20°C for up to 6 weeks and freeze–thawed up to eight times.
4. Prior to total RNA sample/T7-oligo(dT) primer/Poly-A control mix preparation, the total RNA sample extracted from LMD-isolated cells is concentrated to ~2 μL using a speed vac.
5. To prevent condensation that may result from water bath-style incubators, incubations are best performed in a thermocycler.

6. Cooling the sample at 4°C is required before proceeding to the next step. Adding the First-Cycle, Second-Strand Master Mix directly to solutions that are at 70°C compromises enzyme activity.
7. The heated lid function is turned off or the thermocycler lid is left slightly open for the 16°C incubation step.
8. Samples may be stored at –20°C at this step for later use.
9. IVT cRNA wash buffer is supplied as a concentrate. 20 mL of 100% ethanol is added before use. It is important to work without interruption during the cleanup steps.
10. Collection tubes are labeled with sample names since caps may break off during the spin.
11. Samples may be stored at –20°C at this step for later use.
12. The samples may be immediately cleaned up or stored at –20°C for later use. The samples should not be left at 4°C for long periods of time.
13. The mixture should be yellow in color (similar to cDNA Binding Buffer without the cDNA synthesis reaction). If the color is orange/violet, 10 μL of 3 M sodium acetate, pH 5.0 is added to the mix to turn the color yellow.
14. The reaction should NOT be assembled on ice, since spermidine in the 10× T IVT Labeling Buffer can facilitate the precipitation of template cDNA at low temperature.

References

1. Day, R. C., McNoe, L., and Macknight, R. C. (2007) Evaluation of global RNA amplification and its use for high-throughput transcript analysis of laser-microdissected endosperm. *Int J Plant Genomics 2007*, 61028.
2. Bolstad, B. M., Collin, F., Brettschneider, J., Simpson, K., Cope, L., Irizarry, R. A., and Speed, T. P. (2005) Quality Assessment of Affymetrix GeneChip Data, in *Bioinformatics and Computational Biology Solutions Using R and Bioconductor* (R. Gentleman, V. Carey, W. Huber, R. A. Irizarry and S. Dudoit, Eds.) pp 33–47, Springer, New York.
3. Kerk, N. M., Ceserani, T., Tausta, S. L., Sussex, I. M., and Nelson, T. M. (2003) Laser capture microdissection of cells from plant tissues. *Plant Physiol 132*, 27–35.
4. Chandran, D., Inada, N., Hather, G., Kleindt, C. K., and Wildermuth, M. C. (2010) Laser microdissection of *Arabidopsis* cells at the powdery mildew infection site reveals site-specific processes and regulators. *Proc Natl Acad Sci U S A 107*(1), 460–5.
5. Reimers, M., and Carey, V. J. (2006) Bioconductor: an open source framework for bioinformatics and computational biology. *Methods Enzymol 411*, 119–34.
6. Irizarry, R. A., Bolstad, B. M., Collin, F., Cope, L. M., Hobbs, B., and Speed, T. P. (2003) Summaries of Affymetrix GeneChip probe level data. *Nucleic Acids Res 31*, e15.
7. Irizarry, R. A., Hobbs, B., Collin, F., Beazer-Barclay, Y. D., Antonellis, K. J., Scherf, U., and Speed, T. P. (2003) Exploration, normalization, and summaries of high density oligonucleotide array probe level data. *Biostatistics 4*, 249–64.
8. Bolstad, B. M., Irizarry, R. A., Astrand, M., and Speed, T. P. (2003) A comparison of normalization methods for high density oligonucleotide array data based on variance and bias. *Bioinformatics 19*, 185–93.
9. Archer, K. J., Dumur, C. I., Joel, S. E., and Ramakrishnan, V. (2006) Assessing quality of hybridized RNA in Affymetrix GeneChip experiments using mixed-effects models. *Biostatistics 7*, 198–212.

Chapter 21

Visualizing Cellular Dynamics in Plant–Microbe Interactions Using Fluorescent-Tagged Proteins

William Underwood, Serry Koh, and Shauna C. Somerville

Abstract

Interactions between plant cells and microbial pathogens involve highly dynamic processes of cellular trafficking and reorganization. Substantial advances in imaging technologies, including the discovery and widespread use of fluorescent proteins as tags as well as advances in laser-based confocal microscopy have provided the first glimpses of the dynamic nature of the processes of defense and pathogenicity. Prior to the development of these techniques, high resolution imaging by electron microscopy gave only a static picture of these dynamic events and live cell imaging was significantly limited in resolution as well as the availability of relevant stains and markers. The incorporation of fluorescent protein fusions and laser-based confocal microscopy into studies of plant–microbe interactions has opened the door to fascinating new questions about the cellular response to attempted infection. Additionally, studies of cellular responses to pathogen infection may lead to new knowledge about fundamental processes in plant cells, such as mechanisms underlying subcellular trafficking and targeting of proteins and other molecules.

Key words: Plant, Pathogen, Confocal microscopy, GFP, Powdery mildew, Fluorescent protein, Microbe

1. Introduction

The combined use of fluorescent protein fusions with laser-based confocal imaging has yielded novel insights into the spatial and temporal processes associated with the invasion of plant tissues in numerous pathosystems. Fluorescent-tagged proteins can be used in several ways to study plant–microbe interactions. Tagged proteins can be used as general markers to visualize the spatial and temporal dynamics of the plant cell, the pathogen, or both. In a recent example, live cell imaging was used to track the infection process of yellow-fluorescent protein (YFP)-labeled *Magnaporthe oryzae* in rice. In this study, fungal hyphae were found to constrict remarkably to cross plant cell walls at specific sites, presumably

John M. McDowell (ed.), *Plant Immunity: Methods and Protocols*, Methods in Molecular Biology, vol. 712,
DOI 10.1007/978-1-61737-998-7_21,

plasmodesmata, to infect neighboring cells (1). Tagged proteins can also be used as markers to track the dynamics of specific subcellular compartments or cytoskeletal components. Marker-labeled organelles, including the nucleus, cytoplasm, Golgi, and peroxisomes have been found to migrate to sites of penetration by the powdery mildew fungus *Golovinomyces cichoracearum* in *Arabidopsis* epidermal cells (2). In addition, actin filaments have been found to polarize toward sites of pathogen attack in several pathosystems (3–5). Fluorescent tags can also be fused to specific proteins of interest to determine their subcellular localization upon pathogen invasion. *Arabidopsis* proteins involved in penetration resistance against the barley powdery mildew *Blumeria graminis*, such as the plasma membrane-localized syntaxin, PEN1 (SYP121), and ABC transporter, PEN3 (PDR8), as well as the peroxisome-localized myrosinase, PEN2, all display strong focal accumulation at sites of attempted penetration by fungal appressoria (6–8).

In this chapter, we describe methods for live-cell imaging of plant–microbe interactions using the *Arabidopsis*-powdery mildew pathosystem as an example. We discuss methods of inoculation, specimen preparation and mounting, image acquisition, and image processing along with specific considerations that may help to improve the quality of images obtained. Although we describe methods using powdery mildew as an example, these techniques can be adapted to suit many plant–microbe pathosystems.

2. Materials

2.1. Powdery Mildew Inoculation

1. Powdery mildew working culture, such as *B. graminis* f. sp. *hordei* race CR3 (non-pathogenic on *Arabidopsis*) maintained on barley (*Hordeum vulgare*) variety AlgerianS (CI-16138) or *G. cichoracearum* strain UCSC1 (pathogenic on *Arabidopsis*) maintained on squash (*Cucurbita maxima* cv Kuta; (9, 10)).
2. Two- to three-week old soil-grown *Arabidopsis thaliana* plants expressing the desired fluorescent protein fusion. (see Note 1).
3. Cardboard box large enough to surround the flat of plants to be inoculated and several feet tall or taller when box is broken down and stood upright. Referred to hereafter as a settling tower.
4. Compressed air duster (available at office supply stores).
5. Dew chamber or similar dark chamber with high (near 100%) humidity.

2.2. Preparation and Mounting of Specimens

1. Razor blades.
2. Glass microscope slides, such as VWR (West Chester, PA) VistaVision 75 × 25 × 1 mm.
3. Glass cover slips, such as Fisher (Pittsburgh, PA) 22 × 22 mm #1.5.
4. 0.5 mg/ml (10×) propidium iodide (Sigma, St. Louis, MO) stock solution in H_2O. Store protected from light at 4°C. Can be stored for at least 6 months. Prepare working solution by diluting one part with nine parts H_2O.
5. Vacuum grease or fingernail polish (see Note 2).

2.3. Confocal Imaging

1. Confocal microscope with appropriate laser line(s) and filters (see Note 3).
2. Immersion medium appropriate for selected microscope objective. Typically oil, glycerin, or H_2O (see Note 4).

2.4. Processing of Images and z-Series

1. Image processing software for microscopy. The freely available open source software ImageJ (http://www.rsbweb.nih.gov/ij/) is very useful for processing images from confocal microscopy. In addition to ImageJ, we also frequently use Metamorph imaging software (Molecular Devices, Sunnyvale, CA).

3. Methods

Two primary considerations before carrying out imaging are at what point after pathogen inoculation to begin imaging and the desired duration of observations to be made. These considerations depend on the specific pathosystem to be studied and the type of processes to be observed. Powdery mildew fungi typically initiate appressorium formation and penetration of *Arabidopsis* epidermal cells between 12 and 24 h after inoculation. To observe focal accumulation of the ABC transporter PEN3 at fungal entry sites, we typically conduct imaging at 24 h post inoculation (hpi).

Propidium iodide is frequently used as a counter stain to visualize fungal structures as this dye rapidly incorporates into fungal cell walls and cytosol. However, propidium iodide also incorporates into plant cell walls, albeit at a slower rate, presumably due to slow diffusion through the leaf cuticle. This becomes an important consideration when carrying out time-lapse imaging over a longer period of time (more than ~20 min).

3.1. Powdery Mildew Inoculation

1. Choose an inoculation area with little or no air currents that is isolated from uninfected plants and growth areas.
2. Place the flat of plants to be inoculated on the ground and surround with the cardboard settling tower.

3. If desired, place a clean microscope slide or cover slip on the edge of a pot to assess the inoculation density. The number of conidiospores inoculated per unit area can be determined by counting spores on the microscope slide using a light microscope.
4. To inoculate the barley powdery mildew, *B. graminis*, shake 1–2 pots of infected barley plants (10–15 plants per pot) over the settling tower and allow spores to settle onto *Arabidopsis* leaves (see Note 5). To inoculate the *Arabidopsis* powdery mildew, *G. cichoracearum*, detach 3–4 infected squash leaves and use a compressed air duster to blow conidiospores over the settling tower. Allow spores to settle. We typically aim for a spore density of 20–40/mm^2, although the density of inoculation is not critical for imaging provided that spores are not clumped together.
5. Transfer inoculated *Arabidopsis* plants to a dew chamber or other chamber with high relative humidity and incubate 1 h in darkness (see Note 6).
6. Transfer inoculated plants to a growth space reserved for infected plants.

3.2. Preparation and Mounting of Specimens

1. Detach an inoculated leaf and, using a razor blade, cut small squares (~5 mm^2) of leaf tissue (see Note 7).
2. Place drops of mounting medium (i.e., 0.05 mg/ml propidium iodide or H_2O) on the surface of a cover slip and, using forceps, transfer leaf pieces to the drops of mounting medium ensuring that the adaxial (upper) surface of the leaf is toward the cover slip surface.
3. Pipet a pool of mounting medium onto the center of a glass microscope slide and carefully transfer the cover slip, with leaf pieces, onto the glass slide. Ensure that enough mounting medium is used to fully fill the space between the microscope slide and cover slip. Avoid trapping air bubbles between the tissue sample and the cover slip (see Note 8).

3.3. Confocal Imaging

1. Select an objective for imaging. The choice of objective depends on your available microscope equipment. We typically image *Arabidopsis*-powdery mildew interactions using a 63× H_2O immersion objective (see Note 9).
2. Place a drop of the appropriate immersion medium on the coverslip for upright microscopes, or directly on the objective for inverted microscopes.
3. Secure the specimen slide on the microscope stage.
4. Using brightfield, focus on the surface of the leaf and identify an area with one or more germinated conidiospores. You may wish to use epifluorescence (mercury lamp illumination) to identify germinated conidiospores that are well-stained with propidium iodide through the microscope oculars.

5. After identifying an interaction of interest, collect a *z*-series. Typically, the user inputs the beginning and ending *z*-plane stage positions as well as the step size (*z*-plane distance between each optical section collected in the series). In general, step sizes of 0.2–0.5 μm yield good results. If the *z*-series is to be used for three-dimensional (3D) projection, step sizes of 0.1–0.2 μm generally yield superior 3D projected images (see Note 10).
6. When imaging multiple fluorophores, the user typically has the option of collecting the respective *z*-series for each fluorophore in two distinct ways. One can choose to collect an entire series for a single fluorophore, followed by the collection of an entire series for the second fluorophore, or alternatively, the microscope can collect an image for each fluorophore at a single *z*-plane before progressing to the next *z*-plane. The choice of which method to use depends on the information desired. On the one hand, if both fluorophores label dynamic entities and the temporal localization of each entity relative to the other is important, collection of an image for each fluorophore at a single *z*-plane is preferable, as the images at a single *z*-plane are more temporally comparable. On the other hand, if the temporal localization of each fluorophore-labeled entity is not critical, collection of a full series for a single fluorophore before proceeding to the next fluorophore allows for more rapid image collection and mitigates the potential for fluorophore photobleaching.

3.4. Processing of Images and z-Series

1. One of the most useful methods of processing a *z*-series is to create a two-dimensional (2D) projection of the collected series of images, referred to as a *z*-projection or extended focus image. This manipulation can be performed using ImageJ software or other microscopy imaging software packages. The most commonly used variant of the *z*-projection is a maximum projection, where the value of a particular pixel in the final projected image represents the highest value at that pixel from the collected series of images. An example of a *z*-series and the resulting maximum *z*-projection derived from the series is shown in Fig. 1.
2. When *z*-series are collected for two or more fluorophores, the resulting *z*-projected images can be overlayed or merged to produce a composite of the features visualized by each fluorophore (see Note 11). An example of two *z*-series collected at the same stage positions and their resulting merged image is shown in Fig. 2.
3. Another useful method of visualizing a *z*-series is to generate a 3D reconstruction of the collected images. This manipulation allows for visualization of the 3D spatial aspects of features imaged in the *z*-series and can give additional information that would be absent in a 2D *z*-projection. ImageJ software

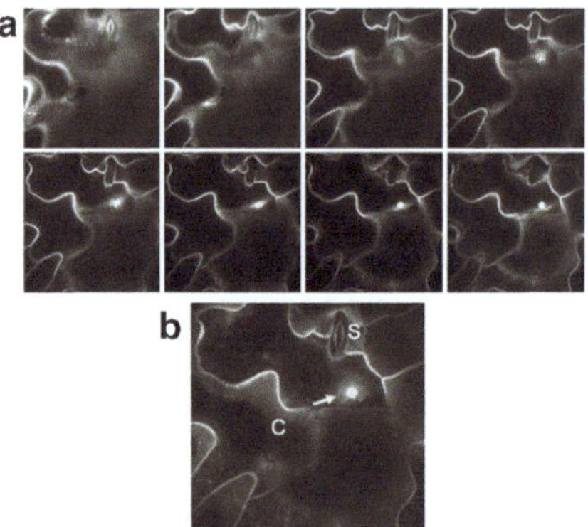

Fig. 1. (**a**) A confocal *z*-series imaging a translational GFP fusion of the *Arabidopsis* plasma membrane-localized ABC transporter PEN3 at 24 h postinoculation (hpi) with *B. graminis*. Images were collected from the epidermal cell layer of an *Arabidopsis* leaf. For this simplified example, eight optical sections were collected with a step size (*z*-distance) of 2 μm. (**b**) A two-dimensional *z*-projected image derived from the series in shown in (**a**), created in ImageJ using maximum projection. PEN3-GFP shows focal accumulation at a site of attempted fungal penetration (*arrow*). *C* conidiospore, *S* stoma.

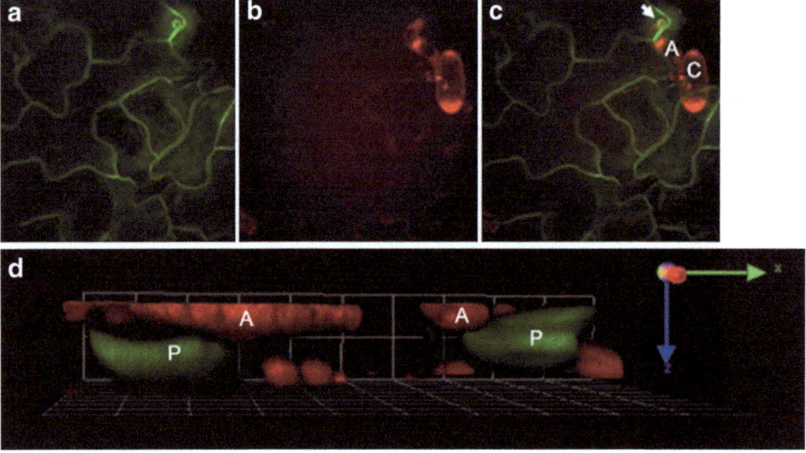

Fig. 2. (**a** and **b**) *z*-Projected images derived from *z*-series collected for PEN3-GFP (**a**) and propidium iodide (**b**). Images were collected from the epidermal cell layer of an *Arabidopsis* leaf 24 hpi with *B. graminis*. The *z*-projected image for PEN3-GFP is derived from a *z*-series consisting of 36 optical sections with a step size of 0.5 μm. The *z*-projected image for propidium iodide is derived from a *z*-series consisting of 14 optical sections with a step size of 0.5 μm. Both *z*-projections were created in ImageJ using maximum projection. (**c**) Overlay of the two *z*-projected images in (**a**) and (**b**), created in ImageJ using the RGB merge function. *Arrow* indicates PEN3-GFP accumulation at the site of attempted fungal penetration. (**d**) A three-dimensional (3D) volume rendering image derived from *z*-series collected for PEN3-GFP (*green channel*) and propidium iodide (*red channel*). Images were collected from the epidermal cell layer of an *Arabidopsis* leaf 25 hpi with *B. graminis*. The 3D rendering was created using Volocity Visualization software and is derived from a *z*-series consisting of 30 optical sections with a step size of 0.5 μm. Note that this rendering is derived from a different *z*-series than the images in panels (**a**)–(**c**). The *green arrow* indicates the *x*-plane direction and the *blue arrow* indicates the *z*-plane direction. One grid unit = 7.6 μm. *A* appressorium, *C* conidiospore, *P* papilla labeled with PEN3-GFP.

has a 3D reconstruction feature; however, we typically achieve superior 3D reconstructions using Metamorph or Volocity (Improvision, Waltham, MA) software. An example of a 3D volume rendering image created using Volocity Visualization software is shown in Fig. 2.

4. Notes

1. Plant growth conditions should be optimized for the specific pathosystem in use. In this case, plants are grown under a 12 h day-length at 22°C day temperature and 20°C night temperature, 70% relative humidity, and a light intensity of 100–150 $\mu E/m^2/s$.
2. Optional. Vacuum grease or nail polish can be used to seal the cover slip to prevent evaporation of mounting medium during prolonged imaging.
3. The following are typical laser lines and emission filters used for imaging common fluorescent protein tags and propidium iodide: GFP – 488 nm laser line, 525/50 nm emission filter; YFP – 488 nm laser line, 535/50 nm emission filter; CFP – 442 nm laser line, 480/40 nm emission filter; RFP – 562 nm laser line, 620/60 nm emission filter; and propidium iodide – 488 nm (or 562 nm) laser line, 620/60 nm emission filter.
4. The choice of immersion medium may be limited by the objectives available. However, if multiple objectives are available with the option of using different immersion media, the choice can have an impact on the quality of the images obtained. The most common immersion media are oil, glycerin, and H_2O. Oil immersion objectives typically have higher numerical apertures. However, when imaging into leaf tissue, particularly into the mesophyll layer, this advantage is negated by the difference in optical density of the immersion oil compared to the mounting medium (H_2O solution). We find that a H_2O immersion objective works well for imaging interactions at the plant epidermal layer and is particularly useful for imaging into the mesophyll layer. In practice, any of the three common immersion media can be used successfully.
5. To avoid clumping of conidiospores and to ensure inoculation with fresh spores, shake infected barley plants 1 day prior to inoculation to discard older, nonviable spores. This allows fresh conidiospores to develop prior to inoculation.

6. We find that maximal conidiospore germination is achieved when plants are inoculated several hours before the beginning of the night (dark) growth cycle (i.e., inoculated in late afternoon).
7. Cut leaf pieces can be larger than 5 mm^2; however, for best imaging results, the pieces should be small enough to allow an even distribution of mounting medium between the microscope slide and cover slip.
8. For longer duration (>30 min) imaging, it is necessary to seal the cover slip to prevent evaporation of the mounting medium as this causes shifting of the specimen over time. The cover slip can be sealed by running a bead of vacuum grease in a square on the microscope slide, pipetting mounting medium within the square of vacuum grease, and affixing the cover slip to the slide using the applied grease. Alternatively, after placing the cover slip onto the slide, the cover slip can be sealed by the application of nail polish around the edges of the cover slip.
9. For longer duration imaging (>30 min), glycerin or oil immersion media are more suitable as H_2O is likely to evaporate during the imaging procedure.
10. Best results are typically achieved by beginning the *z*-series in the focal plane in which the top of the fungal conidiospore is visible and collecting images into the plane of the leaf until just after the desired labeled feature is imaged.
11. Extraneous *z*-planes that do not contain relevant information for a particular fluorophore can be removed when compiling a *z*-projected image by specifying the optical sections to be used for projection.

Acknowledgments

We thank Candice Cherk, Shundai Li, Charlie Anderson, and Ian Wallace for critical reading of the manuscript. This work was supported in part by an NSF grant (Award # 01519898) and funding from the Carnegie Institution of Science to S.C.S. and by a NIH postdoctoral fellowship (F32-GM0834393) to W.U. The content is solely the responsibility of the authors and does not necessarily represent the official views of the National Institute of General Medical Sciences or the National Institutes of Health.

References

1. Kankanala, P., Czymmek, K., and Valent, B. (2007) Roles for rice membrane dynamics and plasmodesmata during biotrophic invasion by the blast fungus. *Plant Cell* **19**, 706–724.
2. Koh, S., André, A., Edwards, H., Ehrhardt, D., and Somerville, S. (2005) *Arabidopsis thaliana* subcellular responses to compatible *Erysiphe cichoracearum* infections. *Plant J.* **44**, 516–529.
3. Opalski, K.S., Schultheiss, H., Kogel, K.-H., and Hückelhoven, R. (2005) The receptor-like MLO protein and the RAC/ROP family G-protein RACB modulate actin reorganization in barley attacked by the biotrophic powdery mildew fungus *Blumeria graminis* f. sp. *hordei*. *Plant J.* **41**, 291–303.
4. Shimada, C., Lipka, V., O'Connell, R.O., Okuno, T., Schulze-Lefert, P., and Takano, Y. (2006) Nonhost resistance in *Arabidopsis-Colletotrichum* interactions acts at the cell periphery and requires actin filament function. *Mol. Plant Microbe Interact.* **19**, 270–279.
5. Takemoto, D., Jones, D.A., and Hardham, A.R. (2006) Re-organization of the cytoskeleton and endoplasmic reticulum in the *Arabidopsis pen1-1* mutant inoculated with the non-adapted powdery mildew pathogen, *Blumeria graminis* f. sp. *hordei*. *Mol. Plant Pathol.* **7**, 553–563.
6. Assaad, F.F., Qiu, J.-L., Youngs, H., Ehrhardt, D., Zimmerli, L., Kalde, M., Wanner, G., Peck, S.C., Edwards, H., Ramonell, K., Somerville, C.R., and Thordal-Christensen, H. (2004) The PEN1 syntaxin defines a novel cellular compartment upon fungal attack and is required for the timely assembly of papillae. *Mol. Biol. Cell* **15**, 5118–5129.
7. Stein, M., Dittgen, J., Sánchez-Rodriquez, C., Hou, B.-H., Molina, A., Schulze-Lefert, P., Lipka, V., and Somerville, S. (2006) *Arabidopsis* PEN3/PDR8, an ATP binding cassette transporter, contributes to nonhost resistance to inappropriate pathogens that enter by direct penetration. *Plant Cell* **18**, 731–746.
8. Lipka, V., Dittgen, J., Bednarek, P., Bhat, R., Wiermer, M., Stein, M., Landtag, J., Brandt, W., Rosahl, S., Scheel, D., Llorente, F., Molina, A., Parker, J., Somerville, S., and Schulze-Lefert, P. (2005) Pre- and postinvasion defenses both contribute to nonhost resistance in *Arabidopsis*. *Science* **310**, 1180–1183.
9. Moseman, J.G. (1968) Reactions of barley to *Erysiphe graminis* f. sp. *hordei* from North America, England, Ireland, and Japan. *Plant Dis. Rep.* **52**, 463–476.
10. Adam, L., Ellwood, S., Wilson, I., Saenz, G., Xiao, S., Oliver, R.P., Turner, J.G., and Somerville, S. (1999) Comparison of *Erysiphe cichoracearum* and *E. cruciferarum* and a survey of 360 *Arabidopsis thaliana* accessions for resistance to these two powdery mildew pathogens. *Mol. Plant Microbe Interact.* **12**, 1031–1043.

Index

John M. McDowell (ed.), *Plant Immunity: Methods and Protocols*, Methods in Molecular Biology, vol. 712,
DOI 10.1007/978-1-61737-998-7, © Springer Science+Business Media, LLC 2011

S

T

W

Y

Z

MIX
Papier aus verantwortungsvollen Quellen
Paper from responsible sources
FSC® C105338

If you have any concerns about our products,
you can contact us on
ProductSafety@springernature.com

In case Publisher is established outside the EU,
the EU authorized representative is:
Springer Nature Customer Service Center GmbH
Europaplatz 3, 69115 Heidelberg, Germany

Printed by Libri Plureos GmbH
in Hamburg, Germany